The BIBLE

on the Lost Dutchman Gold Mine and Jacob Waltz

The BIBLE
on the
Lost Dutchman Gold Mine
and Jacob Waltz

A Pioneer History of the Gold Rush

by

Helen M. Corbin

Wolfe Publishing Co.
2625 Stearman Road, Suite A
Prescott, Arizona 86301

THE BIBLE ON THE LOST DUTCHMAN GOLD MINE AND JACOB WALTZ

(Leather-bound edition limited to 100 copies.)

Printed in the United States of America

06 05 04 03 5 4 3 2 1

Published June 2002

Library of Congress Cataloging-in-Publication Data

Corbin, Helen.

The bible on the lost Dutchman Gold Mine and Jacob Waltz: a pioneer history of the Gold Rush / Helen M. Corbin.

p. cm.

Includes bibliographical references and index.

ISBN: 1-879356-59-7

1. Superstition Mountains (Ariz.)—Gold discoveries. 2. Waltz, Jacob. 3. Pioneers—Arizona—Superstition Mountains—Biography. 4. Miners—Arizona—Superstition Mountains—Biography. 5. Gold mines and mining—Arizona—Superstition Mountains—History. 6. Frontier and pioneeer life—Arizona—Superstition Mountains. 7. Treasure-trove—Arizona—Superstition Mountains—History. 8. Waltz family. 9. Superstition Mountains (Ariz.)—Biography. 10. Superstition Mountains (Ariz.)—History. I. Title.

F817.S9 C66 2002

979.1'75—dc21

2002006802

CIP

To Arizona's pioneers and historian Bertie Roberts, a direct descendant, who knew Jacob Waltz and kept an accurate record. Her persistence for historical truth opened the doors into the past.

Invictus[1]

by William Ernest Henley

Out of the night that covers me
Black as the pit from pole to pole
I thank whatever Gods may be
For my unconquerable soul

In the fell clutch of circumstance
I have not winced nor cried aloud
Under the bludgeonings of chance
My head is bloody, but unbowed

Beyond this place of wrath and tears
Looms but the Horror of the shade,
And yet the menace of the years
Finds and shall find me unafraid

It matters not how strait the gate,
How charged with punishment the scroll,
I am master of my fate:
I am the captain of my soul.

The Scotsman who authored this extraordinary work was himself in constant pain, living in the Victorian Age when medicine was in its infancy. This man suffered greatly and lost a foot from an incurable bone disease. The poem reminded this author of the desperate days of the pioneers who, like the poet, never complained. They just pressed on.

[1] Latin for unconquered.

Contents

Illustrations

Documents

Maps

Photographs

Chronology of Jacob Waltz's Life

Born Württemberg, Prussia 1810

Conscripted into German army 1830

Sailed to New Orleans 1839

Applied for citizenship, Mississippi .1848

Left Castroville, Texas, with wagon train bound for California 1850

Arrived in Grass Valley, California .1852

Went to mine in Kernville, California 1861

Received U.S. citizenship 1861

Left for Arizona Territory 1863

Filed claims at Turkey Creek, Arizona (near Prescott) . . .1864

Arrived in Phoenix to file on Salt River homestead 1868

Filed claim jumping affidavit 1872

Seen buying supplies at Mormon store with gold ore 1884

Sent $7,000 draft (paid with gold ore) to sister in Lawrence, Kansas 1887

Jacob invited Julia and Reiney to Christmas dinner at his farm 1890

Flood wiped out Jacob's farm (February) 1891

Jacob Waltz died (October 25) 1891

Introduction

In July 2000 *U.S. News and World Report* published a special issue on the "Mysteries of History." Among such greats as the Holy Grail, King Solomon's Mines, the Lost Tribes of Israel, Noah's Ark, Oak Island and the Ark of the Covenant was the Lost Dutchman Mine of Arizona.

The missing treasure has been actively hunted by thousands of people since the death of its namesake in October 1891. German immigrant Jacob Waltz met his Maker with the secret to the mine's location on his lips. The treasure lies surreptitiously beneath the formidable canyons of the Superstition Mountains, a 247 square mile range of rugged, impassable terrain 50 miles due east of Phoenix, Arizona. Once home to the fierce Apache warriors, those mountains seduced soldiers, Mexicans, prospectors, Spaniards, foreigners and newcomers to the Arizona Territory with their promise of riches. Regardless of the unspeakable tortures forced on those caught within its boundaries by the savages, men continued to come. They had succumbed to **gold fever**.

The Spanish named the range Sierra de la Espuma because of a limestone formation running parallel for miles to the crest of the main mountain and reminiscent of the Great Flood. The name translates to Mountains of Foam. Later, astute Pima medicine men, cognizant of the dreaded Apaches, whose name means enemy, told tribesmen of an evil spirit who rarely slept and whose wrath would be imposed on all who went there. The mountains became the Superstitions. The name held because of ominous foreboding and death.

In 1990 this author's first book on the subject, *The Curse of the Dutchman's Gold*, was published and subsequently became the leading book on the Dutchman's mine. It is now in its sixth printing. The research for that book came from notes written by people who had lived when Jacob lived and some who came directly after. Newsman Sims Ely's book *The Lost Dutchman Mine*, published in 1953, was the first book on the subject. It is now out of print and a collector's item.

Sims and his friend, cattle rancher and civic leader Jim Bark, had spent years hunting for the mine, and while they located and filed on many claims, they never found the Dutchman. Brownie Holmes left behind a diary of notes from his father, Richard (Dick) Holmes, one of the men who was with Jacob Waltz on the night he died. Holmes claimed Jacob told him where the mine was, but he and his son spent the rest of their lives searching unsuccessfully for the elusive treasure.

Two other major characters in this scenario are Julia Thomas, a mulatto woman who ran a bakery and ice cream parlor in Phoenix, and her adopted son Rheinhart (Reiney) Petrasch. Julia and Waltz were friends; in fact, he died in the storage shed behind her store in Phoenix, where he had been recuperating from pneumonia, which he contracted when a major flood swept down into the valley. Waltz's farm under water posed a threat, and Reiney had been sent out to fetch Waltz to be brought to the safety of Julia's care. He lingered long but never fully recovered.

During that time Jacob tried desperately to tell his two friends where the gold was. He told them mining would be too arduous, but he had three caches they would round up as soon as spring came and he was well enough to travel. Death cheated them. Jacob's promise ruined the rest of their lives and the lives of everyone else connected to this story.

The gold that was under Jacob's bed disappeared after his death. Dick Holmes claimed it was Jacob's gift to him. A feud erupted between the Holmes faction and the Petrasch faction, which is ongoing now over 100 years.

Hordes of people have died: There were a few suicides, many were murdered and some were injured searching the shale-covered slopes where a fall means injury and/or certain death. Some succumbed to rattlesnake bite, scorpion stings and, due to unbelievable heat, the excruciating torture of dying of thirst. Many were never found after animals stripped the bodies of flesh then scattered their bones. The real tragedy is the land still clutches its treasure. It is a long-held theory that the mine is cursed, and only the Dutchman was allowed to own the gold.

Jacob Waltz is now world famous. His story is fascinating and bears retelling one more time because of new information that will confirm the many theories, facts and sometimes half-truths connecting Jacob's life to the mystery.

In the year 2000, this author became privy to manuscripts and notes from new sources – the amazing diaries of pioneer families, whose members included historians of persistent dedication and whose words have never before been exposed, and from hand-written notes of soldiers, prospectors and mountain men who were the first into the Arizona Territory. Now for the first time there is documented proof that Jacob Waltz was among them. Their stories are incredible. As the reader becomes aware of the treachery, terror and pain the pioneers endured, one can only marvel at their indomitable spirit and have new respect for those who came before us. While exposing the truth of their torment, we will have new insight into the reasons for the Apache's mutilation and torture, but more importantly, we will come to know Jacob Waltz as never before and will come closer to his infamous treasure.

* * *

A Brief Review of the Pioneers

George R. Roberts (1828–1884) was born in Virginia. (His grandfather fought in the Revolutionary War and was with George Washington at Valley Forge; he was killed by Indians near Middlesboro in 1835.) At age 15, George and brothers William, John and Thomas left for Mississippi where other family members were settled and owned a farm.

Jones County, Mississippi – George R. Roberts met his cousin Abraham H. Peeples (1826–1892). The pair became fast friends and allies.

In 1848 Peeples and Roberts, along with other family members, went to Cairo, Illinois, to join with Peeples's brother who ran a trade station there. No sooner had they arrived than word of the California Gold Rush reached them. The group decided to go.

George R. Roberts, his brothers, along with cousins Charles, Cyrus and Return left Illinois with A.H. Peeples and headed for

Fort Hondo outside Castroville, Medina County, Texas, where a wagon train was assembling to join the Gold Rush.

Fort Hondo was about five miles from Castroville and was considered the edge of civilization. It was also the jumping off point for the wagon trains.

At Castroville the group was joined by three other family members – Daniel, Mose and Gideon Roberts – who hailed from near Beaumont, Texas. The three were cousins of George R. Roberts.

The wagon train consisted of 60 or 70 people, a good portion of whom were German immigrants, including a man named Jacob Waltz, who traveled to California with Roberts and Peeples.

Once in California and after reaching Los Angeles, the train split up and people went their separate ways. Roberts, Peeples and several Germans, including Waltz, headed for Sacramento. They settled about 20 miles northeast of there at a place called Grass Valley. They located a mine, which ultimately flooded, then moved on to Whiskey Flat, later known as Kernville. By 1861 the mines were playing out, and the Indians were on the rampage in that area.

In 1863 Peeples put together an expedition that would head into the dangerous, wild Arizona Territory. He hired famous mountain man Pauline Weaver to lead the expedition. The party, formed at Fort Yuma, had supplies for 30 men. They left Olive City, Arizona Territory, on April 1 of that year.

Noted members of that now-famous expedition were Henry Wickenburg, Joe Green, Oscar Hutton, John Montgomery and our protagonist, Jacob Waltz. A complete list of the Peeples party is on page 26.

From the Matthew Cavaness memoirs: Jeremiah Cavaness and Kissie Huckaby were married in 1837. Their union produced 10 children, 6 daughters and 4 sons. Matthew, the eldest and most famous of the children, was born in Arkansas in 1843.

The family moved to Texas shortly thereafter and lived alone on the frontier where they were prey to attacking savages. Matt grew up shooting Indians. He was tough and capable and would go on to become one of the great Indian fighters of all times.

Matt Cavaness had six sisters, the last two being Rebecca and Sarah. These two women would marry Abraham H. Peeples and

George R. Roberts, respectively. Sarah and George's daughter Sarah "Bertie" was born in 1877. She lived until 1969. This young woman is the historian for whom we dedicated this book and the woman who devoted her life to collecting much of the information we have used to compile this book. There were many other sources of information used, but Bertie's collection was acute to proving the trail Jacob Waltz took.

Captain Joseph Walker, who had charge of a corps of fur trappers for the American Fur Company, was approached by "Pathfinder" John C. Fremont to guide his party to California for $5,000. Walker asked for $8,000, but Fremont declined and went looking for Kit Carson to fill his needs. Carson and Walker were friends and had earlier gone West together, roaming the Rocky Mountains for years. Walker was, at this juncture, between 60 and 70 years of age, weighed approximately 200 pounds, stood over 6 feet, had a white beard, broad shoulders, was nimble, strong and apparently very comfortable in nature. The stories of their travels to Arizona are remarkable.

The Walker Expedition, Arizona Territory 1863: Daniel E. Conner, a member of the Walker party, kept copious notes on the expedition that entered central Arizona in 1863, exploring and opening the country where the city of Prescott stands today. A copy of the members of the Walker expedition is on page 54.

William Edwards and Joe Green: In 1866 a column of the 1st Arizona Volunteers was racing north toward the Salt River, chasing a band of renegade Apaches, who had murdered some miners near Pinal. They stopped to rest their horses at Brackett's water. While the troop rested, the lieutenant sent two of his men, Sgt. William Edwards and scout Joseph M. Green, up to higher ground to watch for a pair of Indian scouts who were overdue in reporting back to the troop.

The pair rode out along the northwest base of what is known today as the Superstition Mountains. After climbing a series of rises, they came onto a hilltop where they discovered the skeletal remains of 25 people. There was no clothing nor personal effects of any kind; the arrows that had killed them were strewn about – the distinct work of the Apache. The soldiers assumed the bodies had been there for around two years, maybe more, and they believed the dead were from a rival Indian tribe. Their story is further documented in the chapter on the "Massacre Site."

Chapter 1

Jacob's Early Years

Discussing Jacob Waltz inevitably leads to questions that have gone unanswered for over 100 years, at least, since Jacob's death in 1891. Many authors, including this one, took hearsay information and speculated, as did everyone else, about where and when Jacob came to America and with whom? Where did they land and why? What drove these indomitable souls to the New World? But, the even larger question was how did they manage it?

Jacob Waltz was born in Germany in a barony named for Prince William of Württemberg that in 1810 (the year of Waltz's birth) was part of Prussia. This was not a good time to be born in that part of the world. Napoleon's army had penetrated the country, engaging in fierce battles that devastated part of Prussia until 1815. Farming communities were hard hit as the conquerors swept across the region foraging what little food and supplies were available. In 1812 entire communities were burned out, and there were no resources available to rebuild. The winters were fierce. Hordes became refugees seeking work in larger cities that turned into cesspools of poverty, sickness and death. Those who could immigrated to America and Australia.

Most of the populace were slaves. An average daily wage was about 10 silver Groshen (approximately $.15 American). There was also an unjust royal tax placed on the wage, which if not paid, allowed the government to strip the worker of all his possessions. While women and little children begged in the streets, children of nine years were sent away as apprentices under deplorable conditions only to learn as experienced journeymen they would receive 1 (one) Thaler a week. That amount was $.65

American. Children would work an entire year for 14 Thaler ($9.10).

In 1830 Germany, a boy who had reached 20 was conscripted into the army and by law would serve three years, after which he was part of a *Landwehr* (much the same as the National Guard). Payment for this service was a loaf of black bread every five days, and payment every 10 days equaled $.65. Soldiers lived in misery, and execution by firing squad was common, often for infractions such as stealing food. This revelation explains why Jacob Waltz and his friend Jacob C. Weiss both spoke of being soldiers at various times. It was not the American army they referred to; the pair were never in the Civil War.

Both men immigrated from Nagold, a larger industrial city in Württemberg, although they lived in a settlement called Horb. When they sailed for the New World in 1839 with only one bag apiece, Waltz and Weiss were 28 and 29 years old, respectively, and both marked their occupations as laborers. (This information is listed on the manifest that was translated by the International Ship Transcriber's Guild. A partial copy of the document is in this chapter.) The fare to sail from Bremen to New Orleans was 120 to 240 Thaler, approximately $75 to $150 American – a king's ransom.

There was, however, an alternative: "redemptioners." This allowed the passenger an opportunity to sail free and upon arrival pay the debt plus 12 percent in one month's time. If the debt was not paid, the passenger was sold into servitude, usually as a farm laborer, until the debt was paid. Women and children did the same and worked as household help. Although the German government frowned on these practices, ships' captains, especially those going into New Orleans, found the outcome very lucrative. Farm labor along the Mississippi was in short supply, and Germans, who were hard workers, were very welcome. Often if a family was in particularly bad straits, a strong family member went ahead, alone, and would work until he had saved enough to bring other members over.

Jacob Waltz applied for citizenship in Adams County, Mississippi, a heavily rural farm community on the river. It is a good bet he was working off his passage.

The manifest provided by the International Ship Transcriber's Guild shows Jacob Waltz and Jacob C. Weiss left Bremen, Germany, on the ship *Olbers*, captained by H.W. Exter on October 1,

District of Mississippi - Port of New Orleans

Ship Olbers
H.W. Exter Captain

Departed the Port of Bremen October 1, 1839
Arrived the Port of New Orleans November 17, 1839

I, H.W. Exter, Master of the Bremen Ship, Olbers, do solemnly, sincerely and truly swear, that the within list, signed by myself, and now delivered to the Collector of this District, contains the names of all Passengers taken on board this said Ship at the Port of Bremen or at any time since, and that all matters therin set forth are, according to the best of my knowledge and belief, just and true. I do further swear that three of the Passengers have died on this voyage. (Signed) H.W. Exter, Captain and Master of the Ship Olbers, sworn before me this 17th Day of November 1839. C.H. Fechter, Collector, for the Port of New Orleans, District of Mississippi.

DISTRICT of MISSISSIPPI - PORT of NEW ORLEANS

SHIP OLBERS

Departed the Port of Bremen ; October the First 1839
Arrived the Port of New Orleans ; November the Seventeenth 1839

I, H.W. Exter, Master of the Bremen Ship, Olbers, do solemnly, sincerely and truly swear, that the within list, signed by myself, and now delivered to the Collector of this District, contains the names of all Passengers taken on board the Ship Olbers, at the Port of Bremen or at any time since, and that all matters therin set forth are, according to the best of my knowledge and belief, just and true. I do further swear that Three of the Passengers have died on this voyage. (Signed) H.W. Exter, Captain of the Ship Olbers , sworn before me this Seventeenth day of November, 1839.

No.	Name	Age	Settlement	Occupation	
1.	Joseph C. Middendorf	41	Nabern, Bayern	stone mason	
2.	Elise Middendorf	39	Nabern, Bayern	wife	
3.	Joseph A. Middendorf	11	Nabern, Bayern	son	
4.	Michel Middendorf	7	Nabern, Bayern	son	
5.	Jacob J. Middendorf	4	Nabern, Bayern	son	
6.	Gottlieb Dannemann	24	Penzwagen, Bayern	weaver	
7.	Christian J. Geisecke	33	Penzwagen, Bayern	backer	
8.	Jacob Grunwald	21	Penzwagen, Bayern	driver	
9.	Christian Unterkirch	23	Penzwagen, Bayern	farmer	
10.	Anton B. Gradwohl	45	Ostinghausen, Prussia	farmer	
11.	Gertrud Gradwohl	46	Ostinghausen, Prussia	wife	
12.	Franz Gradwohl	20	Ostinghausen, Prussia	backer	
13.	Elsie Mueller	20	Bettinghausen, Prussia	servant girl	
14.	Franz O. Kontze	36	Ostinghausen, Prussia	farmer	
15.	Ferdinand Kontze	14	Ostinghausen, Prussia	son	
16.	Margreta Heitmann	27	Ostinghausen, Prussia	laundress	
17.	Stephan Heitmann	4	Ostinghausen, Prussia	son	
18.	Gertrud Heitmann	3	Ostinghausen, Prussia	daughter	
19.	Anson Heitmann *	11mo	Ostinghausen, Prussia	son	deceased
20.	Adolf G. Vogelsang	48	Besingen, Bayern	distiller	
21.	Anna Vogelsang	20	Besingen, Bayern	servant girl	
22.	George Vogelsang	17	Besingen, Bayern	servant	
23.	Hedwig Vogelsang	14	Besingen, Bayern	servant girl	
24.	Joseph Vogelsang	8	Besingen, Bayern	son	
25.	Martin Bertsch	46	Nagold, Wurttemberg	butcher	
26.	Ernest A. Hobert	37	Nagold, Wurttemberg	tailor	
27.	Hanna Hobert	33	Nagold, Wurttemnberg	seamstress	
28.	Marlene Hobert	12	Nagold, Wurttemnberg	daughter	
29.	Joseph Hobert	9	Nagold, Wurttemnberg	son	
30.	Freidrick Zebler	51	Mernech, Sachsen	farmer	

***The manifest provided by the International Ship Transcriber's Guild shows Jacob Waltz and Jacob C. Weiss on the ship* Olbers.**

79.	Caroline Moser	12	Horb, Wurttemberg	daughter
80.	Gretta Moser	10	Horb, Wurttemberg	daughter
81.	Albert C. Linder	43	Bosbach, Bayern	farmer
82.	Marie Linder	37	Bosbach, Bayern	wife
83.	John Linder	9	Bosbach, Bayern	son
84.	Christian V. Noltemeyer	23	Messenkamp, Hannover	painter
85.	Franz Mihlan	36	Neuroden, Schlesien	locksmith
86.	Andres L. Rau	26	Badendorf, Bayern	gardner
87.	Jacob B. Ost	30	Schwenningen, Wurttemberg	tavern keeper
88.	Dorothea Ost	26	Schwenningen, Wurttemberg	cook
89.	Luisa Ost	4	Schwenningen, Wurttemberg	daughter
90.	William H. Schmidt	25	Horb, Wurttemberg	tailor
91.	Frieda Schmidt	46	Horb, Wurttemberg	cook
92.	Lawrenz Schmidt	13	Horb, Wurttemberg	apprentice
93.	Christian Schmidt	10	Horb, Wurttemberg	messenger
94.	Bernhard C. Pfeiffer	30	Pfalzgraffenweier, Wurttemberg	farmer
95.	Elsie Pfeiffer	26	Pfalzgraffenweier, Wurttemberg	wife
96.	Martin Pfeiffer	3	Pfalzgraffenweier, Wurttemberg	son
97.	Jacob Waltz	28	Horb, Wurttemberg	laborer
98.	Peter Schmidt	23	Horb, Wurttemberg	tailor
99.	Gerhard W. Schmidt	21	Horb, Wurttemberg	locksmith
100.	Jacob C. Weiss	29	Horb, Wurttemberg	laborer
101.	Friedrich H. Kreitzer	38	Horb, Wurttemberg	farmer
102.	Dorthe Kreitzer	39	Horb, Wurttemberg	wife
103.	Henry G. Reinner	30	Wildberg, Wurttemberg	farmer
104.	Gertrud Reinner	27	Wildberg, Wurttemberg	wife
105.	Heinrich Reinner	6	Wildberg, Wurttemberg	son
106.	Ruthann Reinner	4	Wildberg, Wurttemberg	daughter
107.	Gretchen Reinner	1y9mo	Wildberg, Wurttemberg	daughter
108.	Joseph M. Steinhoff	41	Machoff, Bayern	harness maker
109.	Henry F. Muller	60	Sost, Prussia	paysan
110.	Gertrud Muller	59	Sost, Prussia	dress maker
111.	Andres Muller	27	Sost, Prussia	servant
112.	Fredrich H. Kahrs	49	Nagold, Wurttemberg	farmer
113.	Helga Kahrs	44	Nagold, Wurttemberg	wife
114.	Elsie Kahrs	69	Nagold, Wurttemberg	grandmother
115.	Peter Kahrs	17	Nagold, Wurttemberg	farmer
116.	Marjore Kahrs	11	Nagold, Wurttemberg	daughter
117.	Sophie Kahrs	9	Nagold, Wurttemberg	daughter
118.	Jacob Kahrs	7	Nagold, Wurttemberg	son
119.	Frederic A. Wolff	25	Nagold, Wurttemberg	cooper
120.	Franz L. Weirhane	30	Nagold, Wurttemberg	weaver
121.	Dorthea Weirhane	65	Nagold, Wurttemberg	widow
122.	Jurgen F. Keane	43	Nagold, Wurttemberg	brick layer
123.	Mary Keane	41	Nagold, Wurttemberg	cook
124.	Alexander Keane	17	Nagold, Wurttemberg	apprentice
125.	Hermann A. Bodemann	55	Nagold, Wurttemberg	miller
126.	Hedwig Bodemann	50	Nagold, Wurttemberg	wife

Jacob Waltz (circa 1862).

1839, bound for New Orleans. The arrival date was November 17, 1839. (This is the first time in over 100 years that anyone can actually document the existence of Jacob C. Weiss. Waltz speaks of him as his friend and is known to have mourned his friend's death until the last moment of his life.)

The ship carried 205 passengers. The listing also included their hometown, settlement, occupation and luggage. Considering the state of deprivation that existed in Württemberg, it is reasonable to assume the pair may have been working in Nagold or nearby to save money for their trip to the United States. There are, however, many other passengers from that same town. In fact, the manifest lists 108 passengers from Württemberg.

The mass exodus was caused by the horrible conditions existing in the area. And, one can be certain that upon arrival in their new land they would, at least, band together in a settlement where life would be more manageable. These people were intelligent but unschooled; most of them could not read or even write their own names correctly and would have to pool their collective resources to begin their new lives in as much comfort as they could garner.

Later, we can document eight of the clan leaving for Bremen late in the 1840s where they boarded the steamer *Heinrich Von Gargaren*, their port of entry also being New Orleans. It was customary for one member of a group traveling together to be named on the manifest: in this case it was Ignote Waltz, presumed to be an older brother, whose occupation was listed as a tailor. (See Medina County, Texas, 1850 census.)

This was before the United States developed an immigration policy, which didn't occur until 1880. New arrivals gingerly stepped down the gangplank wearing a curious grin and melted, undocumented, into their destiny. Everyone who came steerage was eager to be free of life in the crowded, unpleasant stench of the ship's holds. One can only imagine the excitement they felt disembarking. Unfortunately, nothing is recorded about those first days when having little or no money, with an acute language barrier, among total strangers who were more than likely unfriendly – they had to depend on their spirit and each other. However, since Jacob Waltz was already in Mississippi, it is probably fair to assume the new Waltz arrivals were welcomed and given a decent place to settle.

Waltz family records from Missouri showed initially they settled in Medina County, Texas, (Castroville) in 1850. Ignote is recorded as is another Waltz, Sabastian, who wasn't on the federal census but did show up later on land deeds. Jacob's sister, already married in the homeland, bore the name Schmidt. She and her husband eventually migrated to Lawrence, Kansas. Other records show a Mary Ann Waltz, age 1 year 4 months,

CENSUS YR: 1850 STATE or TERRITORY: TX COUNTY: MEDINA DIVISION: CASTROVILLE REEL NO: M432-912 PAGE NO: 395B
REFERENCE: "TAKEN OCT 29, 1850 BY HENRY K. JUDD"

LN	HN	FN	LAST NAME	FIRST NAME	AGE	SEX	RACE	OCCUP.	VAL.	BIRTHPLACE	MRD.	SCH.	R/W	DDB
1	68	68	HYSENGER	MARY	23	F				GERMANY				
2	69	69	OHR	STEPHEN	29	M		BLACKSMITH	50	GERMANY				
3	69	69	ORH	MARY	20	F				GERMANY				
4	69	69	OHR	LAWRENCE	22	M				GERMANY				
5	70	70	MERRIAN	JOHN	21	M		STONE MASON		GERMANY				
6	70	70	BELHART	JOHN B.	11	M				GERMANY		X		
7	70	70	BELHART	VINCENT	9	M				GERMANY		X		
8	71	71	ALBERTI	JOHN B.	38	M		FARMER	200	FRANCE				
9	71	71	ALBERTI	MARYANN	42	F				FRANCE				
10	71	71	ALBERTI	ADRIAN	18	M		LABORER		FRANCE				
11	71	71	ALBERTI	ADELINE	12	F				FRANCE		X		
12	71	71	ALBERTI	CATHERINE	11	F				FRANCE				
13	71	71	ALBERTI	CARMELA	10	F				FRANCE				
14	71	71	ALBERTI	AUGUST	7	M				TEXAS				
15	71	71	ALBERTI	EMILE	6	F				TEXAS				
16	71	71	ALBERTI	EUGENE	2	M				TEXAS				
17	72	72	ANDRES	NICHOLAS	35	M		SHINGLE MAKER	300	SWITZERLAND				
18	72	72	ANDRES	ELIZABETH	44	F				SWITZERLAND				
19	72	72	ANDRES	ELIZABETH	4	F				SWITZERLAND				
20	72	72	ANDRES	NICHOLAS J.	67	M		LABORER		SWITZERLAND				
21	73	73	NEGLE	FERDINAND	30	M		BLACKSMITH	100	SWITZERLAND				
22	73	73	NEGLE	ROSINI	34	F				GERMANY				
23	73	73	NEGLE	ROSINI	10	F				GERMANY		X		
24	73	73	NEGLE	FERDINAND	1-1/2	M				TEXAS				
25	73	73	NEGLE	EMILE	3/12	M				TEXAS				
* 26	74	74	WALTZ	IGNOTE	36	M		TAILOR	80	GERMANY ✓				
27	74	74	WALTZ	MARYANN	1-1/4	F				TEXAS ✓				
28	74	74	SCHMIDT	LAWRENCE	24	M		TAILOR		GERMANY ✓				
29	75	75	LEONARD	JOHN	39	M		FARMER	100	GERMANY ✓				
30	75	75	LEONARD	ELIZABETH	39	F				GERMANY ✓				
31	75	75	LEONARD	GUSTINE	7	M				GERMANY ✓		X		
32	75	75	LEONARD	GENEVA	5	F				TEXAS				
33	75	75	LEONARD	THOMAS	2	M				TEXAS				
34	75	75	LEONARD	GABRIEL	5/12	M				TEXAS				
35	76	76	BOYLES	WILLIAM	54	M		FARMER		SOUTH CAROLINA				
36	76	76	BOYLES	LYDIA	43	F				SOUTH CAROLINA				
37	76	76	BOYLES	LAFAYETTE	18	M		LABORER		GEORGIA				X
			REMARKS:	IDIOTIC										
38	76	76	STANDFEAR	RACHELE	17	F				GEORGIA				
39	76	76	STANDFEAR	ISAAC S.	1-1/2	M				TEXAS				
40	76	76	BOYLES	SARAH	14	F				GEORGIA				
41	76	76	BOYLES	DORAH	5	F				GEORGIA				

1850 Census, Medina County, Texas
**** Ignote Waltz***

born in Texas to Ignote. Her mother's death from cholera is recorded in 1849. Just the year before, Jacob Waltz applied for citizenship in Natchez, Mississippi, just a short boat ride up the Mississippi River from New Orleans. On the request form, Jacob affirmed he had resided in the U.S. for five years, actually it was 10, but no proof was required.

All ships' records from Hamburg, Germany, 1850-1934, are well documented, but the port at Bremen kept careless records in improper storage. Documents found were unreadable, except for one held by the Waltzes that listed the ship, port of exit, entry and departure of the eight former Württemberg residents headed by Ignote. Of course, later the International Ship Transcriber's Guild was formed and painstaking translations were done.

In the 1930s, a great-niece, Dorothy Middaugh, came from Joplin, Missouri, to stay at the King Ranch near the Superstitions in Arizona. She possessed documents from her relatives about the migration to America. It was her intention to search for her great-uncle's lost mine. While questioning those who were left who actually knew Jacob, she traded information with historian Sarah "Bertie" Roberts.

While Jacob never actually speaks to us in his own words, we will learn to know him through records, documents, family members, friends and the threads of his life woven through history as he lived it. But, now for the first time, we can document him from birth to the grave. Along with this information we can tell the story as it unfolds from the diaries of the pioneers with whom he lived and worked. We will also try to dispel the myriad rumors that surround his life story and replace them with as much truth as is available.

During May 1985, through the auspices of Dr. Bertol Koester, the Honorary German Consul for Arizona, we were privy to the Waltz clan via records held in the city hall of Württemberg, Germany. Dr. Koester had arranged through the government to open the necessary church and archival records. The now-divided town had two mayors, Hugo Meroth and Walter Gutekunst. The records, meticulously scripted in long, narrow, blue and red leather-bound books, listed the Waltz family members (births and deaths) one after another. Our hosts took us to the ancient Waltz family home; descendants still occupied it.

In the course of our discussions, which were translated by Dr.

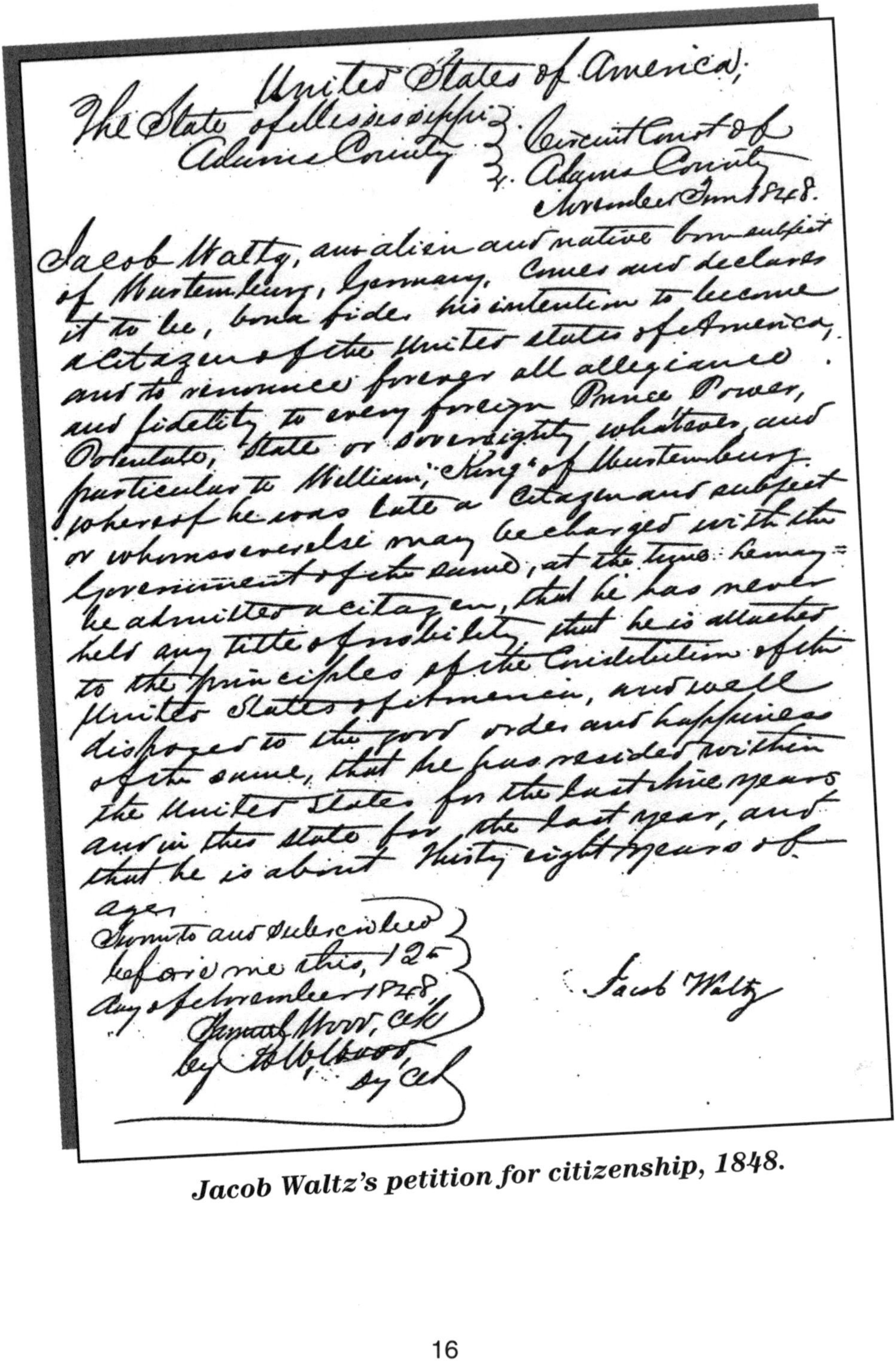

United States of America;
The State of Mississippi } Circuit Court of
Adams County } Adams County
November Term 1848.

Jacob Waltz, an alien and native born subject of Wurtemburg, Germany, Comes and declares it to be, bona fide, his intention to become a citizen of the United States of America, and to renounce forever all allegiance and fidelity to every foreign Prince Power, Potentate, State or sovereignty whatever, and particular to William, "King" of Wurtemburg, whereof he was late a citizen and subject or whomsoever else may be charged with the Government of the same, at the time he may be admitted a citizen, that he has never held any title of nobility, that he is attached to the principles of the Constitution of the United States of America, and well disposed to the good order and happiness of the same, that he has resided within the United States for the last nine years and in this state for the last year, and that he is about thirty eight years of age

Sworn to and subscribed before me this, 12th day of November 1848,
[illegible], Clk
by [illegible], Dy Clk

Jacob Waltz

Jacob Waltz's petition for citizenship, 1848.

Koester, we asked about the rumor that Jacob had sent a large amount of money to his sister in Germany after he located the mine. The mayors laughed. Both men assured us that if a large deposit had come in from America to someone in that town, the gossip mill would have spread the news loudly and rapidly. Now there were two puzzles that were about to be solved. The family name was Waltz. It mattered little what other spellings had been written. Jacob's sister had not been the recipient of a large sum of money; at least, it had not been sent to Württemberg. However, later we will show a document that will prove the money was sent and the story true, but the assumed destination was incorrect.

Rumors of minerals just begging to be snatched from the earth raced like wildfire across the United States starting the mass movement West called the Gold Rush. Soon wagon upon wagon of eager seekers rolled across the plains in search of bonanza, and their stories would become fodder for eastern newspapers. They would be the heroes who opened the new frontier while fighting the bands of savages, roaming at will in search of interlopers to torture and kill.

That time frame – from the 1860s to the turn of the century – was considered the most famous 40 years in American history. The ragtag prospectors would prove themselves as tough, dedicated humans, capable of fighting with the best of them, while becoming experienced miners. Some of them actually became rich, among that group was our protagonist, Jacob Waltz.

For those who continue their search for the mine or who enjoy the adventure through reading about it, we must include new information, some of which has never before been revealed. The diary of Abraham H. Peeples (born July 1826 in Jones County, Mississippi) will be exposed in excerpts verbatim, as he and his cohorts allow a rare window into territorial days. Remembrances of Bertie Roberts, great-aunt of Mathew Roberts, will expose the truth regarding these men, their relatives, their travels to Arizona, their gold mining efforts and the horrors of trying to stay alive during frightening Indian battles.

We are also privy to the diary of Matt Cavaness, who built what would later be known as the Cow House and who was undoubtedly one of the great Indian fighters of all times. And, as the saga unfolds, we'll meet others who are an integral part of Arizona's beginning and whose lives and times involve the Dutchman. Try-

Historian Sarah "Bertie" Roberts (1877-1969)

Rebecca (Cavaness) Peeples (1841-1939)

ing to include all the pioneer players without confusing the reader will be a monumental task. A sincere effort will be made to be concise; however, the stories told by the pioneers in their own words are extremely graphic and unusually important because of their extraordinary courage and determination. To refuse them their own voice would seem unjust.

One matter that probably ought to be cleared up early in this manuscript is the connection between the parties who will speak through their memoirs. Sarah "Bertie" Roberts (1877–1969) was the daughter of George R. Roberts (1828–1884) and Sarah E. (Cavaness) Roberts (1844–1933). She was the niece of Matthew Cavaness (1843–1929) and Rebecca Cavaness (1841–1939). Rebecca Cavaness later became the wife of Abraham H. Peeples (1826–1892). The reader should be aware of the times and the fact that early pioneers were a small, close-knit group. Eventually everyone was connected either by marriage or by place of settlement or by business. That fact will become more evident as the story unfolds.

Rebecca (Cavaness) Peeples and Sarah (Cavaness) Roberts were not only sisters but were also close friends. They conspired to make certain the history was recorded, because a fire had destroyed A.H. Peeples's home in Peeples Valley, near Wickenburg, and the original manuscript of his travels. Fortunately for all concerned and future generations, the sisters convinced Mr. Peeples to allow them to interview him at length when he was willing and able. They recorded the information in longhand during the years of 1890 and 1891. Interestingly enough, 1891 is the year Jacob Waltz died.

Rebecca Peeples died in 1939, and Sarah in 1933. Bertie Roberts, christened Sarah after her mother, was born in 1877 but did not succumb until 1969. This remarkable woman kept a constant record and photographs of the pioneer group. In fact, she initiated the annual Pioneer Picnic in Phoenix, which allowed her to interview the various people who are in this manuscript. And, since Bertie grew up as a neighbor of Jacob Waltz, her words are extremely important.

It would be unconscionable to get ahead of this story, and it will sometimes be necessary to be redundant. The information that follows in Chapters 2 and 3 is verbatim from manuscripts, diaries and memoirs of Bertie Roberts, A.H. Peeples, Matt Cavaness, the Petrasch family and Mathew Roberts.

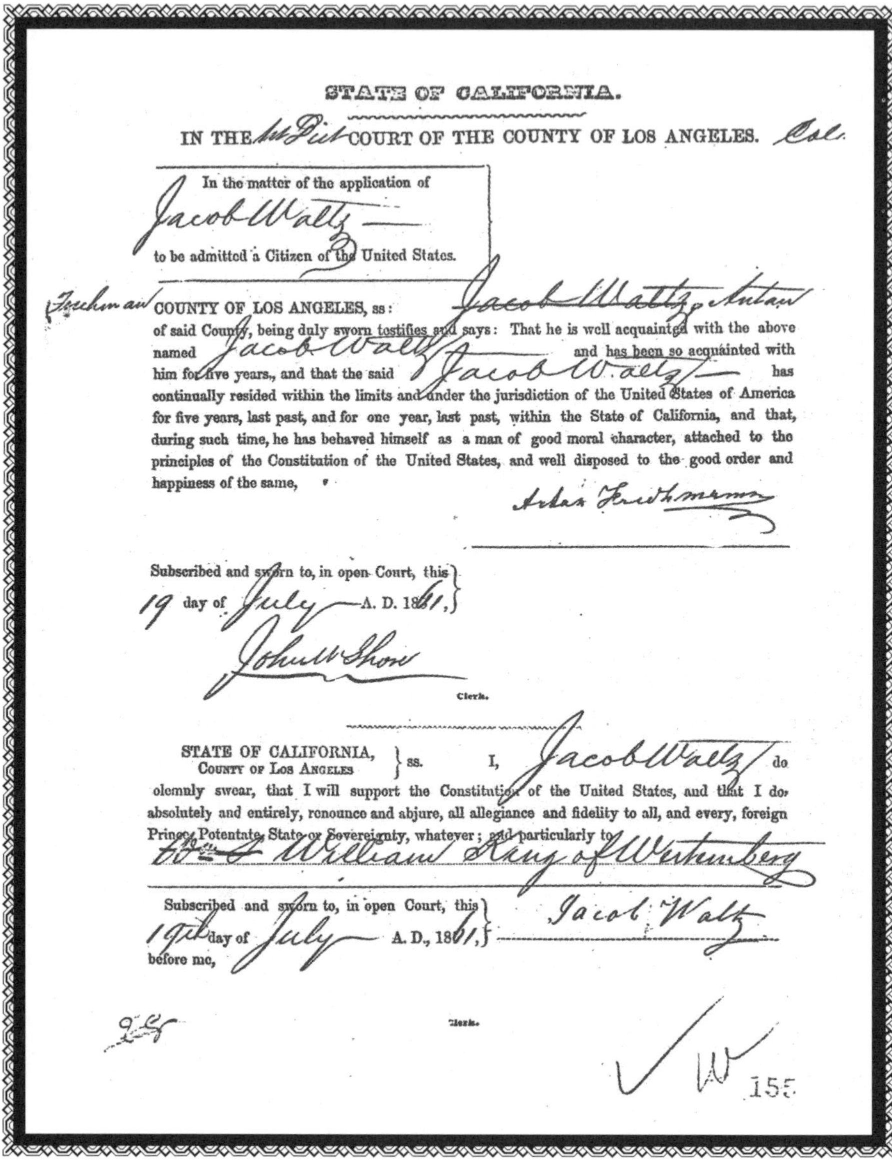

STATE OF CALIFORNIA.

IN THE 1st Dist COURT OF THE COUNTY OF LOS ANGELES. Col.

In the matter of the application of
Jacob Waltz
to be admitted a Citizen of the United States.

COUNTY OF LOS ANGELES, ss: Jacob Waltz, Antan Teuchman of said County, being duly sworn testifies and says: That he is well acquainted with the above named Jacob Waltz and has been so acquainted with him for five years, and that the said Jacob Waltz has continually resided within the limits and under the jurisdiction of the United States of America for five years, last past, and for one year, last past, within the State of California, and that, during such time, he has behaved himself as a man of good moral character, attached to the principles of the Constitution of the United States, and well disposed to the good order and happiness of the same,

Antan Teuchman

Subscribed and sworn to, in open Court, this 19 day of July A. D. 1861,
John W Shore
Clerk.

STATE OF CALIFORNIA, County of Los Angeles } ss. I, Jacob Waltz do olemnly swear, that I will support the Constitution of the United States, and that I do absolutely and entirely, renounce and abjure, all allegiance and fidelity to all, and every, foreign Prince, Potentate, State or Sovereignty, whatever; and particularly to the William King of Wertemberg

Subscribed and sworn to, in open Court, this 19th day of July A. D., 1861, before me,
Jacob Waltz

Clerk.

155

Jacob Waltz's citizenship document.

Chapter 2

The Pioneer Trail
Tales of Their Beginnings

The following are memoirs from personal family records of George Roberts, A.H. Peeples and Matthew Cavaness.

A Brief history of Geo. Roberts and his travels west.

George R. Roberts was born in Virginia in 1828, his grandfather fought in the Revolutionary War and was with Washington at Valley forge. George Roberts mother was Rebecca Peeples the daughter of an Irishman come to America in 1788. George Roberts father was killed by indians near Middlesboro in 1835. At age 15 George and brothers Wm. John andThomas left Virginia for Mississippi where family members settled and owned farm land. At Jones County Mississippi George met his cousiin Abraham H. Peeples and the two became inseperable friends and allies. In 1849 Geo. Roberts and Peeples along with several family members left for Cairo, Illinois where a trade station was being ran along the Mississippi River by an older brother of Peeples. No sooner had the group arrived and word of the California Gold rush reached Cairo. George Roberts, his brothers along with cousins Charles, Cyrus and Return left Illinois with A.H. Peeples for California. The group stopped for supplies and to outfit the trip at a place called Castroville in Medina County Texas. Ft Hondo sat on the edge of civilization before crossing the dangerous and hostile land of west Texas. There at Castroville the group was joined by three other family members, Daniel, Mose and Gideon Roberts from land near todays Beaumont Texas. At Castroville a wagon train of some 60-70 people was assembled for the dangerous journey to California. A good portion of these people were German immigrants. With these immigrants was a man named Jacob

Waltz, he traveled with Roberts and Peeples to California. The wagon train faced untold hardships crossing to California and early in 1850 the group reached Los Angeles and fresh supplies. There the people split up and went their separate ways. Roberts, Peeples and several Germans including Waltz headed for Sacramento and settled about 20 miles N.E. of there at a place called Grass Valley. While there George Roberts was cutting timber for a mine when he noticed a rich deposit of lode gold in a quartz lead. He and the others filed on the claim but groundwater and the steep angle of the vein made it impossible to mine without large machinery. Roberts and the others took what gold they could and sold their claims to the Ophir Mining Company. George Roberts sold his claim for $350.00. Ophir had the same trouble the original miners did and after limited success sold out to a wealthy capatilist Wm. Bourne. Bourne brought in the needed machinery and named the mine the "Empire," it became the richest mine in California history and did not close down until 1969. Roberts and Peeples soon moved on to a place along the Kern River where they discovered wire gold along the river. A town grew up there called Whiskey Flat later to be known as Kernville. Roberts built a hotel and saloon in old Kernville and ran the mine and other placer workings in the area. By 1861 the mines were playing out and the Indians were rampaging all across the Owens Valley. Several German miners showed up in Kernville after a disasterous flood near Los Angeles wiped out their mining camps. They asked George Roberts for work in his mine and Roberts hired them, one of the men was their old friend Jacob Waltz. By early 1863 the Kernville miners had given up and A.H. Peeples gather together a party of men for an expedition into the wild and dangerous Arizona Territory. The expedition was guided by the famous mountain man Pauline Weaver. Peeples outfitted the expedition at Ft. Yuma and secured supplies for 30 men to last 3 months. The group left Olive City, AT on April 1st, 1863. Noted members of that now famous and wildly successful expedition were Henry Wickenberg, Joe Green, Oscar Hutton, John Montgomery and the Dutchman himself Jacob Waltz.

George Riley Roberts
(1828-1884)

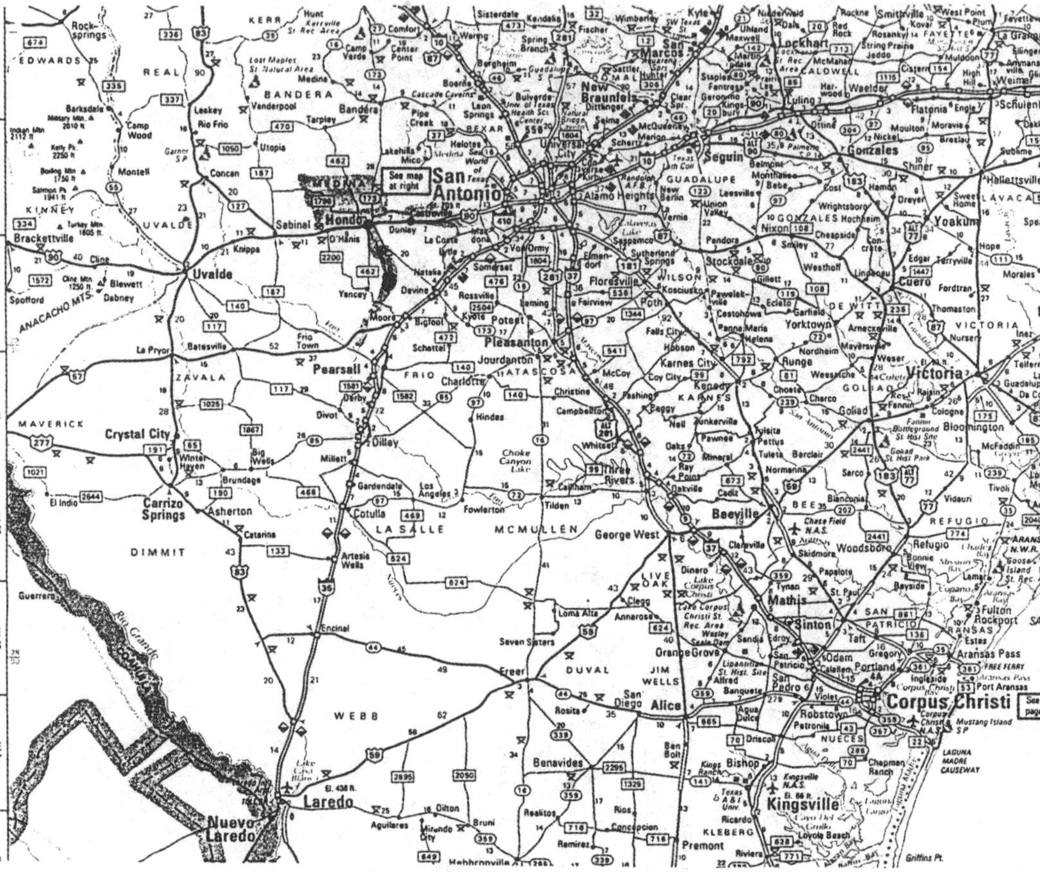

Highlighted: Medina Co. Texas, Castroville, and the city of Hondo which had its beginnings in Fort Hondo which sat along the Honda River in 1848.

Fort Hondo in 1848 was the last outpost of civilization before crossing the barren West Texas desert to El Paso.- 500 miles of bandits, Comanche and Kiowa indians!

Manuscript from Mathew Roberts's family records.

"The first evening found us less than 7 miles from the Plaza at Marysville from whense we began. The next morning we awoke to find one of our very best mules was missing. We spent a good deal of that day looking for him and had to give up the search, Mr. Roberts deciding the mule had been stolen. Sending his load back to Marysville, we again commenced our train towards our destination. . . .

"Our party had been traveling a long distance without water. Mr. Roberts and myself were some distance ahead of the train when near the trailside amid a thick stand of stately pines we saw a beautiful spring flowing freely from under a rocky ledge. Mr. Roberts spurred his mule to a run and it was a race to see which one of us arrived ahead of the other. We drank our fill and waited as the rest of our party came up. Once all the stock and men were well satisfied we rested there in that beautiful spot until the urge to git once again overtook us.

"We began to descend and after an hour or so we arrived at a lookout. At the base of this mountain lay the Grass Valley with the south fork of the Feather River. This beautiful valley lay tucked between the hills, only about a mile wide and six or seven in length. A beautiful green piece of earth surrounded by high red mountains with black outcroppings of slate, almost devoid of trees and brush. A beautiful brook flowing the entire course of the valley falls into a deep canyon at its western extreme. The altitude of this little valley makes it very cold with light frosts common even in the warmest summer weather.

"It was there by that little valley that just a scant year past your father (George Roberts) was cutting timber for the mines down by the American River when he come up on a quartz lead, an outcropping that showed free gold laced through and through. A richer prettier sight I never saw. It was just a setting there waiting to be found. We laid a claim down on that lead and your father (George Roberts) sent me on down to the mining camps along the American River about 30 miles out east of sacramento with the lumber, <u>and I was to fetch back with me Mr. Waltz and Mr. Young as they had come west with us and had endured the same hardships as we had all the way from Fort Hondo</u> (Castroville in Medina County, Texas). <u>Those two gentlemen would claim the sides and one end of the lead as it was difficult to decipher exactly which way the lead dipped</u>." – Mathew Roberts

The PEEPLES / WEAVER *expedition*

Members of the A.H. Peeples expedition to explore the mountains and rivers of central Arizona Territory in APRIL of 1863.
From the diary and journal of A.H.Peeples.

No.	Name	Role	Origin
1.	A.H. (Abraham) Peeples	(Captain)	Mississippi
2.	Pauline Weaver	(Guide)	Tennessee
3.	Joseph M. Green		Tennessee
4.	Henry Wickenberg		Germany
5.	Lennard Pappan		Virginia
6.	Arthur Henry		England
7.	Ben (Mclenden) Munroe		Georgia
8.	Oscar Hutton		Virginia
9.	Mathew Webber		Mississippi
10.	Samuel Groupa		Germany
11.	George R. Roberts		Virginia
12.	Lorenzino Para		New York
13.	James Ansley (Coho) Young		Tennessee
14.	Chas. White		New York
15.	Charles Hahn		Germany
16.	David Porter		Louisiana
17.	George Eiskler		Germany
18.	J.H. Frank		Germany
19.	Christian Kertz		Germany
20.	Jacob Waltz		Germany
21.	Valentine Giegerich		Germany
22.	Gerhart (George) Fifer		Germany
23.	Talbot Freeman		South Carolina
24.	Charles Straid		Germany
25.	Joe Fugett		Missouri
26.	Edward Mason		Kentucky
27.	Gideon Roberts		Texas
28.	Wilson Jones		Virginia
29.	J. (Jim) Miller		Kentucky
30.	Samuel Houten		Germany
31.	Wm. Roberts		Texas
32.	John Montgomery		Illinois

Excerpts from the Memoirs of A.H.Peeples

"Messrs. Paulen Weaver, Joe Green, Coho Young and I secured the needed provisions at Fort Yuma and packed them up along the Colorado River to Olive City where we were met by George Roberts and the other men from California. Mr. Weaver decided the best way to commence was to split the group and travel about a day apart. A smaller party would go ahead searching for some good diggings and marking the trail.
The remainder of the men would follow, packing up the supplies. Our two groups kept in communication by signal fires, a method used by the Indians and one Mr. Weaver understood and used to his advantage. The forward group was led by Mr. Weaver and Len Pappan. Messrs. Joe Green, Matt Webber, Arthur Henry, Ben McLenden, Henry Wickenberg, Oscar Hutton, Samuel Groupa, George Roberts, Lorenzino Para and myself kept a sharp eye out for diggings and Indian signs.
Packing up the supplies and muleteering were Messrs. Chas. White, Coho Young, David Porter, Chas. Hahn, Geo. Eiskler, Gerhart Fifer, Jacob Waltz, Chas. Straid, J.H.Frank, Talbot Freeman, Joe Fugett, Edward Mason, Gideon Roberts, Jim Miller, Wilson Jones, Wm. Roberts, Samuel Houten and John Montgomery.
Mr. Weaver himself had never been to our destination at the San Francisco Mountains but he had passed near to them in the 40's while with a trapping party -------- ."

" The early name for Nevada City, California was the Coyote Diggings. George Roberts had discovered a quartz lode there at Grass Valley in the fall of 50'. It was just he and I and Matt Weber back then, cutting timbers when we made the strike. In a couple months the word had spread to the mining camps down along the American River and the miners came pouring in. The placer mines around the quartz lodes were very rich in coarse gold and money came flooding in as the town sprang up overnight, in six months there was two thousand men in congregation along the banks and in the canons. In the beginning everyone was finding gold, twenty to two hundred dollars a day was common. Supplies were high, a pinch of gold dust was worth a dollar and no measure of supply cost less than a dollar then. Often miners would pay for goods pinching out measures of gold dust on the merchants scale while spilling excess onto and around the scales. No attempt was made by the miner to clean up and recover his spillt riches, haughtily stating, "there's more where that came from." Everybody had money and everybody spent it. The gulches were alive with men, some swinging pick and shovel while others stood out knee deep in the cold icy water washing the sand for the gold. A digging was understood to be 40 feet by 40 feet and each miner respected his neighbors distance, disputes were rare in the beginning but as more men poured in fights and bloodshed erupted. Often miners would dig out the easy spoils and move on, sometimes digging at numerous places in the course of a single week. Newly arrived miners would take up the abandoned ground as fast as the previous occupants would give it up. Coho Young and Jacob Waltz mined out the ground next to George Roberts and myself. At about one hundred feet along the side of the quartz lead we hit water

Gideon Roberts – Arizona pioneer, miner and member of the Weaver/Peeples expedition of 1863. He was with Jacob Waltz at his death.

and it flooded our holes, abruptly ending our rich luck. George Roberts took out about fifteen thousand dollars and I about the same before we abandoned the dig. Every hole dug in that valley met with a similar fate and in a few months all the miners sold out to a mining group that proposed to bring in a steam pump up the river through Sacramento -------."

" Our exploring party consisted of Messrs. Webber, Hutton, Green, Waltz, Hahn, Porter, George Roberts and four others. We came over the pass at San Miguel and crossed the lake slough near the Tulare Lake, and then passed on the north side of King's River to the foot of the Sierra Nevada. Here we were met by a large number of Indians from the mountains who showed us large quantities of gold. They refused to trade with us unless we came with them to their campments. They appeared every mark of perfect friendship for us and we travelled two days into the mountains with them. We had travelled about ten miles on the third day and the number of Indians around us ever increased with each hour and now numbered over one hundred. Some of them had gold which showed quite coarse and to our enquires of where they obtained it, they pointed to the east. Oscar Hutton became warry when there was a stir amongst them and noticed the squaws and children suddenly were missing.
Placeing everyone on guard with rifles at the ready and pistols in our belts we prepared to defend ourselves against an attack that we knew would commence shortly. The Indians had no guns and were cowed by our formidable situation. Their hope was to catch us off guard and fall upon us before we had time to make a show. Nearing a narrow passage between a stream and a bluff Mr. Hutton halted us saying all their signs showed this was the place they chose to do their work on us. Messrs. Webber and Roberts turned to guard our trail while Joe Green and I covered the Indians between us and the creek. Messrs. Waltz and Hahn closed on our left flank while Hutton and the others confronted the chief who spoke in a loud voice. Mr. Hutton deciphered the old chief was calling out orders to his warriors and in an instant Mr. Hutton raised his rifle and shot the chief dead. A shower of arrows filled the air all around us and rained down on us from the bluffs above. The qiuckness of the chiefs death startled and confused the Indians and in the confusion we fired a couple volleys into them, killing many. They fell back but regrouped quickly and charged again. There being little cover every man was down on his knee and ready to sell his life as dearly as possible. "Use your pistols boys", Mr. Hutton shouted, and the charge was met on all sides with a steady fire that did great dammage in their ranks. Seeing us so well prepared for a fight the Indians fell back to the bluff and continued to shoot arrows down on us with little effect. Our only casualty was Charles Hahn who had an arrow sticking out of his right leg. Mr. Waltz had helped him limp to a large group of Sycamore's that showed to provide us some good protection as it was beginning to get dark. With the creek behind us and the Indians above us on the bluff we spent a sleepless night watching their fires and listening to their whoops and shouts. We figured they would try and rush us again at dawn but come morning they had pulled back off the bluff and could not be seen although we felt their presence close by.

Messers. Roberts and Green attended to Mr. Hahn's leg in the night, the arrow having gone almost clean through had to be broken off and pushed through the rest of the way and drawn out. A crutch was cut for him and by morning Mr. Hahn was raised and hobbling along with the help of a man beside him.

At daylight we struck on west for King's river where a mining camp then known as Glade's diggings would afford us some comfort and protection. The Indians were beginning to gather around us again in large numbers but having lost their chief they had no more taste for a fight with our rifles and kept themselves back a tolerable distance. Mr. Hutton said it would do us no good to continue easterly into the Nevada's as these Indians were the Chowchillas, the most treacherous and blood-thirsty tribe of the Tulares. --------- "

The Memoirs of Matt Cavaness

1843 - 1929

Arizona Territorial Pioneer

Typed as written from the longhand manuscript of Matthew Cavaness Jr. Arizona Territorial pioneer.

By: Joseph Miller 1950

The Matt Cavaness Memoirs contain references and information concerning the following Arizona Territorial pioneers :

George R. Roberts
Abraham A.H. Peeples
Joseph M. Green
James Ansley (Coho) Young
Henry Wickenberg
King Woolsey
Abe Sutton
Aaron Mason
Wm. Oury
Elisha Reavis
Ed Peck
George Walker
John Y.T. Smith
James M. Barney
Frank Cosgrove
Chas. Mason
Benjamin W. Regan
George E. Mowry

Matthew Cavaness, Arizona Pioneer (1843-1929)

Matt Cavaness's freight team at the Gila Bend Stage Station in the 1880s.

Memoirs of Matt Cavaness 12

the Overland had built half way across the Plains had water in it. We was the first outfit to cross the Plains since the Overland had pulled off. So we got up to the head of the Concho and laid over one or two days and then started across. We knew if we didn't find water half way we would have a hard time to get across. Our oxen was all fat and would make ten to fifteen miles a day.

I think it was the third night about midnight we got to the reservoir and as luck would have it it was full of water. When the oxen got close to it they smelled it and they just broke and run and over the edge of the reservoir and into it went wagons and all. It was just deep enough to make them swim around in the water all the teams got in and everything in the wagons was wet. We didn't mind that. So we got water, made camp and stopped there three days. Had good rest and our oxen got rested up. Went out and kill us some fine deer. We wasn't uneasy there about Indians as we had nothing they wanted. After we got rested went on down to the Pecos river at the horse head crossing we stopped there one or two days and caught fish, then we went on to what they call Leon holes. There we found several families camp. The first thing they done come out and meet us, says if any of your wagon tires is loose, don't you put them down there in that hole of water. We taken ours down there and put them in and when we went to get them we couldn't find them. We can't find any bottom to it and two of our men has gone back to get some wagon wheels so we can travel. And we stop there until those men come back and they told us all the news. They said the South had seceded from the United States and had declared war and had elected Jeff Davis president and was preparing for war and in a short time they would be drafting men for the war and they said the road was lined with people coming this way.

Now we knew we better have lots of help to go on from where we was to the Rio Grande river was right in the Indian country and at Fort Davis they had moved all the soldiers off and there was no one on the road until we reach the Rio Grande river. So we stopped right there until all the imigrants came up. They came from everyplace, Arkansas, Missouri and eastern Texas and when they all came in there was two hundred and fifty wagons and about three hundred families. They had all kinds of stock. Some of them got off in such a hurry they brought their milk cows and all the stock they had. Some were working oxen, some working horses and mules, anything to get away. So they didn't want any war. When we all got together we had to come to some understanding how to travel. We knew that for us all to do and travel as we please that wouldn't do. We all call meeting and everyone would talk and say what he thought was best way to get along.

So father put up a motion to the people and it was seconded that they would elect a captain to run the outfit. So we had election and elected a man by the name of Joe Curtis for captain. He got up and says to the whole camp I will accept it on this condition only. I must have full charge in every way and if any man does not obey my orders he is to cut out of the train and have to go his way. So that was secon all over. So Mr. Curtis arms his men and put so many in each outfit and he had so many for night guards and so many for day guards. All men under forty-five and over eighteen he put them on night guard and the others on day guard.

Memoirs of Matt Cavaness 13

He had fifty men for day guards, all horseback. There was over one thousand head of cattle and horses loose in the day herd. So when he got everything arranged as he wanted it, says tomorrow we make a start. So everybody was willing to go as we was uneasy about the war. So we all got ready and the time came. He rode up to our outfit and says to me to pull out and all of fathers outfit pull out and he sat there and gave orders to each outfit to follow in behind. Told everybody to keep their places as he directed them and when we all got strung out the train was a mile long, everyone to keep his place. I was wondering how he could camp us all that way, but it was easy after one seen how he did it. His outfit was the last one to follow, so we travel that day. He had the day herd behind so when it came camping time he rode up to me and says Matt you follow me. He rode in circle around and when the last wagon left the road he came right there and he had every man to drive his team right behind the team ahead. He says to everybody turn all your teams loose. Had all hands to drive up close to the wagons ahead, close enough to take the tongue of your wagons and tie it to the hind wheel of the wagon ahead. So they did that and when that was done we had a complete corral made for all the stock. His outfit and father's was at the gate to drive the stock in and everything was driven in that corral at night and the night guard was put outside of this corral. So there was no show for the Indians to steal our stock. At day light he was up and had the day guard up and everything was turned out to graze for two or three hours, then everybody had plenty of time to get breakfast and get ready to travel.

We travel slow, five to ten miles a day only. We went on to the Rio Grande river. Got there about noon one day and stopped for noon and while we was there in came General Sibley with a whole regiment of Southern soldiers and he stopped for noon. Our captain went up to his camp and had a talk with him. He wanted to know where we were going. He says to our captain, you come on up to El Paso and we'll have a powwow. So that settled that. We didn't want any powwow. So the captain came back to camp and told us what the general said. Everybody was wild over it. Didn't know what to do. Father got up and said, gentlemen and ladies, I make a motion that we have our captain to take fifty men with axes; go down to that river and cut a road through the vines and dig down the banks so we can cross the river and get in to Mexico and those soldiers will have no show at us to have a powwow with us. Everybody yell out to go at once. The captain took his men and started and by the time we hook up and started he had the road ready for and down we went into the river. We went across; we went into Mexico. Camp at top of the bank just below a little town called Agnashho. The captain went up and saw the mayor of the city and told him we wanted to go through to California. The mayor said we couldn't go through the old imigrant road as it hadn't been traveled for years. That was the old imigrant road in 1849. So the mayor said he furnish us with a guide for one hundred and fifty dollars in advance. So we paid it and started on our road for California. I think that was on July 1, 1861.

We came to a little Mexican city in about twenty or thirty miles where we could buy provisions and everybody needed something and the mayor of the city put a tax on our stock and every wagon had to pay so much to pass through. So we paid that tax and went on and the next city it was the same way. So we paid, but the next city the mayor call on us, father say that have us all broke if we keep paying every time we come to a city. All hands was work up over paying such a tax, I for one. Father says to have our captain to go up and see the mayor and

Memoirs of Matt Cavaness 14

tell him that we wont pay any more taxes. They all say go and tell the mayor. We wont pay it, that we was buying at every city what we needed. The mayor said you can't go through unless you pay. The captain told him that we wouldn't pay and we would go on in the morning. The old mayor says I'll stop you all right. Here we was camped outside the city about half a mile. So next morning we all got ready to travel and our captain called for seventy-five men, all armed, horseback and rode right ahead of the lead wagon. We got in site of the city. We saw that they had about one hundred Mexican soldiers at the street where the road entered the city and they had the street blockaded with big logs. The captain rode up and stopped. The Mexican soldiers were setting there. They say you can't go through. Captain says, boys, get down and pull them logs out of the way so the teams can go through.

Every man dismounted and you ought to see them logs tore down. We detailed ten men to ride ahead of me as I was driving the lead wagon. He stayed there with the rest of his men until the last wagon passed through. Then came the loose stock and he came through with that, so that ended us paying taxes. Every city we came to they came out and bring us all kinds of things that they had to sell and was real nice every place we came to. In a short time we got out in the Indian country. I don't know what tribe it was but they was sure bad. They had lots of horses and good ones too, and we couldn't travel fast as the road was hard to follow. But there was plenty of grass and water and all our stock was in good fix. So all we had to do was to keep what we had and there was such a big bunch of us that the Indians couldn't do much unless they could get us in a place where we couldn't keep together. But our captain sure was a dandy. He just moved us along slow. Had guards out all the time head and behind and on both sides of the road and when he camp us we had a plain view of the country.

On the road one day there was a child died and we camp to bury that child. Camp on a high mesa right on the edge of a big dry creek that came down out of the mountains. Had all the stock down in the flat and the guards was on guard, horseback. None of the stock was over two hundred yards away. I was helping dig the grave for the child about fifty steps from the camp. I left my rifle at the camp. We was working away on the grave and I looked up this dry wash about half a mile and I saw the Indians coming horseback, about three hundred. They was riding as fast as they could run. I went to the camp and got my rifle and got back to the grave just as the lead Indians got right in front of us. I dropped down on my knee and raised my rifle. The Indian set his horse up and turn him around. As he did that I cut loose and at the crack of the gun, down went Mr. Indian. But he was tied on his horse. Two other Indians ran to him, one on each side and run his horse off over the hill. He lost his bow and arrows and lots of beads and buckskins and moccasins. I ran back to camp and got on father's horse and took after those Indians that was driving this Indian's horse off that I had shot. I ran him about three miles. We was shooting at each other all the time and all at once they ran down into a steep canyon and there was lots of Indians in sight. And up came a man that had been after me trying to catch up so he taken a look at the Indians. Say, Matt, the best thing for us to do is to get back to camp. Just look at them. We can't catch that horse. They got more Indians and let them go.

So we started back. Hadn't gone far and we heard shooting at camp. The Indians had surrounded the camp and all the shooting you ever heard

Memoirs of Matt Cavaness 15

it was going on. We got close. No I says, we got to go right through those Indians to get to camp. He says, if you go ahead, I'll follow. I'm going to camp. If they take that train and take all the women and children prisoners they do that over my dead body. Now I says, leave your rifle in the sling. Take your pistol in hand and we go through shooting. So we did. We got in about one hundred and fifty yards and the Indians saw us coming. Here they come. Here we went meeting them. We got in about thirty steps of the Indians and I cut loose with my pistol and I shot so fast the Indians didn't know what to do. They was running in every direction and I went to camp. This man was right behind me. He says, Matt, if I could use a pistol like you did I wouldn't want a rifle. He says, I just shot to frieghten them off so I could come to camp. You might call that frightening them if you didn't kill three or four Indians.

Well, we got inside the corral and I took my rifle and went to work shooting at the Indians. There was Indians everywhere. During the time we was off the guards had got all the stock inside the corral and those Indians was doing everything they could to get the stock out side of the corral, but they couldn't do that. They stayed around shooting at us with bows and arrows. They didn't have any guns. So we had the best of them. They left about dark and we didn't see them until the next day. We went across a big and beautiful valley and camped on it. Next morning as we was ready to start there was about twenty little boys and girls that had ponies that ride along head of the train and behind and all around every where, but the captain of the train wouldn't let them get more than three hundred yards from the train as we had seen so many Indians the day before he knew we would have to be very careful if we didn't lose our stock and maybe some of the children. So this little bunch of children started out ahead. They had gotten about three or four hundred yards ahead of the train and the Indians was close to the road. They came in the early morning and got in a little low place in the high grass and there they was when the little kids came along. They made a run to cut them off from the trains. The children seen the Indians and here came the kids and all the yelling you ever heard it came off right there. Now if you don't think them little fellows didn't ride and sure run. The captain always had fifty men on his guard, horseback, ready at moment's notice anytime. He was right there ready to move the train. He and his men ran out and saved the children from being captured by the Indians.

The Indians retreated back a short distance, the captain after them. About a mile and he came in sight where they were camped. He stopped a minute and took a look. He say to his men, just look at the Indians. We better get to camp. If they ever get us cut off from the train we have no show at all to save ourselves. So he came back to the train and told everybody that we was at the mercy of the Indians. If they ever get us separated from the train they take every woman and child prisoner. They want children to raise and stock.

When those kids was brung back to camp and the Indians trying to cut them off, the old women like to went crazy. They jusy yell and jump up and down. Now, after that it was no trouble to keep those kids in camp. When the captain told everybody how things looked, he says he saw the whole valley not two miles from here alive with Indians, just as far as he could see, nothing but Indians. The captain says all I want is to have everybody satisfied and take things as they come and obey orders. I'll take this train through all right and not lose any stock either, and no children would be taken off. Now if you don't think we had a hard time and again we had a good time. We didn't make over five miles a day. It was guarding day and night, two hours on; two hours off and it lasted that way for two months. The Indians just travel along close to us. They camp sometimes not more than a mile from us. Have their big fires and dance. We could hear them beating on their drums and could see their fires. So we got us some music and we commence having our dances right out on the ground. All the young people get after supper and clean off a place and when dark came we get out there and dance and yell as loud as those Indians. We had lots of fiddles.

We traveled through a beautiful country for several hundred miles, the finest grass country you ever saw. This Mexican we got at the Rio Grande river that came along as guide, he knew the country all right. It would be days that we never saw the road. We go through pass in the mountains then we see the road in places. One place we came to where we went through a pass and went down quite a box canyon. Well, it looked like we couldn't get through there at all. The Indians was everywhere, in the canyon and on both sides of it. They didn't seem to care for us seeing them. They just managed to keep just far away so we could not shoot them, and they never had any guns, only bows and arrows.

Now to get down through this canyon was a job so we would just move a few hundred yards a day. Now and then we had a chance to take a shot at the Indians and they retreat back and come again. We move down a little farther on in that way we finally got through that box canyon and come out on a large open country and went on and camp right out on a nice level country. This guide told us the country ahead would be all right, so next morning when we got up our guide was gone.

The old man Saunders had two fine mares that he was taking to California to raise him some fine colts. The first thing Mr. Saunders looked for his mares and they was gone. He took his son and a man named Keley went with him. They had taken a trail of the guide. He had taken those two mares and his horse a nd struck right out for a range of mountains that lay north of us about twenty miles away. So they got on three good horses and taking his trail, followed it in to the mountains. He went in to it and stopped at a tank and got water and then he went down a canyon. They followed him down the canyon for miles and overtook him traveling along, leading those two mares the road up behind him, and their guns went off and the guide jumped off. As he did he lost his boots. They brought the mares and the guide horse back and his boots. All they would say that the Mexican got so bad scared that he lost his boots. They never did say they killed him but we guess at the balance. That night they came in with the two mares and Mexican horse and boots.

So we traveled on and came to the head of the Santa Cruz river that ran down by Tucson in Arizona. There was a Mexican town there about two thousand people. Nice settlement there. We stopped there two or three days a nd down came Lieutenant Mowry from the Patagonia mines. He wanted to hire some men and teams to go with him up to the mine and work. So father and uncle Isaac Person and uncle Nathan Huckaby and J. R. Boner went up to the mines to work for Lieut. Mowry. So we all went up there to work. So we work there a short time and Mr. Mowry wanted uncle Nathan to take his hat off when he spoke to Mr. Mowry. So that didn't set good with us and we settle up with Mr. Mowry. He paid us all he owed us and we pull out.

Went on over to the Sonoita river and there was nice farms on it. All had nice crops in all kinds of vegetables. They had moved all the soldiers out of the country and all the farmers had to leave their home and get out of the country as there was nothing to stay in that part of the country in Arizona. Went on down to old Fort Back where they used to keep soldiers at. It was a lonely looking place. Everybody gone, citizens and all.

So we went on to Tucson and when we got there we found one white man in the city, balance all Mexicans, and not very many of them. We pull right into plaza and camp there that night. The Indians came right into town and taken out one hundred head of horses. Next morning the Mexicans got about fifty in number and followed the Indians. We lay over that day to see how they came out. About midnight they came back packing a man on a horse. They had overtaken the Indians in the Santa Catalina mountains about thirty miles from Tucson, the Indians killed the man that lost the horses. So we was on the lookout again. The white man in Tucson told us as soon as we got to the Pimo reservation that we would be all right as the Pimo Indians never was at war against the whites. And in fact they was at war with the Apache Indians. And any time the Apache Indians would slip in on the reservation and steal their stock and clean down the Gila river close to Yuma. So the Indians told us to look out. The Indians no good. He steal horses and cows.

We went on down the Gila river and just this side of Oatman Flat is where the Oatman family was killed. There had been a party there after some Indians. They killed one or captured him, we don't know which. But anyway they had hung him up to a tree right over the road. His feet was over two feet above the ground. We left him hanging there and went on our road. We saw where the Oatman family was killed and the Indians taken off one or two girls and one of them raised a family by Indian chief. So I read in history. That was right about them.

We went on down to Yuma. We found out that part of our train was just ahead of us as we left the big train at Santa Cruz where went up to the Patagonia mine to work for Lieutenant Mowry.

So we cross the Colorado river and went down about thirty miles and found lots of families camping there resting up before they start across that desert. Go ninety miles without water. Some had pull out. Some time in the night two men rode into camp. Just passed by. Never let themselves be known. We didn't think anything of it. But about sunrise, in came two more men. They ask if they could get a cup of coffee. Of course they did and one says, by the way, did you see two men or hear them pass during the night. Father said yes I seen when they passed by. They went down the river. They sure got the coffee and away they went on the trail of those two men.

That evening they came back leading two big mules. They said those two men were Mexicans and had stole the two mules from them out on the desert. We asked them if they arrested the Mexicans. They said the Mexicans was all right. They didn't think they would steal any more mules. They had the Mexicans horses and saddles. So we all knew that they had killed the two Mexicans. That was nothing to do out of the way. So they went on and overtook their outfit.

There was quite a lot of imigrant camps there. We commenced pulling out to cross this ninety miles desert. We pull out and get across to water, a place called Carrizo creek. There we found quite a lot of imigrant camps. One man camped there by the name of Buck and he had a brother-in-law named Roland. They married sisters and was dead immages and has caused the whole outfit a lot of trouble to keep those two men apart. And I went up to the station where the store was and there was

Memoirs of Matt Cavaness 18

Mr. Buck. Him and I was standing inside by the counter and in came Mr. Roland. Mr. Roland step in the door with a double-barreled shotgun in his hands, raised his gun and and Mr. Buck drew his pistol and they both fired at the same time. I stepped to one side and they both fell in side, one on top of the other. We picked them up and laid them outside beside each other. I ran down to the train and told the two sisters what had happened. They both came up and their father and mot[illegible] Their father said to the girls, say girls, those two men has caused us lots of trouble and now they are gone. We have no way to give them a coffin. We got to bury them the best we can. The old man says we dig a grave large enough to put them both in to it, side by side and we did it. The girls took some quilts and put them around each one and we laid them in the grave and covered them up. Both of the girls and the whole family said they was glad it turned out that way. They rather see them both go than to one go. We all pull out for Los Angeles, Cal.

We had a nice trip all the way in. Struck a game country and out of the Indian country. We were relieved when we got close to Los Angeles. The plains was covered with food for our oxen and wild geese by the thousand. So we had all kinds of fresh meat. When we got to Los Angeles we expected to see a city, but we found it to be a few old adobe houses with dirt roofs, hardly one house in a dozen had shingles roofs on them.

We stopped about three weeks and traded our oxen for work horses. We went on up to Tulare county this side of Visalia in the mining camp. All the rest of kin folks went on up to King river where mother had two brothers living that had come to California in 1849 and had made quite a raise in the early day of California.

We stopped there in December , 1861 and in the spring of 1862 we went up to King river and stayed there that year, then we came back to the mines on White river and went to work in the mines for a man named George R. Roberts. He married my sister Sarah. I bought six yoke of oxen and wagon and went to hauling ore for the quartz mills. Mr. George R. Roberts had a partner named A. H. Peeples. They would hear some talk once in a whiles about Arizona. So Mr. Peeples made the proposal that he come to Arizona. So he and Joe Green and Coho Young, they pull out for Arizona.

In the meantime we left White river and moved over Greenhorn mountains to Kern river at a mining camp. Mr. Roberts built himself a hotel. I haul lumber off Greenhorn mountains for the mines and city and that year the Owens river Indians broke out east of us about eighty miles. That was a mining country then the Indians was killing everybody they could, so all the people was worked up over it. Most of the supplies for that section came in from Los Angeles with oxen teams. There was five teams on the road to our mining camp, called Kernville. The teams got in about twenty miles of Kernville and the Owen river Indians had come down close to that part of the country and they saw them teams and waylaid them and killed four of them. One man got away and came in and reported the killing. Of course we rustled all the men we could, cowboys and miners. We got eighty men together and away we went after the Indians. Didn't take any grub as we thought we catch them in very short time as the Indians didn't have but few horses. The start the Indians had broke all the good boxes open and took all of the drygoods they could pack and from the time they made we estimated we made. Was three hundred in big and little. We taken the trail and out in the rough mountains they went, us after them. We followed them the first day in the most rugged mountains as one ever saw, but when night

Memoirs of Matt Cavaness 19

came of course, we stopped as we couldn't trail them after dark. When I left the house mother ran out to me and says, Matt, here is some dried meat and a loaf of bread. Tie it on your saddle. It might come in good play as you are going after Indians you don't know where you be and when you get back. Well, I didn't show up my loaf of bread and meat that night but next morning I did. I got it out and say, boys, here's what I got. We eat it and followed them Indians for three days but never did find them. They knew the country better than we did and kept in the rough country where it was impossible making any headway trailing them.

We went in to the Agua Caliente rest station. There was a man and his family there. He was the Indian agent there. The captain got down and went in and told the agent that we was out trying to catch those Indians that killed and robbed that train and we didn't have anything to eat for three days. The agent replied and said, he wouldn't give us anything to eat. So all the men yell out, we hang him to a tree and started out with a rope to the to the horse. The agent yell out to the captain to stop his men and tell them to come in and they try and feed us something. His wife and daughter went to cooking and the agent had some of his Indians to get a beef and kill it for us. So they did and gave us horse feed too. We had plenty to eat for one night and day. When we left he gave us meat and bread to do us home. He treated us very nice after he seen that we wouldn't take no for an answer. We didn't get the Indians.

We buried the men right where they was killed. We gave up the chase and went home. In a few days in came a man from Visalia. He had eighty head of pack mules all loaded for Owens river merchants and the mines, and when he got there all his men quit him and told him they wouldn't go out to Owens river for all his outfit. So there he was in bad fix. Couldn't move his outfit. He tried everybody to get some help but everybody was worked up over the Indians outbreak. I got sorry for this man and I said, if you give me five dollars a day I go with you to Owens river so you can deliver your freight to Owens river. He was glad to do it. So next morning we loaded up the eighty head of mules and pulled out for Owens river. The first day we made about fifteen miles and camped on south fork of Kern river. Next morning packed up and pulled out again. Didn't go but a short way until we miss one mule. The boss went back to get that mule and I went on with the outfit, seventy-nine head of mules all alone. Traveled all day. Didn't know what moment the Indians would tackle me and take the outfit. That evening I got to water and camp. No boss came, just I camp my outfit right on top of a hill where I had good view all around, so if the Indians did come I had a good chance to keep them off. About midnight I heard someone yelling out at the top of his voice. I got up and answered him. Who was it but the boss. He had found his mule and come on.

Next morning he got up and looked all around and says to me, say kid, down there in the flat would have been a nice place to unpack and make good camp. That's so boss, but if the Indians had come they would have been all around the camp and you have no show to kill one. At this place they can't come anywhere close but you have a good shot at them. The boss say, kid you are all right. After this you got to pick out our camp. So after that he never had anything to say about camping.

We went on to Owens river and delivered our freight. The whole country was all worked up over the Indians outbreak and they was doing lots of killing and robbing and it looked as though we couldn't get

Memoirs of Matt Cavaness 20

back home. No one would dare to leave camp. Well, the boss was worked up over it. He got uneasy. Was afraid he would lose his mules and that was all he had. He said, I lose those mules I be ruined. That's all I got, every dollar is in this outfit. Well I say, if you be on the lookout I think we can make it all right. My God, kid, I never had any experience with the Indians and I don't know what to do. Well I says to him the best way for us to do is to change our way of traveling. He says, I don't know how to do that. Now kid, if you know I be awfully glad if you will tell me how you think is the best way to get out of here. I says, that's easy. He says, right now kid, you got this outfit on your hands. I turn it over to you right here. I ain't going to have a word to say. So I knew that the Indians wouldn't or didn't travel of a night. We was camped right in town, so just after dark, well, I says to the boss, we better pack up and get out of here and make the next water about fifteen miles. He never said one word, just went to pack up the mules. All we had to do was to put aparajós (pack saddles) on the mules and we was ready to travel. I says to the boss, I go ahead with the bell horse and the mules would follow. You come on behind and see that no mules don't stop on the road. We made camp at the water all right. We got there about three in the morning. I says to the boss, now you lay down and take a little sleep and I look out for the mules. So he went to bed and went to sleep. When daylight came I seen the Indians was camped off about a mile. They had come in there and was expecting us to come in there in the evening, then they take us in. But as we came in the night, that fool them. So they stayed in sight of us all day. We walked around our mules with our rifles in one hand so they could see us. I knew that wouldn t tackle us as long as we keep on the lookout, and Indians won't make a fight unless he gets the drop, and gets the first shot.

So we stayed in there all day and the Indians stayed where they could see us and I knew just what they was planning on, to take us in next morning at day light, but we fool them. After night, probably about ten or eleven o'clock we pack up and pull out, taking the bell off the old mare that we used for the mules to follow. Next morning we was twenty miles on home road. So we done that every night for several nights and the Indians finally gave us up and we got in to home all right, and I stop at home and the boss had no trouble to get men to go to Visalia with him, but he wouldn't take any more freight for Owens river.

The people out there had all kinds of trouble to get their supplies in. The Indians came in from Owen river and make their headquarters upon the head of south fork of Owens river. Stayed there quite a while and done lots of devilment. Finally the citizens thought it was a good idea to go up there and clean them out . The cowmen and miners and farmers all got together and talk it over. This same captain that was out on that other raid said if he had one hundred and fifty men he go up there on the head of the south fork and clean these Indians out. Lots of cowmen knew the country and knew about where the Indians was at. We all turns out and raise one hundred and fifty men.

This captain had been a general in the English army. He thought he could lick all the Indians in the country. There was one caw man that knew all that country. He told all of us that if the Indians was in a certain box canyon that therewas only one way to get in it, and was an Indian trail. Where they would go up and down in the canyon it was about forty miles to the head of the river. So we all got ready. Took lots of grub and ammunition and we go to clean those Indians out. We went up the river one day and camped. Next morning this cowman said to the captain, have all your men now to clean up their guns and be ready for the clean up. So we pull out up the mountains. It was awfully rough and the trail was very narrow and we had to ride single file one behind the other. This cowman led the way until he got to the trail that went down in the canyons. The cowman says to the captain, now you go ahead about fifty yards and you can see down in the canyons and if those Indians is here you can see the camp. So the captain says I want to see the Indians. He made a start and went about one hundred yards and came in plain view of the Indian's camp. You bet he stopped and call all hands to get down and come forward and take a look at the camp. So we did and the whole canyon was full of Indians. They saw us up on the rim of the canyon. The Indians made all kinds of signs for us to come on down to their camp. We view it out as well as we could but we couldn't see any possible way to get there, only go right down this trail. So we stopped up there that evening and night and done everything we could to find a way to get to their camp, but we couldn't find any place close enough to shoot into their camp. We only had short range guns those days and we could see that the Indians had the best of us. We couldn't go down that trail as the Indians would have us out in plain view to shoot us and we wouldn't have any show to kill an Indian until we got down into the canyons.

We stayed around on top a few days trying to find a place to get close enough to shoot in their camp but we couldn't find it. If we had such guns as we have now we could have killed the whole tribe of Indians. So finally we pull off and let the Indians alone and didn't make any fight. It was no use to make a fight unless you had some show to win it. So we went back home and gave it up as a bad job to clean them up. We went home and didn't hunt any more Indians after that.

We waited until 1864 when we got a letter from Mr. A. H. Peeples to come to Arizona. We went to work and got ready to move out in May, 1864, we pull out for Arizona. Father's family and George R. Robert's family. We had good outfit, about eighty head of fine horses and brood mares. We came on down until we struck the Mohave river and travel down that until we came to where it give out and went across a sand desert twenty miles and came to Soda lake. Went on until we came to the Colorado river and Fort Mohave. There we cross the river.

Old Man Hardy had a trading post there. He looked at our outfit and said to us, you can't go on as the Hualpai Indians has broke out and are killing everybody they find traveling on the road. Now you people lay over one or two days and I see if I can find some men that will go with you to Prescott, Arizona. So we did. We found six men to go with us to Prescott.

So we pull out. He gave us some maps of the road showing all the water on the road. Water was short on the road and someplace twenty miles apart. The first camp was twenty miles out. We made that the first day and camp in a flat close to a hill side that night. The Indians came along on top of this hill and rolled a big boulder down the hill right in camp. They thought it would scare our horses and they would stampede and run off, but we had them all tied up and hobbled so they couldn't run, and besides we was on guard too.

Mext morning I got on my saddle horse and told father I go ahead and find the next water and find a good camp to camp in. I took a small canteen of water with me. I rode to where I thought the water was and couldn't find it and I went on all day and I didn't find any water, and just before sundown I found some nice green grass. I knew my horse was

Memoirs of Matt Cavaness 22

awfully dry and tired as I had rode about seventy miles that day and it was hot in July. I got down and let him eat some of that green grass. After dark I started back and a short distance I seen an old Indian trail crossing the road going right into the Hualpai mountains. So I says to myself, I take that trail. It will go to water some where. So I did take it. I rode about a mile and I hear someone talking. I listen to them and I soon learn that was Indians. They didn't seem to be more than two or three hundred yards away. It was a nice clear night, bright stars ashining. I rode a short distance and in came another trail. I stopped and heard them talking. I knew then that I was ahead of them. So I went back a short distance and came to a tree. I got behind it and stopped and could hear them talking, coming right toward me. I sat there on my horse until they got close to the tree and I rode out right in front of them and I yelled out to them, friend and water. They stopped and both had rifles and threw down on me and I yell out friend. I taken my canteen off my saddle and threw it down to them and told them in Spanish, water. They still had their rifles pointed at me. I got down and walked over to them and talk all the time that I want water. They was talking Indian all the time to each other. Don't know what they said but suppose they was talking whether they should kill me or not. I taken my rifle off my saddle and I laid it down on the ground by their feet; taken my pistol off and laid it down by their feet, but they still had their rifle on me. When I did all that they taken their rifles down and commenced talking and I made them understand that I wanted water.

They had no water. I made them understand that I wanted to go with them to water. I was awful dry. My tongue had commenced swelling and I could hardly speak. I tried to give them my rifle and pistol but they wouldn't take them. We made motions and finally they made me to understand they wanted me to get on my horse and they take me to water. I got on my horse and they taken the trail and into the Hualpai mountains we went. We came to a big canyon. The trail went up that canyon about six or eight miles. All at once they turn out up on a little fault and they stopped and turned around and said, lots of water.

They threw down their load they had on their back. I got down and pull off my saddle. One of them caught me by the arm and motion me to water. We walk about twenty steps and there was a nice running stream of water and big hole of water. My horse just run in to it and I laid down and went to drinking. I was all in. I sure was drinking and all at once one of the Indians caught me by the back of my head and pull me up and the other Indian got hold of my horse and pull him out and taken us back to camp. I tied my horse to a tree and set my gun up against a tree and by that time the Indians had a fire. I taken my saddle and laid it down on one side of the fire and taken my saddle blanket and laid down and I don't think it was twenty minutes before I was asleep It must have been about midnight and I rode all day close to seventy or eighty miles. I was sleeping fine and one of the Indians came to me and pull me up and motion to me, come and eat. I got up and they had some kind of meat and had some kind of bread, I don't know what it was made of, but anyway I I eat it. They had lots of mescal. That was good. I sleep there that night. Next morning just as day was breaking one of the Indians came over to me and pull me out to eat breakfast. I got up and eat with them. They fix their load and I got my horse and water him and fill my canteen, saddle my horse and shake hands with the Indians and they went on up the canyon and I pull out to find the train and see if they had found water. I went back on the road about ten miles and meet them. The first thing they ask, Matt, have you any water. There was several small children crying for water. They stopped right here and unharnessed the teams and part of us stayed with the families and me and brother Williams taken all the stock and canteens we had and went back to the water. We sent two men back with water and we stayed all night at the water. Next morning went to the place where we left the families and they was all right. Had a good sleep and plenty of water. All hands felt good.

We pull out for the next water place called Willow springs. We didn't get there that day but the next day about noon we got close to the water and the Indians was all around the place and it look like we couldn't get there. But when we made fight the Indians have back and we went in to the water. The Indians those days didn't have but few guns and they were afraid of a gun. So it was no trouble to route a bunch of Indians, but whenever they kill a man or taken in a train they generally got lots of ammunition and some rifles. So it wasn't very long until the Indians had lots of guns and ammunition and sure knew how to use them.

We went on to Prescott. Got there the first day of August, 1864. We went on down to Wickenburg and on the way at Kirkland valley father and his family all stop there except myself and brother William went on with Mr. Roberts and family. We got there, found Mr. A. H. Peeples waiting for us. He told us the Indians was awful bad, killing every chance they got. In about one week in comes news that the Indians had killed two men at Bell canyon. That was about a mile from where father had stopped. So Mr. Roberts and Mr. Peeples said that we better go up there and get father and the family to come down to Wickenburg. So me and my brother William went up to father's place and they was more than willing to leave that place. Mr. Bell and his partner stayed all night with father and got on their horses and started for Wickenburg. Got off about a mile and the Indians waylaid the road and as they came up the Indians fired on them killing both of them. Mother and all the children heard the shots, so father say we go to Wickenburg. Down we come and got there. Built a house and father was running a eating house for the public. Doing very well. Mr. Roberts and brother and myself went to putting up arrastras to work some of the Vulture mine ore. Mr. Peeples had a interest in the Vulture mine and he was one of the discoverers of the mine. We had a Mexican herding all this stock that we brought from California. We had lots of stock to work. So we thought we go at mining instead of putting this stock on the road hauling freight as the Indians was so bad it wasn't safe to go any place unless about ten to twenty men in a bunch. We got up one raster (arrastra) and one morning I got up early and taken one horse and put him in the arrastra and went to work. Turned the other stock out and told the Mexican to herd them there in the flat right in front of the house. So he did. We sat down to breakfast and the Mexican and the stock was over two hundred yards away. We called the Mexican in to eat his breakfast, and he came in and sat down to eat. Was there about twenty minutes and the Indians must to have been just under the bank of the river. They came out and run around the horses and taken them across the river and stop about five minutes and all of them caught horses and mounted them and away they went. We all got our guns and ran over there but didn't get close enough to take a shot at them.

There was four and five year old horses in the bunch that never had a rope on since he was branded. The Indians would just get on them wild horses, two and three on a horse and from where we could see

Memoirs of Matt Cavaness 24

the horses just went off like an old saddle horse, never buck a bit from what we could see. There must have been three hundred Indians in the bunch.

Well we went back to the house and had a powwow. What to do. All our stock gone. Left afoot. Had one horse left. The Indians taken off about eighty head of horses. Father there with a big family and Mr. Roberts with his family. Mr. Peeples said to us, I know where I can buy some ox teams down at the pass on the Colorado river, and I take that horse that's left and I go down there and buy them teams and bring them up here and we haul ore in here for the miners here.

Mr. Roberts can run the saloon while I gone. so he taken the horse and pull out for the Pass, one hundred and twenty miles away. So he did, right through the Indians country and alone. He got there and bought three teams, six yoke to the team. Got a load of freight for Wickenburg for the merchants there. Me and my brother went to work in the Vulture mine until he came back. He got in with the teams. I taken one to drive and brother taken one. Went to hauling ore and by that time there was several families came in and all had ox teams so we made up quite a lot of teams there. In all there was ten ox teams hauling ore. There was one hundred and twenty-five arrastras running. Probably three hundred men in camp. All those people came in had quite a lot of stock. So they all join in and had a man or two herding this loose stock. The Indians was killing people every few days. We all joined in and hired one extra man to the team. There was ten teams of us hauling ore. All them was was running together. It takes us three days to the trip. To haul water for our teams we go out half way the first day and camp there, leave our water and go next day to the Vulture mine and come back to our water that evening and water our cattle and next morning come in to Wickenburg and load and fill our barrel and pull out next day at noon for one-half way camp. So that the way we done.

We had to stand guard around our oxen all the time day and night and one of our trips we went out and camped. Next morning pull out for the mine. In about two miles of the Vulture we had to go through a box canyon about two hundred yards long. I was driving the lead team. I saw where a bunch of Indians had crossed the road that morning. I stop and had all the teams to drive up and make a corral with the wagons. I says to the men, those Indians is down there at the box and they are waiting for us to get in that box and then fire on us. Then all the men was uneasy. Time now, I say to the men, we got to do something right now. All the men say, you name it and we be with you. I say to them, now I take half the men and the other half stay here with the teams and go down to the box and give the Indians a fight if they want it. They all say, take your men and go, we stay here and if the Indians come we sure do the best we can. If you hear us shooting you know that the Indians has attacked us and you can come back. I say I sure do that, so I divided the men. Five of us went on one side of the box and when the Indians saw us coming that way they pulled out over the hills. We could see them running everywhere. We went on down to the box and right on top of the box the Indians had laid rock down so they could put their guns on and shoot right down in the box. They would be right over our heads. I knew if the Indians didn't get the first shot, he r run sure. I knew that if we run after them a short distance, could get close enough to get a shot at them.

It look like there was about three to five hundred Indians. We went back to the teams and told the men what we seen. Went on to the mine, loaded up and came back to our camp, water our oxen as usual.

Memoirs of Matt Cavaness 25

Next morning we pull out for Wickenburg. Got in about three miles of town. I had put two men ahead of my team as I was driving the lead teams and the other men on each side of the teams so the Indians couldn't get close to us. Uncle Jo Blackwell was one a head about thirty steps ahead of the team. We was on a ridge, a little ravine on each side of us. All at once bang went two guns. Down went uncle Joe and his partner. I signal to the other teamsters to run their teams up and make a corral and as soon as we got that done we went to shooting at the Indians. They kill uncle Joe's partner dead. They shot uncle Joe just above the knee and it paralysed his leg and he fell and rolled down in the ravine and laid there. He couldn't get up and stand but where he was the Indians couldn't see him, but he could watch and whenever he seen an Indian stick his head up uncle Joe would take a shot at him. We fought them there until late in the evening before they give it up. We had the advantage of the Indians. We had our wagon corral and we could get behind this. They shot lots of arrows in our cattle but didn't kill any. When the Indians pull off we went to uncle Joe and he says, boys, I can't walk but I can shoot. Just when one of the Indians stuck his head up and taken several shots at the train and I could hear him yelling out something to his men, I laid my gun right where I could take a shot at him. I laid there and it wasn't long until I see the black head come in sight. I waited until I could see his eyes come in sight. I cut loose. Now I want some of you to go over there and see if you can find some hair or blood, so I says to uncle Joe, I go. One other man went with me and when we got there it wasn't blood but I do think there was a tea cup of brains on the rock. Of course, there no Indians. They always pack off the dead whenever they can do it.

We taken uncle Jow up and put him on the wagon, pick up our dead man and put him on the wagon and pull out what arrows we could out of the cattle and went on to Wickenburg. Everybody was all excited as they knew we was having a fight with the Indians. We buried our dead man and uncle Joe got well and we just kept on hauling ore. It was only few days after that the Indians came right in the edge of the city and cut off that day herd about one hundred head of cattle and horses right through you might say the city. There was a man named Ball ahd his son had gone out after hay. They was in four hundred yards of home. They heard the Indians yelling. They shot the herder down and was yelling at the cattle to make good time. They saw Mr. Ball and Mr. Ball saw them. He yelled out to his son to jump and run for the house. The son didn't lose any time but Mr. Ball cut the haim strings of three horses, pull the bridlesoff and started the horses down the road for home and while Mr. Ball was trying to get his fourth horse loose, Mr. Indian saw him and made at him with a long spear. Run right up to him and push it at Mr. Ball. He dodged the spear and pulled out his pistol and taken one shot at the Indians but didn't hit the Indian.

We had just got in with our teams that was hauling ore from the Vulture mine and of course everybody was riled up. Seeing the Indians going up the Hassayampa river with all their stock. Everybody that had a gun got it and we taken after the Indians. They went up the river about one mile and left the river, taken up a small canyon. We seen them as we left the river. There was about eighty to one hundred of us. We went on up this little box canyon about one mile from the river. We hadn't gone far up the canyon until the Indians sure stop us. They had got at the head of the box canyon and had stopped, kill one of the milk cows. They knew we was after them. They decided they was able to lick

Memoirs of Matt Cavaness 26

us so they came back on each side of this box canyon as we came along they fired on us. They killed one man and wounded several others. They shot from both sides of the canyon and it kept us busy dodging arrows and you couldn't get behind anything but what the Indians could have a good shot at you. I done everything I could to hide behind something so I could take a good shot at one but I couldn't get time to take a bead on Mr. Indian. Here come arrow and I have to dodge that. So we fought them there about one hour and the men that was wounded yell out to us for God sakes boys to take them out of here. They didn't want to be left here for the Indians to kill us. There was one man that had his leg broken just above his ankle. He couldn't walk. He put his arm around my neck and up came one of the other boys and we started down the canyon with him. The other boys picked up the dead man and we retreated down the canyons and was glad to get out alive. We got home and all the extra stock was gone once more. There was one hundred and twenty arrastras working and all the stock that was working in the arrastras was all we had left.

Joe Green and I bought six yoke of oxen and went to hauling ore from the Vulture mine and everything went on as though nothing happened.

About two weeks after that one early morning we happen to see two men coming down the river. We watch them and when they got in about half mile of the city we saw them get down under a big cottonwood tree and pull off their saddles and lay down to sleep. As they had left Prescott the evening before and had rode all night to make Wickenburg and as they was so close to the city they felt perfectly safe. They hadn't been asleep over one hour. We was working around our arrastras and all at once we heard a gun shot and we looked up and seen the Indians all around those two men. We ran up there but when we got there the Indians had taken everything off, even taken the men's clothes off. We buried the men and went on working. We didn't have force enough to go out after the Indians and leave enough to care for the women and children. So we had to do the best we could.

Now I tell you how we got our supplies. The government had a flour mill at Casa Blanco on the Pime reservation, on the Gila river about seventy or eighty miles from Wickenburg. The Indian agent at Casa Blanco, he put on the road six, ten-mule teams to haul us flour and wheat to Wickenburg. Now we was doing well mining if the Indians would let us alone. It was a lively camp. J. B. Allen, of Tucson, he heard of the camp. Here he came. Brought lots of mules to work in the arrastras. He put something like one hundred arrastras and Mr. White, the Indian agent at Casa Blanco, had a big train and besides his help the Mexicans from Sonora, Mexico they came loaded with beans and penoche that was a kind of sugar. And besides that they bring penole, that was roasted wheat ground up very course, make a fine meal if you had dried meat to eat with it. So everything look good for the camp. Every one says we got to hold them teams that hauling ore from the Vulture. It won't do to let the Indians take that in. So the men that was running arrastras they all join in and furnish us with extra help so we wouldn't be taken by the Indians.

Everybody was making money. There was one man that had quite a stock of drygoods, but he couldn't come any way near keeping enough groceries for the camp as he couldn't get teams to haul them in from old La Paz as it was called then. It Ehrenberg now, on the Colorado river but we was doing fine. Those six teams of the Indian agent would come in once a month with flour and wheat. The Mexicans in Sonora they got on to it that the agent would make that trip so they made it a point to be at Casa Blanca and they travel along with the mule teams and come in with them with their beans and pinoche, penole and in fact we bought everything they bring us. We was in great hopes that we make all kinds of money. We knew the very day that the teams would be in. Well, the teams will be here tomorrow. That was on everybody's mind, but instead of the teams coming in it was a Mexican girl about sixteen years old. She said the teams and her outfit camp together out on the desert close to White Tanks. She said just as they was pulling out camp at daylight the Indians fired into the camp, She says she just jump up and ran down the road with two Indians after her. She said she just outrun them so bad that they stop and give it up. She was about twelve miles from the river and when she got there she said, I was so tired I had to stop and rest a little before I started for Wickenburg. She said I don't know who was killed. My father and mother was along and several kin folks. I am so uneasy I want to go back and see if any of my folks is alive. So we stop all the arrastras and taken the animals and rode out to where she left the outfit. We had to let her go so we gave her a horse and away we went. You never saw such a sight. They had killed the whole outfit, taken all their clothes off and cut the bodies all in pieces. Emptied out the flour and wheat by the wagon and cut the harness all up and set fire to the wagons and Mexican carts. Taken the Mexicans oxen and all the mules and was gone when we got there. All the mules collars was cut out a piece that would make a moccasin sole. They cut everything they didn't take away. That poor girl. Every man in the bunch couldn't help shedding tears for her. We buried all the bodies right there and went back home as it was no use for us to think of following the Indians. They lick us if we found them. We take that girl back with us and she stayed there with a family for about one year and married a good man. He was a blacksmith by trade. Years after that him and I was partner in the blacksmith business in Phoenix, 1872. They raise a big family. The girl she made a fine woman. We oft times talk about the way she got away from the Indians. She says I just pull my dress up around my waste and went to running. Two Indians taken after me and I just got down to running and they sure done their best to catch me. I soon seen that I was leaving them then that gave me more courage and I slowed up a little, but after I got off two or three miles I couldn't see them any more. I slowed up a little and when I got to the river I was awfully dry for water and I rested a little while only, then I hit the road for Wickenburg. I didn't know how far it was there. It was twelve miles. She made awful good time anyway, and when they was attacked they was about thirty-two miles to Wickenburg as well as I remember. She got in about ten a.m. She must have made that run in about four hours or less.

We never got anything from either outfit. Well, that cut off our supply trains so we had a hard time to get in supplies again. There was one man volunteered to go to Casa Blanca and report it to the agent there. He went through all right and made his report. The agent called in the Pima Indians and told them what had happened. The Pima Indians turned out about five hundred strong and came out and took the trail, followed it out by Frog Tanks and the Apaches stop there and kill a mule or oxen and had a dance. The Pima Indians followed them up on the Verde river. The Apaches stopped there and when the Pima Indians showed up, they two to one Apache. So the Apache just give them Pima a good licking and sent them back home to their reservation. The man that volunteered to go and report, he got back in about six miles of Wickenburg and the Indians killed him and took his horse and rifle and two six shooters and all his ammunition. That was the way they done. They knew all

Memoirs of Matt Cavaness 28

Americans always had good arms and if they caught one, they do that to get the gun and they sure done all the killing they possibly could. That was in 1865.

There was a man in the country by the name of King Woolsey. He had been in the country since 1858. He took up what is called now Agua Caliente ranch, one hundred miles below Phoenix. He lived there during the time that the old Butterfield stage line was run through Arizona, and when Prescott was laid out and the Joe Walker party came in and Mr. Peeples party came in and other miners, Mr. Woolsey go up to Prescott. He got about twenty men before he started from the Agua Caliente ranch. He took lots of pack animals. He said if he could raise one hundred men he go out and see if he couldn't clean those Indians out. I think that was in 1862 (1864). He got to Prescott and raise ninety men. Joe. Green and A. H. Peeples went with him. They was gone three months, all through Tonto Basin and on to the head of Salt River and down Salt River to a place they call now Wheatfields, just below now called Globe. There they had a big fight with the Indians. Had several fights on their trip. Woolsey one day was off from the party some distance and he found a little Indian girl about five or six years old, all alone. He took the girl up on his mule and fetch into camp and when he got back to Prescott he took this girl to John Dickson and wife and told them he pay them to raise that child and send her to school and when she get to be about sixteen I come and get her. So they did and when the time came he went and got his girl and took her down to the Agua Caliente ranch and raise a family. I think the mother and two girls live in Phoenix now.

Well, in 1865 I get back to where I was hauling ore from the Vulture mine. Joe Green and I had bought us a team and was hauling ore. They sent us some U.S. soldiers in the country. General Stoneman was in command. He established his headquarters in Los Angeles, Cal. He had established Fort Whipple at Prescott, Arizona. He had heard of Woolsey making this trip after Indians. So he wanted a guide for the soldiers. So he wrote a letter to King Woolsey as he was call. He said to King Woolsey, you go to Prescott and report to the commanding officer at Whipple and go as guide.

Chapter 3

Diaries and Memoirs

On the following pages are excerpts from the memoirs of A.H. Peeples. Peeples relates the early days of the Arizona Territory (1864-1866) and the terrible problems the pioneers faced from Indian depredations.

Photo of Abraham H. Peeples (Visalia, California 1858). Solid gold frame made from ore dug up at Rich Hill.

In those years before the Territory became settled the threat of sudden death from indian attack was always with us everywhere we traversed. The many Apache bands were the worst, no atrocity or fiendish devilment was beyond their imagination. Those who came after the Apache wars were settled could ever dream of the horrors and depredations we witnessed. No mere words could describe the terrible and heart rendering scenes that befell us time after time. Old friends and neighbors, whole families, little children and all were killed, butchered and mutilated beyond recognition by the heartless savages.

A Mexican woman lives today in this city (Phoenix) that was the sole survivor of an Apache attack about 25 miles north of Wickenberg on the old emmigrant road to Prescott. She was just a young girl then, eleven years old, traveling with her family and some fifteen other Mexicans for Prescott city. They had crossed the California deserts without being molested and had crossed the river into the Arizona Territory at Ehrenberg. There at Ehrenberg they were warned not to attempt the dangerous journey to Prescott without an escort of soldiers. But having no troubles up until then the leaders decided an escort wasn't necessary. They pushed on and got almost to the headwater of the Hassayamp when they were set upon by about 150 Apache what had been laying in wait for wagons along that old emmigrant road. At first the Mexicans made a good show being well heeled and the Apaches moved off but quickly resumed the fight anew. The Apache then abruptly quit the attack and made signs to the Mexicans that they wanted to parlee. Several Apache leaders met the Mexicans and stated they only wanted half the mules attached to the wagons and the Mexicans arms and they would let the party continue on their journey unmolested.
The Mexicans being inexperienced in the ways of the Apache and seeing they were sorely outnumbered, agreed to the terms.

No sooner had the Mexicans given up their weapons when the Apache devils fell upon them and began killing the entire party. The little Mexican girl ran and hid in a hollow log about a hundred yards away. In the struggle she was able to slip away undetected and witnessed the horrible scene from her hiding place.

The Apache took all the men and stripped them naked, then tied them to the wagon wheels upside down so their heads were about two feet off the ground. Then they build fires under them and proceeded to roast them in that position until their brains burst from their heads. Some of the children were clubbed down dead in front of their families and a tiny baby was thrown into the air and caught on the end of a butcher knife as it fell through the air, all the while the savages was laughing and whooping at their devilment. Two women were stripped and ropes tied around their feet and drug through a stand of cactus over and over again. When they were almost dead from the unbearable pain the savages poured coal oil from the wagons over them and set them afire. Some of the older children were found later, most of their skin had been cut from them and their bodies filled with dozens of arrows. The little girl said only one or two of the party were taken captive and never seen again.

That little Mexican girl walked back into Wickenberg two nights later and told her terrifying story. I was among those that went out and found that horrible scene exactly as the little girl had told. Every man there shed a tear for that little girl and what those savages had done to her whole family. Most of the men put in and hired a family in Wickenberg to look after that little girl and raise her up. She later married Frank Cosgrove here at Phoenix and they had several children. Cosgrove ran a blacksmith shop on Jefferson and whenever I meet that woman on the streets she always give me a hug. She never forgot that horrible time and has nightmare's to this day. I guess none of us can ever forget those days and the lasting memories of the Apache atrocities.

– A.H. Peeples

The Walker Party

Arizona Territory 1863

Copy of a manuscript written by Daniel E. Connor who was a member of the Captain Joseph R. Walker party which entered Central Arizona in 1863, exploring and opening up the country where the city of Prescott stands today.

This manuscript is a verbatim copy of the manuscript.

Typed by Mulford Windsor in 1909 from the longhand papers of Mr. Connor.

Members of the Captain Joseph R. Walker party explorers of central Arizona Territory June 1863. From the manuscript of D.E. Conner of Wilmington Ca. who himself was a member of that expedition.

1.	**(Captain) Joseph R. Walker**	**Ca.**
2.	Joseph R. Walker Jr.	Ca.
3.	Martin Lewis	Ca.
4.	Jacob Lynn	Ca.
5.	George Blosser	Pa.
6.	Alford Shupp	Pa.
7.	John J. Miller	Il.
8.	Jacob Miller	Il.
9.	Samuel C. Miller	Il.
10.	Solomon Shoup	Il.
11.	Hyrum Cummings	NY
12.	Hyrum Mealman	NY
13.	Harry Wheelhouse	NY
14.	Dr. (General) Coulter	NY
15.	Johnny Bull	Canada
16.	George Lount	Canada
17.	Roderick McKinney	Canada
18.	Wm. Williams	Conn.
19.	Mr. Benedict	Conn.
20.	A.H. French	Vt.
21.	Jacob Snider	Ger.
22.	J.H. Dixon	Ms.
23.	Mr. Young	La.
24.	Frank Finney	Ks.
25.	Jackson McCracken	SC
26.	J.W. Swilling	Ga.
27.	Mr. Chase (James)	Ohio
28.	Chas. Noble	Mo.
29.	Henry Miller	Mo.
30.	F.G. Gillilan	Ky.
31.	Felix Burton	Ky.
32.	Tom Johnson	Ky.
33.	Charles Taylor	Ky.
34.	D.E. Conner	Ky.

In September, 1858, Daniel E. Conner, a Kentuckian, while visiting friends and relatives in Platte county, Missouri, paid a visit on horseback to Leavenworth City, Kansas. Attracted by some unusual excitement, he learned that parties who had participated in the recent gold discoveries on Cherry Creek, a tributary of the Platte, at the point where the city of Denver now stands, were in town, and the people were electrified by their wonderful reports.

The excitement culminated in a public out-door meeting, addressed by many speakers, mounted upon an improvised rostrum. These speakers described the gold fields, and it was declared that by taking time by the forelock, Leavenworth would become the key to the great Western plains lying between that point and the Rocky Mountains, and as naturally to this further land of marvelous wealth.

Conner returned to Platte county, imbued with the Pike's Peak idea.

The following Spring saw him and fifteen companions, with five covered wagons well laden with supplies and drawn by oxen, off for the gold fields. They crossed the river at Leavenworth on April 7, 1859, and by way of Lawrence, Council Grove (the border settlement of Kansas), Diamond Spring, Lost Spring, across the Little Cottonwood, Cow Creek and other tributaries of the Arkansas river, "Big Bend" of the Arkansas, up that stream past Walnut Creek, the Santa Fe Crossing, Ask Creek, Pawnee Fork, Bent's Fort, Fountaine-qui-Bouille (emptying into the Arkansas where Pueblo now stands), up that stream to Soda Springs at the foot of Pike's Peak, across South Park (which lies between Pike's Peak and the main range of the Rockies), arrived, forty-five days after leaving Leavenworth, at the point where Hamilton now stands, on the west border of South Park, a favorite hunting ground of the Ute Indians. From here a difficult trip of sixteen miles took them over the summit of the Rockies, the almost snow-bound pass traversing what was known as "Snowy Range," to Blue River, the scene of their ElDorado.

A Reminiscence of Fremont and Walker.

"Fremont made one of his paths across this mountain, but he didn't take a wagon with him. It appears that he was in search of a guide, and came across Capt. Joe Walker, who had charge of a corps of trappers in the interest of the American Fur Company. The great Pathfinder offered Walker $5,000 to pilot him and his party to California. Walker asked $8,000 and Fremont refused to give it, and was given directions by Walker where to find Kit Carson, who was also in charge of a company of trappers and encamped on Green river, a stream to which Blue river is a tributary. Fremont found Carson, who accepted the task for $5,000.

"The reader knows the rest. Kit Carson was published into the greatest mountaineer of the American continent. These two men, Walker and Carson, went to California together in 1832, from whence they roamed the Rocky Mountains at pleasure, long before this region was noticed by the outside world.

"These little jottings are put here, not as my own information, for they were obtained from Capt. Walker himself, after I had traveled with him a year in these mountains, and became well acquainted with him. He told me many other things connected with the different expeditions made in the Rocky Mountains, including those of Gen. Fremont, but as the latter has given his own account of them I shall leave the subject."

Movements Prior to Joining Walker Party.

The following two years were spent by Conner in prospecting expeditions and mining experiences on Blue River; Gold Run, eight miles north; Georgia Gulch, Humbug Gulch; by way of Pike's Peak, toward New Mexico, at the foot of the Rockies south of the Arkansas; back to the new towns of Jo Buckskin, Fairplay and Jefferson, the latter being in South Park, six miles from the town of Hamilton; to American Gulch; expeditions for distances toward the Dakotas and the Great Salt Lake; up Swan river, to explore Bald river; to Washington Gulch, the present site of Leadville, and in August of 1862 he was back at Jefferson.

Joins Rebel Recruits.

By way of Canyon City he left Jefferson for Mace's Hole, where a Confederate regiment was being recruited by Col. John Heffner. His object was by this means to secure means of returning to the States, without joining either Federal or Southern armies. He came to be known as a Captain, and as of that rank was searched for by the Federals after the gathering at Mace's Hole had been broken up and dispersed.

Devoting himself to eluding the pursuit of the Federals he worked in October, 1862, at Hicklin's Station, disguised as a common laborer. From that point he started on horseback to join a party of rebels said to be at Apishapa Canyon. After a forty-mile night ride he came to a tributary of the Arkansas called "Picket-Ware," (afterward Las Animas), where there was a Federal mail station. Federal soldiers were there, but he got into communication with the station keeper, a Southern sympathizer.

Joins Walker Party.

Acting upon the advice of this friend he determined to join a party, consisting of thirty men and fifty pack animals, under Capt. Joe Walker, which had passed this station three days before, bound for Arizona.

Following no trails, after four days of almost constant travel without food, he overtook the Walker party, and found that two of his Rocky Mountain friends, members of the original Missouri party, were among the number. One of these was Frank G. Gillilan.

Description of the Captain.

"Captain Joseph R. Walker, who had been chosen as the leader of this expedition, was a native of West Tennessee--a contemporary and friend of Kit Carson. He went to the Rocky Mountains, and thence to California, in 1832, and had roamed and lived in some part of these mountains and the Pacific country continuously ever since. He was between sixty and seventy years of age and still a fine specimen of physical manhood. He weighed about 200 pounds and stood over six feet in height; had broad shoulders, black eyes and a luxuriant growth of hair and unshaven beard. His profusion of hair and beard was perfectly white. He stood erect, and was well preserved for his age, walking and moving as easily and nimbly, apparently, as a boy. Taken in all ways, he was still a dignified and reticent man, of strong, unobtrusive will and deliberate judgment, a good judge of human nature, and always apparently cool and fearless.

"Walker and Carson were associates up to the date of Fremont's crossing of the mountains, when he took Kit Carson with him to California as a guide.

"I traveled with Capt. Walker through the Rocky Mountains for more than a year, and in the midst of and during all the hardships and dangers of mountain, desert and vale I never saw him excited or agitated but once, and that was when his orders were disobeyed during an important and critical situation. And during my whole experience with him I never heard him dwell upon any of his own exploits nor tell any wonderful story for the edification of those present. Yet he had roamed the Rocky Mountains continuously for thirty years, and that too during a period when this hillimitable romantic wilderness was almost totally unknown by the masses of the people of our country. * * * *

"This party had no tents, but took the weather as it came. But the preparations for the protection of the necessary supplies were efficient, and quite sufficient. Each pack-mule was provided with two boxes, or pockets made of rawhide, with the hair outward, and attached thereto, near the back upper edge, two loups of the same material by which to hang the box on one side to the pack saddle. The other box was similarly hung on the other side to balance the load. The sides of these two boxes which lay next and against the animal extended high enough to be drawn over the top of the box and tied down to keep out the rain. These boxes were made large enough to contain seventy-five pounds of flour to each, sack and all, and were constructed too strong and tight to endanger the safety of anything put into them. Each mule carried two of these pockets, besides the cooking utensils placed on the top and covered with blankets, before binding the whole with pack-ropes. One of the mules thus caparisoned was loaded with mule shoes and nails and hammers with which to put them on. The party was prepared to live for an indefinite time in the wilderness, by all conveniences except tents. We had to take the weather as it came."

Arrive at Taos.

Out of the roughest mountains, by the shortest route to Santa Fe (where they desired to secure from Gen. Carleton passports to Arizona), the party arrived at Taos. Several days were spent here before a military detachment from Santa Fe arrived, and compelled them to take the oath of allegiance. Capt. Walker notified the commanding officer that if he had known Gen. Carleton, with whom he had played many games of euchre, was not coming in person, he would not have tarried, as his men could go anywhere without passports, but he had thought to secure passports out of respect for his old friend.

Passing to the left of Albuquerque, the party struck for the Rio Grande at Peralta, the river crossing. On this leg of the journey, Conner saw a Mexican grist-mill, which he describes as follows:

A Mexican Grist Mill.

"This mill stood in a little hollow on four legs, and covered a square surface of about eight feet. A floor was built on the top of those legs, and upon this floor was erected a little room about six feet high. One mill stone was laid on the floor and made fast, and the other was placed on the first one. A rude wooden shaft, with wooden gudgeons, stood perpendicularly under the centre of the floor, in a socket in a rock, and the top of the shaft extended through the floor and the first stone and was made fast to the second one. The lower end of the shaft was bristled with numerous rude paddles, which were driven into it horizontally and were about three feet in length. A little stream of water was flowing down the hill through an artificial ditch, and was guided against the paddles by a small trough. The corn was poured by hand into a hole in the centre of the upper stone, and the meal came out from between the two stones all around their circumference, and fell on the floor ready to be scraped up for use. If there was a particle of iron about this mill, even as much as a nail, it was not visible.

The Mexican Vehicle.

"Their vehicles, like their mills, were built after the same pattern, from one end of the Territory to the other. This contrivance consisted of two wheels, an axle, a pair of thills and a rude bed. The wheels were about four inches thick and about two and a half feet in height, and each was composed of three cottonwood slabs fastened together into one piece, out of which the wheel was cut round, or nearly so. It was amusing to see a little Mexican ox hitched to one of those carts and pulling a load. The vehicle was connected to a board, which was first firmly bound to the animal's horns. Having to draw the load entirely by the horns and forehead the little animal would project his nose to the front with his forehead horizontal, pull his best as he traveled, and keep constantly licking out his tongue and rolling his eyes with great efforts to see the ground, and would patiently prolong these exertions until I began to feel tired myself from looking on at such desperately hard work."

The Mexican Plow.

"The plows used by the Mexicans for cultivating the soil were nearly like, in construction, those manufactured in the United States. They were of the most simple, primitive order, consisting of and formed from the fork of a tree. One fork was cut short and shrpened to go into the ground, while the other extended horizontally for abeam, by which it was drawn. The main trunk of the tree, from which the model seemed to have been taken, extended back a sufficient length to support a handle by which to guide the machine. The handle was usually a straight stick, bound on firmly with a rawhide thong cut for the purpose. There was no iron used in the construction of the Mexican plow."

Route to Arizona.

"The Rio Grande was crossed by the Walker party a few miles above Peralta, and the route lay down that stream past Peralta, Valencia, Tome, Limitar and numerous other little towns on both sides of the river, one camp being made under the cottonwood trees, on the battle ground of Valverde, where Gen. Canby planted his artillery in the fight with the Confederate Gen. Sibley, in 1861. Passing up a creek coming in from the right they camped beyond Fort Craig, a couple of days, in a bend of the river. The commandant at Fort Craig sent over a five-gallon keg of whiskey, and the large portion of the party enjoyed a "whiz." That night several random shots were fired into the camp, and the next morning an elaborate Indian head-dress was found near by.

At this point the party turned due west, into the wilds of Arizona (as the southwestern part of New Mexico was then called by some). Within two or three days the Apache signal smokes were seen, and from this time on eternal vigilance was the watchword. During the following two weeks the effort was to pierce to Central Arizona direct, but the way was continually beset with savages, and there were very many narrow escapes. Capt. Walker declared that it was useless to attempt to force a way through this mountainous country, so completely overrun with Indians and in which it was so difficult to find water, so he changed the course of the party to the southwest. His first objective point was "Cook's Springs," a watering place at the foot of a perpendicular mountain, which he had found some thirty years previously, and the point was reached after several bouts with the alert foe. This was about December, 1862, though days and dates had been lost sight of.

Traveling westward, the party came in sight of Stein's Peak, and pushing on to find water, what was thought to be an Indian signal smoke was found to be the last camp of three white men who had attempted the "Southern route" to California. They were hanging by their ankles, all in a row, to a pinon limb, hands tied behind them and their headshanging to within a foot of the ground. Under each head was a little fire, and that was the "signal smoke" the explorers had seen. The hair and skin were completely burned off of the skulls, giving the bodies a ghastly appearance as they swung there perfectly naked.

Several weeks more were spent in an attempt to get through the mountains to Central Arizona, but the Apaches baffled every attempt. The time was spent in a series of maneuvers on the part of both reds and whites to secure an advantage, but the cunning traps of each were successively checkmated.

Turning again to the southwest a camp was finally reached within a mile of the old "Southern route," and the next day the road was struck at a point where were eight dead bodies, mute testimonials to an Apache visit. One of the victims had been fastened to a clump of cactus and burned, a portion of the body remaining to tell the tale..

Continuing on a southwest course a stop was made, within the next few days, at a hot-water crater, eight feet in diameter, in the centre of a knoll. The water was blue, clear and very hot, and although the weather was cold steam from the water could be seen for a long distance. Some twenty feet down the side of the hill a hot water spring also trickled out of the ground and coursed toward its foot. Near the hot water crater was a cold water spring, which supplied water for camp purposes.

Another day near and along the mountains, still in a southwesterly direction, broughtthe party to the ruins of an old fort built of pine logs, some five or ten miles from the edge of the mountain woodlands. This was Fort McLean, about eight miles from the "Southern route." Twenty miles north of it lay Pinos Altos, where headed the canyon, or hollow, that debouched at Fort McLean.

Near Pinos Altos was the headquarters of Mangus Colorado, head chief of the Apaches of all this section of the country, and Capt. Walker conceived the idea, while making a stop of some weeks at Fort McLean, that if Mangus could be captured and held as a hostage, the passage of the Indian country through which they were struggling might be the easier effected.

Here is Connor's narrative:

Capture and Death of Mangus Colorado, in February, 1863.

"Jack W. Swilling was the chosen director to accomplish the task. In the meanwhile the advance guard of some California troops, enroute for the war in the States, arrived at our encampment under the command of Capt. Sherland of the United States army. He said that he had espied our fresh trail and tracked us up to see who we were. He had thirty or forty soldiers with him, and informed us that several hundred more were behind, under the command of Gen. West, Major McClave and others. We were glad to see them, being the only white men that we had seen for months.

"Our pro tem captain, J. W. Swilling, was ready now to proceed after Mangus, and invited the Captain and his soldiers to go with us to Pinos Altos. Some of them accepted the invitation, and we started, leaving some of each party at the old fort, in camp.

"We arrived at Pinos Altos before night, and remained until morning, and succeeded by hoisting the white flag, in drawing some Indians into our sight. The proposition to make a treaty was entertained by the Indians at a distance. The negotiations were conducted in broken Spanish on both sides, and after a long and tedious indulgence of prudential precautions by both parties they approached to a point within easy talking distance. The professions of friendship here so extremely indulged by both of these high contracting parties lent at once a patent impunity to a common treachery that froze their words of love into such hostile motives as might offer the best advantage to be taken in a deadly conflict.

"Mangus and three or four of his followers came up, but the main body of his menremained back on the hillside, amongst a profusion of boulders, which are the natural breastworks of 'Poor Lo.' Our men now presented their guns and ordered Mangus to stand still and surrender, which he did in surprise. He turned to his companions, after he found that they were to be released, and warned them that they were not fooling with Mexicans now, and not to look for him until they saw him.

"We asked him to tell his people that his safety depended on their good conduct toward us, and that if they would let us travel in peace for 'ten moons' we would send him back in safety to them. He talked to his people in gutturals, toward the last, so that we would not understand him, and his face wore an air of care and perplexity.

"His dress consisted of a broad-brimmed, small-crowned chip or straw hat of Mexican manufacture, a checked cotton shirt, breech-cloth and a pair of high moccasins, or moccasins with legs to them, like boots, only that they fit the legs closely. Mangus was apparently fifty years of age, and a large, athletic man, considerably over six feet in height, with a large, broad head covered with a tremendous growth of long hair that reached his waist. His shoulders were broad and his chest full and muscular. He stood erect and his step was proud, and altogether he presented quite a model of physical manhood. If Mangus ever had any, or many peers, among his people, in personal appearance, I never saw them during a five years experience in their country.

"We hurried off to our camp at old Fort McLean, and arrived in time to see Gen. West come up with his command. The General walked out to where Mangus was in custody, to see him, and looked like a pigmy beside the old chief, who also towered above everybody about him in stature. He looked careworn and refused to talk, and evidently felt that he had made a great mistake in trusting the pale-face on this occasion.

"Our men kept Mangus during that night, and on the following day the General concluded that he would take charge of Mangus until the old 'brave' accounted for the loss of two plundered government wagons which were lost on the Rio Grande some months previously.

"Two soldier sentinels now took charge of Mangus, and kept himall the evening at a fire built of the old cabin logs at the fort, which were lying promiscuously around the camp. Our party had been there long enough to form a regularly beaten path a hundred and fifty paces in length, on the west side of the camp, by walking sentinel day and night. The soldiery formed their adjoining camp and the sentinel beat at right angles with that of ours, so that the two beats came together, forming a right angle at the fire where old Mangus was lying on his blanket, a prisoner.

"One soldier walked their beat and one citizen walked our beat, and two extra sentinels were placed over the person of Mangus. This was the situation when night came on, and it was a bitter cold night on that bleak prairie. I was on guard during the fore part of the night, which was exceedingly dark. Being cold and disagreeable, all but the guard retired to their blankets early, and soon left the camp wrapped in profound silence. No fire was kep burning, except the one at the junction of the two guard-beats, where Mangus lay upon his blanket, with his trinkets under his head for a pillow.

"A while before midnight I noticed Mangus moving, now and then drawing up his feet restlessly, and tucking the lower end of the blanket with which he was covered, over one foot with the other. I would walk up to the fire and stand for a moment with the two sentinels, to enjoy the liberal warmth of a bright fire, and then walk off in the dark on my beat, to which I had become so accustomed that I could follow it without difficulty this dark night. When I arrived at the far and lower end and turned to come back to the fire, I noticed that the soldiers were annoying Mangus in some way, and they would become quiet and silent when I was about approaching the fire, and keep so until I again walked off in the dark on my beat. But my curiosity as to what they were doing to the old Indian became aroused, and as soon as I departed into the dark far enough to get beyond the reflection of the firelight I walked rapidly to the lower end of my beat, then turned and walked leisurely back and observed the sentinels' pranks. I could see them plainly by the fire light, as they were engaged in heating their bayonets in the fire and putting them to the feet and naked legs of Mangus, who was from time to time trying to shield his limbs from the hot steel. When I came up to the fire each time they would become innocent, and sleepy, and remain so until I departed in the dark on my beat again, when they would arouse themselves into the decided spirit of indulging this barbarous pastime. I didn't appreciate this conduct one particle, but said nothing to them at the time, and really I had some curiosity to see to what extent they would indulge it.

"I was surprised at their ultimate intentions; just before midnight, when I was about midway of my beat and approaching the firelight. Just then Mangus raised himself upon his left elbow and began to expostulate in a vigorous way, by telling the sentinels in Spanish that he was no child to be played with. But his expostulations were cut short, for he had hardly begun his exclamation when both sentinels promptly brought down their minie muskets to bear on him and fired, nearly at the same time, through his body. The chief fell back, off of his elbow, into the same position in which he had been lying all the forepart of the night. This was quickly followed by two shots through the head, by each sentinel's six-shooter, making in all six shots fired in rapid succession. The old chief died without a struggle, in the precise position that he had occupied continuously since dark. The force of the musket balls marked the frozen ground beyond the body for a distance of five or six feet, with an unbroken line of blood, plainly indicating the respective directions of the bullets after passing though the body. The four six-shooter balls likewise streamed the old Indian's long hair in strands over his face and straight out upon the ground its full length, indicating the direction of the pistol bullets as they left the victim.

"The shooting aroused the whole camp, which soon settled into silence again when the cause was ascertained, and the two sentinels retired with the rest of the camp members.

"George Lount, who was to relieve me of guard duty at midnight, now promptly took my place, stating that it was not quite midnight but that it was so near it that he would stay up and go on duty immediately.

"I retired to my blankets and slept soundly until morning, within less than ten paces of the dead body of Mangus, and arose to find the weather extremely cold, and found the body of the Indian in the same position and untouched. I raised his head and took from under it quite a number of curious trinkets, one of which was a rectangular wooden block made of oak. It was about four inches in length by about two and a half inches in width and three-quarters of an inch thick. It had a hole through one end, by which it was fastened to the other trinkets temporarily. A stout string passed through the hole, with its ends permanently tied together, forming a loup large enough to pass the hand through, and I though that it was to be worn on the wrist, as it was apparently worn smooth by much handling. This block was marked by hyeroglyphics, which were burned deeply into the wood with some hot instrument and quite neatly executed. The hole had been burned through the block. I gave them to a lieutenant.

"Quite a number of soldiers came to where I was standing near the corpse, and amongst them there was one who called himself John T. Wright, of California, who asked the loan of my butcher knife with which to scalp the chief. I declined, upon the ground that my knife was the only cutlery I possessed with which to prepare my food. He then applied to the soldiers' cook, Wm. Lallier, who furnished a large Bowie knife with which the soldier took off the scalp of Mangus. He wrapped the long hair around the scalp and put it in his pocket. I thought that I had never seen the skin about the head of a buffalo much thicker than this scalp of Mangus.

"It was, as before stated, very cold, and therefore the operation of taking this scalp was a bloodless one, leaving the inner skin as white as the skull from which it was taken. The boy was lef t where it lay until nearly noon, when it was aarried on a blanket and dumped into a gulley, blanket and all, and covered up.

"A few nights after this some soldiers dug Mangus's body out again and took his head and boiled it during the night, and prepared the skull to send to send to the museum in New York. I afterward saw the skull frequently before the soldiery departed with it.

"Thus ended the career of the most notorious chief which the Apaches ever boasted of, since the United States has owned Arizona and New Mexico.

More Apaches Killed.

"An hour after Mangus was killed a portion of the soldiers were started for Pinos Altos, where they arrived about daylight and attacked the Indians from an old log cabin, where they were concealed after having the Indians decoyed to within gun-shot, and succeeded in killing two of them. This act confirmed the savages as to the fate of Mangus.

"They returned in a day or two to our camp, and some of our party accompanied them into the mountains north of Pinos Altos, and surprising a rancheria killed ten Indians."

Explanation of the Killing.

"The two sentinels who killed Mangus reported to their superiors that he was attempting to make his escape, and that report was never questioned while this camp was occupied. However that may be, Mangus went to his happy hunting ground without any living friend to tell the tale of the manner of starting; and it was useless for any of the members of that camp to affect to believe that the chief was attempting an escape, for they all knew better, officers and all, before they left that encampment."

Progress Westward.

Still harrassed by the Indians, in the latter part of March, 1863, the Walker party, from which the soldiers had previously separated by taking up their quarters at Pinos Altos, resumed the march westward. They passed through the gap between the Burro mountains and the range to the right of them, and descended the Pacific side of the mountains at a place called Mangus's Rancheria. This place bore no evidence of a shelter, but looked, from the existence of overgrown irrigating ditches, to have been cultivated. It covered about forty acres, was covered with grass, and was altogether an inviting spot. A plow leaning against the projecting rocks bordering the arable land, was in a state of decay.

Camped on the Gila.

"In a few days we pitched camp on a high, treeless hill overlooking the Gila. This hill was covered with a fine growth of grama grass, and was located in the right angle formed by the junction of a little valley on the north and the river on the west. The Gila flowed close to the foot of this hill, and hence its side next to the river was very steep. This hill, like the greater part of all prominent points overlooking a stream of water in Central Arizona, was fortified on its top and around its brow by ridges of volcanic rock, thrown up to about the height of three feet. The whole space thus inclosed was subdivided into rooms by the same character of ridges, like a chess-board, but without its regularity. Some of the rooms were probably twenty feet square, while some were greater and less in size, and a few others were triangular, rectangular and circular. But they all joined from one side of the hill across its summit to the other. I call them rooms for want of a better term. This hill had only about a dozen rooms within its outer wall, or rather ridge. Around this hill for a distance of from two to four miles there was a rolling country, studded promiscuously by small clumps of hills, divided by sharp ravines and all covered with a fair growth of grama grass. But little timber was in sight, either on this scope just described or on the distant mountain ranges. A few scattering clumps of scrub oak marked the ravines and hillsides, with perhaps enough cottonwood trees to indicate the direction of the Gila, growing upon the Gila, for there was but little bottom land upon this river so high up. Just across the Gila from our hill there was a small knoll which was fortified by one circular wall around its centre, about ten feet in diameter.

Expedition to the San Francisco River.

At this point the Walker party was again visited by the military, commanded by Major McClave, and the old Indian fortifications were made the scene of their cmap and named Fort West. A trip to the San Francisco river, about 150 miles in the direct line to Central Arizona, was proposed by Capt. Walker, and Major McClave, with some of his soldiers, were induced to join the expedition. About thirty soldiers, with the mules and cattle belonging to the command, and Conner and three other members of the Walker party, were left to hold the fort.

The expedition was gone five weeks, returning without any serious mishap, several brushes with the Apaches having resulted in nothing worse than a few minor wounds. At the camp all was well, beyond the fact that an Indian stampede had resulted in the killing of all of the cattle belong to the military at Fort West. This disaster was added to but a few minutes after the escort guarding the extra cavalry horses of the expedition came into camp a short diatance ahead of the command. A sudden attack by the Indians resulted in the loss of seventy-eight animals.

After a perilous fifty-mile trip to Pinos Átos, by Conner and four others of the Walker party, for the purpose of gainigg information from a Mexican supposed to be there regarding the fate of one C. P. O., who had attempted to make his way from California to New Mexico three years before, with a large amount of gold dust, and had been murdered by Indians at the instances of Mexicans, the explorers were ready to proceed on their journey to Arizona.

A Typical Indian Stampede.

Before leaving, however, another Indian stampede of the cavalry horses occurred, and as it was typical of the Apache system, is here given in Conner's words:

"The horses were lazily strolling up the river bottom, upon a cloudless, bright morning, followed by five soldiers who had their muskets slung over their shoulders. They were within gunshot of the tents on the hill and were passing now and then the arroyos which opened onto the level bottom land over which the herd was passing, up stream, all quiet and happy. An Indian rode out of the mouth of an arroyo behind the soldiers and the herd, put spurs to his horse, passed within twenty paces of the surprised soldiery on a full run and entered the herd of stock, passed through to the foremost of the herd and rode on in a hurry. But this Indian had been followed by two others, who dashed at the soldiers and shot one apiece with arrows, delivered a terrific war-whoop which stampeded the horses directly after the foremost Indian, who now led the herd his own direction on a wild race. The two hindmost Indians followed the flying herd closely, lying flat upon their horses to escape the musket balls of the soldiers. The soldiers were so completely surprised, and the whole transaction executed so suddenly, that they never got ready to shoot until the Indians, herd and all, were too far away to accompliah anything."

Camped Near "Doubtful Pass."

Some time in April, 1863, the party made its departure from this camp a short distance below Fort West, proceeding again to the "Southern Route," or Overland road, the first camp at all definitely described being indicated as follows:

"We arrived at and encamped near the entrance of what was known as 'Doubtful Pass.' It was a deep, narrow and meandering gorge through high mountains, and about one and a half miles in length. The walls were about perpendicular, and in some places very high, and in others not over thirty or forty feet in height. It was a fine place in which a few Indians might kill a hundred men by throwing rocks down upon them, without danger to the assailants. About midnight the party were surprised by our Captain's order to 'catch up.' Our animals were soon packed and again strung out in the usual single file through this ugly canyon. It was a quiet and silent march, as well as a dark one, but we succeeded without difficulty, and came out on the other end, onto a real desert, upon which we encamped for the remainder of the night. * * * * *

"We steadily pursued our course, without unnecessary halt or delay, over a desert country for many days, and until we arrived at Tucson, which had lately had the honor of becoming the capital of Arizona." (This of course refers to the provisional government of the <u>proposed</u> Arizona.)

Didn't Find Many Inhabitants.

This was about the beginning of May, 1863, and Conner says that if there were 150 inhabitants they were not at home, while of this number he saw only three white men.

A Snake Incident.

Stopping at Tucson a night and day, the party proceeded toward the Pima Villages on the Gila. When about half of the distance had been covered they made a noon-day stop in a grove of thinly-scattered mesquites.

"While we were at this camp we ate dinner, and one of the party, George Lount, who was on guard the night previously, had gone a little distance from camp and out of the way of camp bustle, to refresh himself with a short nap. He stretched himself, therefore, flat upon his bank on the sand and in the shade of a lone mesquite tree, and fell asleep. When dinner was ready, his messmate (Who, according to a foot-note, was Conner himself) missed him, and on failing to find him about the camp, began to suspect something wrong. Gathering up his gun, he went about from tree to tree, in search of his lost mess-mate. He at last came to the tree where Lount was lying on his back sound asleep, and discovered a huge rattlesnake creeping up to the sleeper on the opposite side, already within four or five feet of him, evidently thinking to creep under an old log. But when the ugly monster saw the messmate approaching Lount it began slowly to coil itself preparatory for a leap, if found to be necessary for defense against the intruder. The snake's head was raised high off of the ground viciously, as he began to shoot out his ugly black tongue and to rattle his warning. Our friend found himself in a quandary. If he wakened Lount he would lazily stir himself up, unconscious of the snake's presence, already in striking distance, and receive its deadly fangs before explanation could be given him. The snake was already coiled into the requisite attitude to strike. The messmate, like any other clear headed man who was inured to hardships and sudden dangers, quietly and cautiously knelt upon one knee in the sand and very slowly presented his gun across the body of the sleeping man, and as near to the reptile as he dare risk, took deliberate aim at the snake's neck as it arose glaring at him, and fired. The sleeper sprang to his feet in surprise * while the messmate simply pointed to the snake, with its head shot off, which Lount had not yet seen."

Pima Villages and Maricopa Wells.

A flag of truce was necessary before the party was able to pass through the Pima Villages, as the Indians were laboring under the impression that the white men were Texans, against whom they had been warned by the United States forces. A pow-wow cleared the atmosphere, however, and passing through the villages a camp was made at Maricopa Wells. Here two days wewe spent, and Conner takes occasion to describe the Apaches:

Something of the Apaches.

"The great Apache nation was a federal institution, subdivided into several lesser nations, each having its own chief. I could record several of these branches if I knew how to spell their names, but they are made up from broken Spanish and Indian lingo, and have not yet appeared in any history at my command. But I shall, however, attempt some of them, as directed by their sound: Mesquilaros, Tontos, Kiotaros, Pinals, Hualipais, Apache-mohaves and a White Mountain tribe, and others. The Apache-mojaves are of a recent tribe and took the Mojave part of their name from the necessity of seeking the protection of the regular Mojave tribe against the main body of the Apache nation, from which they had recently seceded. They are all Apaches, and so call themselves. They fight together, live together and go on the warpath together, recognizing one common, great chief, to command all of them when necessary. They have different customs in many ways, from each other, all trifling in effect. Some are known by having two barbs of feathers on their arrows, by which the arrow is made to fly straight, and others have three for the same purpose, while others again fix the feathers cylindrically like the threads of a screw,

Maricopa Wells Stage Station, Pinal County, Arizona (circa 1863).

to give the arrow a rotary motion like the rifle does its ball. Some of them form their moccasins so as to turn up at the toe. The Tontos cut their hair short, and differ in this respect from any other Indians I ever saw. They cut it square in front, so that it comes just to the eyes, thus concealing their forehead entirely and giving the wearer a hideous appearance. * * *

"The Maricopa men have aquiline noses and the women have retrousses. They surely are of the same tribe, and this peculiar difference is very noticeable."

By Way of Sacaton.

"On a pretty morning," though "the weather was becoming exceedingly hot, we were ready to start," and their progress was recorded thus:

"We left the Pimas, crossed the Gila and followed down the river, on the opposit side, some ten or twelve miles, and encamped for the night in the sight of an old monument built by surveyors to mark an original boundary between the United States and Mexico. This camp we called the Sacaton station, after the grass of that name which we found in abundance at this place. A little further down, on the next day, we came to two graves which were covered with rocks and prickly-pear plants and other specimens of the cactus family. These graves contained some of the bodies of the Oatman family, whose history was a sad one."

Departing for the Desert.

"Before departing for the desert, and on the eve of starting, we were joined by the chief of the Mojaves, Irotaba; a black warrior, a squaw, and Charley, a Mojave boy perhaps twenty years of age, who had loafed about Fort Yuma, in California, long enough to speak a little broken English. They were enroute for their homes on the Colorado river, beyond our destination in Central Arizona, and the chief proffered his services as guide, which were accepted of course. In company with this cortege we started out boldly upon the desert for new and unexplored lands. * * *

"Irotaba could not talk one word of English, and if he knew any Spanish he kept it to himself. He seemed to be a kind of protector of a little disaffected branch of the Apache nation, which roamed some of the country into which we were going. He could speak their language, but Charley could not, so our information had to be drawn through several channels before its final reception.

A Desert Water Hole.

"We learned from the chief, through Charley, that there was

some water in a little slump of hills, about midway of the desert, unknown to the palefaces, and that he was directing his course to it. We traveled this desert in the night, as much as was convenient, on account of the heat. * * * We arrived at the clump of hills about noon on the following day, and halted for rest and water. The old Indian pointed up to a cliff to indicate the place where the water was to be found, and Charley, followed by some of the men, departed through some gorges out of sight, but they were presently seen scaling a precarious wall, upon top of which they finally found the water. They had taken a rope off of one of the pack-mules, with which they let the water down by suspending to it the canteens, gourds, etc. This water was alive with little flat, scaly bugs, about the dimensions of hungry chintzes, which are called in the Pacific country water lice. * * *

"There was no spring here, but some tanks, naturally worn under some cliffs, held this fragrant water, all of which we used before we left it.

"We were soon in single file again on a northerly course and traveling toward the mountains bordering the desert in the distance."

Reach the Haviamp.

Traveling much of both day and night, and suffering considerably for want of water, in a few days they made camp at a point which sun-rise showed them was "within five miles of what could be easily seen to be barren mountains.

"We were off again by daylight, and in due time reached the dry mouth of the creek that we were destined to follow, pronounced by the Indians 'Haviyamp,' with a labored grunt on the 'p'. This creek emptied its high waters upon this desert, but as this was a low-water time we had to follow its dry, sandy bed quite a distance into the hills, through a gorge-like channel, before we reached the damp sand at a point to which the water had flown the day previously. This, I thought, was a singular freak, but I learned since that it is quite common in this dry country for the water to gradually cease to flow at night, and then extend its distance as the sun rises.

"We kept up this creek, and the sand became more damp as we proceeded, and finally met the water coming, and encamped just out of the banks, and in half an hour had quite a flood of water running at our feet. And we all did drink. At night the water again restrained itself, to the extent that there was none to be had, and again came afresh on the following morning and lasted all day. * * *

Camped Close to the Vulture.

"We moved up stream, and encamped within a few hundred yards of the subsequently discovered and famous gold lode known as the Vulture. For several hundred yards the croppings of this mine stood up several feet high out of the ground, within plain view of our party, who only observed it as we did any other ledge of rough rocks. But, being black from the weather and time, it was calculated to deceive anyone as to its real character. * * *

"Day by day we continued our journey up this stream, which I thought in my soul was the roughest country, to be composed of ordinary mountains and hills, that I had yet seen in all my Rocky Mountain experiences. The creek was a narrow, gorge-like canyon, that could not be traced, either direction, with any degree of certainty, as to its general course, and it wound its crooked way amongst the mountains, apparently, down deep beneath the general level of the balance of the country. Numerous obstacles in its bed, such as falls, accumulated boulders, etc., over which it was impossible to take the mules, frequently turned us out of the steep banks, where our route would become either up or down, continuously, on account of the numberless arroyos, gulches andgorge-like hollows found to be pitching down the mountains into the main creek. This kept us in continual search for places to get up, down or across these precipitous barriers. The only trees about were the sparse clumps of cottonwood, in the bed of the stream, and the pines in the distant heads of hollows, against the summits of the main ranges on either side. This extensive uneven scope of ridges and hollows was liberally clothed with a slow, desert growth of stunted chapparal, mostly of oak and manzanito, three or four feet in height. * * * *

Found Gold in the Haviamp.

"We found a liberal prospect of gold all along this creek and its tributary hollows. We slowly proceeded up stream until we came to quite a stretch where the water in the creek had sunk beneath the surface, leaving the bed dry, and at the instance of Irotaba we encamped under one of the half-dozen large cottonwood trees, there on the sand in the creek.

A Pow-wow Withthe Apache-Mojaves.

"The chief told us, through our interpreter, Charley, that he had been signaled for a pow-wow, by the natives. We were at a loss to know how he got his signal, for he and his associates were all the while with us, but as the slang phrase goes, 'he got

Prospectors working "washer" at the Vulture Mine (circa 1935).

The Vulture Mine office (circa 1935).

it all the same.'

"Irotaba was the chief of the Mojaves, whose territory was nearly confined to the Colorado river and southern border of California. His people had fought the Californians until they had become about helpless, and they were friendly with the white man rather from necessity than from choice. Irotaba hesitated not to say so. He was a tall and large man, with a careworn cast of countenance, knitted brow, and of a quiet, observant and thoughtful demeanor. He was dressed, upon this occasion, in a cotton, checked shirt, thin blue cotton pants, and a chip, or straw hat of Mexican manufacture; was barefooted part of the time and wore mocassins part of the time. None of the savage features that so distinctly stamped those of Mangus, whom we captured, were to be traced in the face of Irotaba. As before stated, he was acknowledged as a kind of protector over a small band of Apaches, who inhabited this immediate locality, who called themselves Apache-mojaves in consideration of this alliance with the Mojaves, but all the Apaches roamed this whole country at pleasure. Irotaba has since visited Washington under the supervision of Capt. John Moss, and was said to have learned to speak English while on his trip thence, * but at the time he was with us he knew no English.

"However, when we were all settled in camp, in the shade of the cottonwood trees, on the Haviamp, Irotaba delivered, in a yell, a string of gutturals that resounded all through these canyons, gorges and craggy mountain sides, and then stood looking up the rocky defiles perfectly quiet. After quite a pause he raised one arm aloft and steadily held it there, while his eyes were fixed upon a high object up in the mountains. He stood there for some moments, like some crazy practitioner in witchcraft, and then quietly resumed his seat upon a large rock in the edge of the creek, on the same side and to the left hand of Captain Walker, who was seated upon his blanket and saddle. Charley was lying in the sand in front of them, and the black warrior with hair on his legs, and the hump-backed squaw, took rather a back seat. The members of our party who were not on guard, lay scattered around in the sand patiently awaiting the result of Irotaba's vocal hieroglyphics. Although we were silently awaiting the arrival of the savages, whom we had not seen or heard anywhere in this apparently habitless country, we were all surprised to see thirty or forty of them, black and nearly naked, watchfully and stealthily emerge from the nearest rocks around our position, nearly within ten paces, and trepidatiously halt, as though they doubted the propriety of coming nearer. This was a sudden apparition to me, for a low hum of interjections from some of the boys caused me to look up and find this crowd of Indians all in sight, without knowing how they came. If Irotaba had never been with us we should never have known, perhaps, that one of these savages were anywhere about us, unless we had been attacked by them. * * *

"A few gutturals soon had them arranged around in a semi-circle, and the treaty-making began. The Indians stood like they were at a funeral, in silent attention. Capt. Walker opened services by saying, 'Charley, tell Irotaba to tell these people that we are strangers, and come to them as friends, and don't want to hurt or harm them.'

"Charley shot off some gutturals at Iretaba, after which the latter arose from his seat and in a standing position delivered a little, earnest speech to those naked creatures, who seemed not so well pleased with it as they might have been. Irotaba evidently did not want us to go further up that creek. He resumed his seat at the end of his speech with his brows more knitted than ever, and Capt. Walker did not believe that he had told the Indians what he had been directed.

"The Captain continued: 'Charley, tell Irotaba to ask these people whether or not they are going to be our friends.' Charley got off some more of his Mojave gutturals, while he was evidently embarrassed. He employed himself, as the lay on his elbow in the sand, by picking little straws to pieces as he talked. Charley evidently felt that he was in a tight place.

"Iretaba arose again and made a long speech with the gutturals, he also seemed to make, as he proceeded, and a number of savage glances by the natives, one to another, was the consequence. Irotaba again resumed his seat, evidently unable to restrain his feelings, and was plainly dissatisfied about something.

"Our side made several similar attempts at learning their feelings toward us, as well as their intentions, but without success. Irotaba grew more vigorous and loud in his speeches, and looked more savage when he took his seat. The audèènce also showed great displeasure by their gesticulations, either at our language or that of Irotaba, the import of which we had no means of understanding. The surest means we had of understanding them was through their actions, which indicated more hostility than friendship, and therefore we had no confidence in the treacherous, half friendly replies which they gave us. We were probably trying to make the first treaty of peace with these natives ever attempted.

"Iretaba told us that we were the first white men ever so near the source of this creek, and further informed us that one other party attempted to ascend it and perished. * * *

"Our Captain made his last observation by telling Charley to tell Iretaba to tell his people that we came to hunt gold, of which they knew nothing and for which they cared nothing, and that we wanted to be friends and would not hurt them. 'But,' he continued, ' if they begin to steal our mules and to shoot our men we will quit hunting gold and go to hunting Indians.'

"Charley began his rigmarole as he mechanically picked at his straws, to all of which Irotaba listened attentively to the end, and then arose a little excitedly, and made a furious speech, which drew from his audience a few bursts of indignation. The savages would frequently, during this last speech, turn toward one another, overflowed with gutturals.

"Captain Walker asked Charley what Irotaba was saying to the Indians. Charley replied, 'Me don't know what he say when he talk Apache.'

" 'Well, what does he say in your language, to you?'

" 'He say he no fight if you no go up Haviamp.'

"That settled the matter, for we did intend to go on up the creek and therefore could no longer entertain the conditions of our friendly enemies.

"But Iretaba did not know that Charley had negligently given that little piece of information, and still thought that he was cheating us as to his own real views. At the end of Irotaba's last speech the savages would look at the nearest friend and lean up toward him, and deliver a number of 'Ugh, ugh, ugh's.

"Irotaba dismissed his Indians, who, excepting those who came with him, proceeded to scatter off in the rocks and disappeared as easily as they had come. They were unarmed, but of course their arms were not far off.

"The old chief tried hard to make us believe that there was no water above this point, and that we would perish if we persisted in going further into the country, and offered his services to pilot us across to the Colorado river, to the border of California, all of which we respectfully declined. He now became morose and silent, and after we all retired to sleep he and his associates of the Royal Family took French leave of us, and also took all the smoking tobacco that the party possessed with them."

Sickness Overtakes Them.

"We moved up stream about fifteen miles without finding water in the bed of the creek or anywhere else, but encamped at the end of the dry stretch.

"But now the continuous hardships of desert travel began to tell their story. One man after another began to get sick. F. G. Gillilan became delirious, and had to be supported upon his mule for hours, until we reached a camping place. Others became unable to travel, and there were but few who escaped. It always started by a sudden attack of vertigo, which would rapidly increase to unconsciousness.

Camped Below the "Big Canyon."

"My turn came just after eating a hearty breakfast, early on a pretty morning, when feeling as hearty as I ever did in my life.

We had encamped the night before on a little upland flat, just below what was afterward known as the "Big Canyon' in the Haviamp. The party had packed up and was in a line, going to the lower end of the canyon, in which there were green cottonwood trees within its walls for probably a mile of its length. Before descending the river bank to get into the mouth of the canyon I felt the attack of vertigo, which rapidly increased to violence and vomiting. Then a cold perspiration and deathly sickness followed, and by the time we were fairly into the canyon, winding a devious way between the cool walls and under the shady trees, delirious sensations came flowing over my senses, like an irresistible wave, to engulf me. I just had sense enough left to remember the unavoidable rough handling which some of the others had received while in the same situation, and therefore, without warning anyone of my condition, I dismounted, threw the bridle reins over the saddle and allowed my mule to keep its place in the line, as though I would walk awhile. As soon as the party had passed me I turned out into the shade of a large cottonwood, and stretched myself upon the cool sand, which had drifted deep around, and under this tree. A little later and I would not have had time to make any arrangements whatever, for I immediately swooned away into a delirious, horrible dream, which seemed to last an age. I would now and then awake into a semi-consciousness, just to again sleep off into unconsciousness. This flighty condition seemed to have lasted an indefinite length of time, as though I had always been there. But it began to subside as the cool sand allayed the fever, and a clear-headed recognition of my surroundings confronted me. I did not remember anything whatever of what I did with my arms, but found my gun sitting against an old fallen trunk some distance away, and my belt and two six-shooters lying loose and wallowed in the sand.

"I quickly started on upon the fresh trail of the party * (and met Sam Miller and M. Lewis in search of him, with a mule. The party was soon overtaken.)

Camped at and Named 'Walnut Grove.'

"The mules were ready packed and only awaiting for our arrival, when the line of march was taken up and continued until we reached a grove of little black walnut trees, where we again halted for supper. This is one of the prettiest points in Arizona, and has since retained the name we gave it, 'Walnut Grove.' * * * Here the Indians were hovering the summits and ridges of the neighboring mountains. The 'signs' were not quite right, and our captain ordered another 'catch up' only a few minutes after supper.

"Now came the tug-of-war. Bordering the upper edge of this pretty little valley, on both sides of the creek, lay a terribly broken and chaparral country, with a few scattering pines commingled. Just where the creek came out of this rugged country into the valley were a number of impassable falls, above which the creek was but little else than a gorge clear on to the distant pine woods which we had seen in the distance for several days, and were trying to reach. A woodland country began to look right pretty and inviting, since we had seen so little of it for several months. About a half mile to our right hand, facing up stream, and therefore on the left of the stream, was a tall, sharp, chaparral mountain peak. We left our little grove of walnut trees and climbed that mountain and encamped upon its treeless summit, among the chaparral, all night.

Named "The Tanks."

"Upon examining those falls we found deep holes worn into the rocks, where the water fell, and therefore we called the place 'The Tanks,' which name it afterward retained, and I presume it still does.

"I am more particular in describing this country than usual, because it was the place of our destination, where we expected to plant settlements, and I might state here that we succeeded.

"We directed our course up-stream quite a half-mile from the creek, and therefore in crossing the mountain ridges and their intervening hollows that pitched into the creek we had a constant up and down hill route. * * Upon arriving on top of the next ridge we turned at right-angles and followed this sharp backbone down to the creek above the falls. Here we found the immediate bank of this gorge so precipitous as to have to push the mules down upon their haunches to the bed of the creek.

Camped at Mouth of Copper Creek.

"We now turned up this narrow gorge a few hundred yards and encamped about dark at the mouth of another little creek, which flowed, when it ran at all, in from the opposite side of the Haviamp, on the west. Upon making tea with the water from this creek it was found to be so strongly impregnated with copper that it could not be used, and we therefore named this stream Copper Creek, which name it retained. * * *

"When we left the mouth of Copper Creek on the following morning we went up the steep, sharp and high spur formed by the junction of the two creeks. * * Upon the top of this hill we had an open view of the large woodland tract of country, which looked beautiful and inviting. We descended into the creek again, down from this steep mountain through dense chaparral, and left the bed of the creek only once more, to pass a blocked place, and came into the edge of the woods, and followed the creek a few hundred yards into the woods, and encamped for the night upon a pretty and comparatively level situation near the creek. There was no undergrowth of any consequence, and therefore the woods were just open enough to look pretty.

Made a Permanent Camp.

"We considered our long journey at an end, for we were at last in the unexplored regions of Central Arizona, the place of our destination, and located in the finest and largest woodland country by far to be found within the extensive limits of the whole Territory.

"This, our first camp in these woods, has always since been a prominent place, and will be alluded to frequently hereafter. It was within a few miles only of the extreme head of the Haviamp, whose mouth we had to cross deserts to find and whose whole course we had patiently followed as we might follow a gorge through an uninviting, craggy and hot country, clothed by chaparral only.

"These woods we found to lay in a strip, extending about twenty-five miles from southeast to nortwest, and about six or eight miles in width. Our camp was located in the south edge of it. Out of the woods, in all directions, lay a desert country, dried and parched. Toward the north was a grassy plain, perhaps over twenty miles in extent, but it was dry and almost waterless. These woods were on the highest country between the Rocky Mountains and the Colorado river, except perhaps the San Francisco mountains, which stood toward the northeast, about eighty miles away, but were distinctly seen from our 'big woods.' Therefore many creeks, of which accounts will be herafter given, have their sources in these woods and dry up in the distant deserts as they take their different courses north, south, east and west. This fact, to some extent, explains the reason that this country was so long unknown and unvisited. The water courses having their rise in it dried up before crossing the desert country which lies extensively all around it. There never was created a finer location for a headquarters of a lawless banditti than these woods previous to our visit there, so well were they fortified by formidable obstacles in the way of waterless deserts."

Satisfied With the Gold Prospect.

"We spent the following day in a more thorough examination of the bed of the Haviamp, to determine the character and extent of its gold production, and were pleased with the prospect. One day after another added confidence and resulted in the final decision that there was gold in the creek in paying quantities, and there was plenty of deer and other game in the woods.

"This being settled, the next thing to be done was the building of a permanent corral with which to secure the mules at night and to do away with the nexessity of standing around them all night continuously, as we had been doing constantly for the last nige months.

Took Turns With the Axe.

"We had but one axe, and by first one and another it was kept in constant use felling large pine trees, one after another, in the same direction, upon each other, until we had a large rectangular corral constructed with whole trees which were too large to be moved or broken through. This was the stoutest corral ever built in Central Arizona, and when the stock was once in it they might be killed but there was no such thing as storming them out of it.

The First Mining District Organized.

"The next move made was the holding of a miner's meeting. Being only twenty-six of us (some, as will be later seen, had deserted before crossing the desert. M. W.) there was not much discussion. The district was first defined, by bounding it by the distant ridges and primary ranges of the country, without the trouble of any measurement. A mining claim was defined and declared to be 100 yards of an ore vein or quartz lode, and the same extent on a creek bed in surface or placer diggings. The claimant was to hold all the ground on either side of the creek, in placer diggings, and on either side of the vein, if it was a quartz ledge, for a distance of fifty yards, thereby constituting a claim of 100 yards square, lying upon the 'pay streak,' whether of quartz or surface mines. The claimant was to own the possessory right to 'all of the earth and minerals therein contained.' It was further provided that each member of our party should be entitled to two claims upon all discoveries of any character of mines whatever--one by the right of discovery and the other by virtue of pre-emption.

"On permanent organization, Captain Walker was declared by acclamation to be the Governor of the District, and declined upon the grounds that a young man should be appointed. Whereupon the humble instrument of these notes, being the youngest of the party, was appointed, and also declined on the grounds that he would not be there any longer than the first opportunity to get away presented itself. Then Thomas Johnson was appointed, and accepted. I was then elected secretary and recorder of claims for the district, and declined, when Mr. Wheelhouse was appointed, and accepted and acted. I was one of the committee appointed to draft by-laws, and the writer of only a few resolutions, which were adopted and the meeting adjourned. Thus ended the first miners' meeting ever held in Central Arizona.

"One of the resolutions referred to provided for a committee to be appointed by the Governor, to measure and stake off and number fifty-two claims upon this creek (Haviamp) which ran by our camp, and to begin with No. 1, which was provided to be laid off just opposite the camp, and to number each way up and down the creek an equal distance. Then it was further provided that the secretary should mark on bits of paper the numbers, corresponding to the numbers on the stakes, both up and down stream, indicated 'No. 2 Up,' 'No. 5 Down,' etc. These scraps of paper, after due preparation as provided by the resolution of the meeting, were put into the soundest hat to be found in the party, from which each member drew one ticket twice around, and thus settled the personal ownership of the respective mining claims, indicated by the stakes, to the number of two apiece. * Each man now began to hunt his claim, by the numbers of his tickets, and to mark them on the trees. This humble miner drew No. 1, opposite the camp, which proved to be the most worthless claim on the creek.

Created the Name of "Hassayampa."

"Mr. Wheelhouse, while arranging his records, remarked that he intended to give this creek a Spanish pronunciation, and therefore entered the flat 'Haviyamp,' as pronounced by the Indians, with a strong accent upon the last syllable, upon his records, spelled 'Hassayampa,' which has been sustained by our geographies, thereby rendering a pure Indian word, by the assistance of a Spanish dialect, obscure and uncertain.

" I shall hereafter call this creek 'Hassayampa,' but it is not the pronunciation of the Indians. Irotaba, while indicating the passage around a great bend in this creek, after Indian fashion in such explanations, would raise his hand high as he pointed at arms length, and with his finger pointing toward the distant bend slowly and steadily follow its course around, repeating fast all the while the name of the creek--'Haviamp, Haviamp, Haviamp,' and 'Haviamp' to the end of the course indicated.

Indian Name of Lynx Creek.

"There was another creek that headed in these woods which he called 'Cum-yam-muck,' with a strong accent on the second syllable, and whose course he indicated in the same way as he did 'Haviamp.' I believe it to be the creek known now as Walker, or Lynx Creek."

Return to Maricopa Wells for Supplies.

Return to Maricopa Wells for Supplies.

"We had now but one more duty to perform which appeared to me a hard and doubtful one. That was to return over the deserts to the Pima Indian village, to obtain some of their home-made flour and frijoles, as we would soon be out.

"This was in June, 1863, but we knew not what day of the month it was, and we really dated our records as of that month at a venture.

"We dug a deep hole, off from the corral on the opposite side of the creek, in which we buried all of the superfluous plunder, and then built a fire on the ground to conceal the fresh dirt, and took our departure down the Hassayampa, bound for the Pimas again.

" * * * We again reached the Maricopa Wells, and encamped there, and did our trading, and left letters with the Pimas to send to New Mexico and California by the first soldiery which might pass either way. These letters, of course, gave a full account of the flattering prospects for the future of our new El Dorado. We were gone to the Pimas for supplies just twenty days.

Encountered the Pauline Weaver Party.

" * * * We packed up and started across the desert just before night, and about midnight we were surprised to learn that we were about meeting a body of mounted men on our trail, and stopped under preparations for a conflict with the Apaches.

"But the strangers proved to be a party of men from California, who had heard of the Walker party strolling around through Arizona for the last year, by messages to the military department from the soldiers whom we had met before. These men had come into the desert from the Colorado river under the leadership of Pauline Weaver, an old trapper. They continued on to the Pimas, after receiving the description of our New El Dorado, and immediately returned on the trail as far as the mountains, and began to prospect on their own hook forty or fifty miles southwest of our woods.

Back at Home.

"We arrived a second time in the woods safely * * and found our camp just as we had left it. We again settled back into our old and respective quarters around the corral and devoted our time to business.

Built the First House in Central Arizona.

" * * * My three messmates and myself concluded to

build a stout log house, which would be convenient for a fort in case of necessity for self-defense. We finished one in about two weeks, built of logs three feet in diameter and covered with stout timbers, upon which we put pine boughs, and lastly a foot of dirt, which rendered it fireproof, but if it ever became waterproof it was after I left the country. This was the first house built by any of our party, and consequently the first ever built in Central Arizona, that we know of, by the Angl-Saxons.

"We now devoted our time to hunting and prospecting, when not at work upon our claims. Trees were felled, out of which the ordinary rocker was constructed for washing gold. Rude planks, which were split out with an axe, had to answer our purpose, and such implements as could be made out of such materials for mining purposes were pretty soon in operation.

"The diggings did not yield near our expectations, but it was enough to convince usthat we were in a gold region, where the best of mines might be found.

Forerunner of Apache Troubles.

"The Apaches began to visit us. They seemed not to know anything of our efforts to make a treaty with Irotaba. They importuned us by long and senseless speeches to go with them upon a 'big' bear hunt. They were troublesome only for a short time, for they went to war with us, and that ended their visits. One of our party improvidently kicked one of them out of his way while cooking his supper, after a tiresome day's tramp, and the whole batch of them immediately left the camp in great indignation and never returned again."

Discovery of Lynx Creek.

"It began to get late in the fall, when the members of the party were prospecting about the different tributaries nearest our camp, in small squads, to find better diggings. Pursuant to this object a number of us followed the Hassayampa to its source and passed over a range lying at right angles to it and out of our organized district. On reaching the eastern foot of this range we found another creek and encamped near its head. We foundnhere the best diggings yet discovered. While prospecting about a mile upon its course Sam C. Miller came across a lynx, and shot and wounded it. Instead of trying to escape the vicious little animal made a charge upon its assailant, who had to use a six-shooter vigorously before it was subdued. This incident gave this creek the name of Lynx Creek, which it retained.

Discovery of Turkey Creek.

"We now passed out of the head of Lynx Creek toward the southeast and over another range, and thence descended into another creek whose course lay about east. Here we found some gold, but the prospect was not flattering. After traversing this creek a few miles down its course we came nearly to the edge of the woods which bordered on the rough chapparal country, where we found a large covey of wild turkeys, some of which we shot and ate for supper. This incident gave this the name of Turkey Creek, on which many important scenes were subsequently enacted.

Discovery of Wolf Creek.

"We turned at right-angles with this creek, and passed out of it and over the adjoining range north into a chapparal country, and thence into another creek nearly parallel with Turkey Creek. We were now about twenty-five or thirty miles from our camp. We camped in this creek and dispersed the individual members of a very large convention of wolves. For what purpose all of these wolves were assembled on this creek I could never surmise, but there were enough of them to give this creek the name of Wolf Creek, which it ever after retained. We found no gold here at all.

Discovery of Big Bug Creek.

"On the following morning we continued our course north and found another creek, in a chaparral country but not far from our woods, and it really headed in the edge of them and ran east, as did the other creek just passed over. This creek subsequently became the centre of a fine mining district, whose mines will continue to be worked probably for fifty years to come. Here we found gold in what we supposed paying quantities. Before we left this creek, J. H. Dickson, a large, good natured, noisy fellow, came into our temporary camp carrying between two sticks the largest bug that ever grew there and the largest that I ever saw. He laughed and yelled over it until every one of our party had to examine the wonder. This creek became Big Bug Creek in consequence, and a Big Bug District has since been organized here.

Discovery of Granite Creek.

"We left this creek and passed northwest over a range and into Lynx Creek lower down than where we passed it in going east. We kept on our northwest course for probably ten or twelve miles along the edge of the woods and the border of the grassy prairie lying north of these woods, already referred to, until we reached another creek which flows out of the woods and takes its course across this grassy plain for thirty miles in a northern direction. Here we encamped for about two days on the bank of this creek, in the edge of the woods. We called it Granite Creek, in consideration of the amount of granite boulders found upon it and scattered in the woods near it.

Site of Prescott City.

"I will here remark that our camp ground here is covered now by the city of Prescott, which has been the capital of Arizona. We encamped on the right side of the creek, in the mouth of a little valley, probably twenty paces wide and about 300 or 400 yards above the present public square of the city. Upon the opposite side of Granite Creek, in the flat woods, was located an Indian rancheria. We had a little conflict with them before we were through, * *

"We decamped, and traveled across the woods south for about the distance of six miles to our place of beginning, and found our house, corral, etc., just as we had left them. * * *

"Captain Walker, who had been roaming the mountains for thirty years, remarked that this was the only unknown section of the United States, and that this was probably his last expedition, and therefore we had better stay here, as it might be the best region now to be found. The only fears he had were that we might fail to hold out against the Apaches until we had assistance from immigration. And it was feared that we would have to hold out for nearly a year before we were found by enterprising gold hunters."

Moved to Lynx Creek.

"We had occupied our encampment with only one house and one corral for several months, and until we found that our claims on this creek were not going to pan out as liberally as we supposed, and therefore began to prepare for a change. We made preparations over on Lynx Creek for a new home. The heads of these two creeks lay within less than a mile of each other, but flowed at right angles with each other near their sources. We bade farewell to our old camp and soon had a new headquarters established on the head of Lynx Creek, in a new field for activity. The placer diggings opened up flatteringly, and proved to be worth all of the trouble it cost in the exchange. We now considered the question of residence finally settled. * * *

"This creek was nearly entirely dry, and we therefore used rockers of home-made manufacture, with which some of the men would wash out as high as $50 in a day, some days.

"I again made another raid upon the deer, with better success, and on returning to camp I crossed the creek quite a distance from camp, at a lonely and lonesome place, where the bed-rock bottom was bare and clean. I stooped and chipped off a small fragment of the shattered bed-rock with my knife, under which I found a particle of gold of about the amount of a dollar. I then laid aside my gun and searched the little crevices until I had obtained an ounce of gold, and all in less than an hour. I began to feel that I was giving too close attention to gold-gathering to be safe, and that I might be shot without knowing how it was done. I therefore located a claim here and left for camp.

"We began to find gold enough for the present use, but could buy nothing with it, and it therefore appeared worthless at last. No one but he who has experience this sort of a situation can properly appreciate how little riches are coveted. Food appeared to be of immense value beside that of gold. When I returned to camp on this day my dinner consisted of hard bread, made of unbolted flour, alone. * * * But it is not intended to be understood that we lived this way all of the time on this creek, for when we captured a deer or a bear we generally made up for these fasting spells."

Conner Tried to Leave.

Eager to get back to civilization, but unable to induce any of his comrades to accompany him, Conner started alone for the Colorado river, intending to cross the desert to the settlements in California. After having several adventures with the Indians he reached the river, where his horse was stolen from him, and he was compelled to return to the camp of the Walker party, afoot. This trip occupied twenty days or more.

Found by General Clark.

"It was not long after my return when some of the men who were out hunting came in at night and reported that they had seen from the mountain top some white objects, like tents or lodges, off toward the north, on the grassy prairie, some ten miles away. Captain Walker said that he guessed that the imagination had helped the picture, which he rather supposed were probably rocks in some of the departing creeks from our woods. But they were confident that it was somebody, and therefore arrangements were made to investigate the strangers on the following morning. When morning came all were impatient about 'yesterday's' phenomenon, but Captain Walker, whose prudence was always on guard, suggested that care and caution should have the first consideration. He said the Navahoes roamed in that direction, and that they were to be feared. He sent a little party up the mountain with the telescope, who came back and reported that the objects were in the same position that they were the day before, and that they were United States government mule teams. They appeared to be swamped from some cause or other. We took our usual dinner of unbolted flour and gravel, and had hardly finished when we discovered a half company of United States soldiers come riding down the creek, single file, from the opposite direction of that of the wagons. They were within a hundred paces of us before we saw them. Because their clothes were not patched all over I thought that they were the finest dressed set of men I ever saw. We had not seen any white people for so long that we were delighted to see them. They had passed our camp and gone entirely through the woods to the south edge, and turned back and found our encampment on their return. That accounted for them coming to our camp from the south. The squad was under the command of the Surveyor-General of New Mexico, Gen. Clark, He informed us that our letters written from the Pima Indian villages on the Rio Gila had found their way to New Mexico, and upon their receipt Gen. Carleton, commanding that District, immediately sent this command to search for us until we were found. They had been out some months, and found us by accident at last. While on this prairie one of their wagons had broken down, and while delayed there they saw the smoke arising from our woods, differing altogether from the Indian signal smokes seen on their route. They also heard the distant reports of fire arms more regularly in these woods than elsewhere, and therefore determined that they would investigate our Eden. They said that the locality had been so well indicated in the letters that they traveled 525 miles without varying from the true course but twenty-five miles, and that was to the west of our position.

"The General examined our mines, panned out some gold out of some rich dirt prepared for him, and pronounced the diggings rich. He said that we had a romantic situation and pretty woods, if the lower region of desert country did effectually corral it. The soldiers stayed with us and rested a few days, and took their leave of us back upon the same route which they came. Their route lay by the Zuni village, thence to Albuquerque in New Mexico. They deserted their wagons, which they left standing in the plain north of our woods, where we had previously discovered them.

Weaver's Discoveries.

"Soon after the departure of General Clark, the old trapper, Pauline Weaver and party, already referred to, discovered rich diggings about forty miles southwest of us. The creek in which the gold was found took the name of Weaver, after the name of the old trapper, and retains it.

Peeple's Valley.

"There was a pretty, grassy prairie near this creek, of several hundred acres, 160 of which was located by one Pee-ples, of Weaver's party, and subsequently bore the sobriquet of 'Peeples' Ranch.'

Naming of Antelope Hill.

"The owner thereof, upon the arrival of the party into an encampment there, chanced to kill a couple of antelope on this ranch, which gave the large hill standing in the angle of the creek and ranch the name of Antelope Hill.

Discovery of Rich Hill Diggings.

"In a little arroyo on the summit of this hill was discov-ered the richest placer diggings found in Central Arizona.

"This little party located at Antelope, or Weaver, or both, and maintained themselves as possessors until immigration came, and made a permanent settlement of it.

Arrival of Van Smith and Party.

"The first immigration to arrive and come to our assistance was Van Smith and party, from California. It was late in the fall of 1863. They located on Granite Creek, at the north edge of our woods, upon the present town site of the city of Prescott.

First House in Prescott.

"Geo. Lount, of our party, built the first house ever erected there. Smith located a ranch there, built a cabin and corral, and accepted the stock of immigrants to graze and care for at $1.50 per head per month, and the place thus became a general headquarters for the incoming immigration.

"Smith's party was followed by another, and another, until arrivals were numerous. Strangers were seen daily coming, in large and lesser parties. Parties of men soon took possession of all the mines which we had previously deserted on the Hassa-yampa, including my old cabin and the old corral. Our old camp on the Hassayampa, the first and oldest of all the camps, became the headquarters of all the immigration that came from the Gila, on our old original trail.

"Lynx Creek was soon claimed up, from one end to the other, and cabins built upon its course every hundred yards upon that portion which lay in the woods. Every prospect for a speedy population of the country about these woods was considered good.

First Killing of White Men.

"The Indians discovered that there was not a Capt. Walker at the head of all this immigration, and profited by that discovery, for they began to succeed at killing and capturing stock on every hand. They killed seven men on the Hassayampa at one swoop and drove off their horses. This was the first killing of white men about these woods, but the example was followed up promptly, and every day saw some unlucky one bite the earth. Men located claims one day, and abandoned them the next. Cabins would be built, and deserted in four or five days. Parties of men would arrive, and take their leave for the States without investigating the cause of the excitement, as though they had expected on their arrival to load up with gold and return to their homes. Many expressions of surprise were indulged as to the merits of what was called 'this God-forsaken country, where the grass even had thorns on it.' I heard one fellow say that he had no doubt but there was plenty of gold in this New El Dorado, but that it had too much dirt mixed with it, and he therefore considered life too short to begin to pan out a desert to obtain it. Another declared that if he could load his pack animal with gold and start immediately, it would not bear transportation over the route to California. Many of these new immigrants lost their animals, and many lost their lives and animals both on the route, while others who were more lucky would sell out their surplus outfit and immediately take their departure for home, whether they ever reached it or not. Of course many of them failed, for the Apaches were having a glorious picnic.

"The first inquiry always made by the immigrant on his arrival, was, 'Where is the Walker party?' They seemed to be deceived, for they had expected to find the Walker party and failed. Of course the identity of the old party, as a body, soon became lost, in the tide of immigration, and the individual membership thereof scattered."

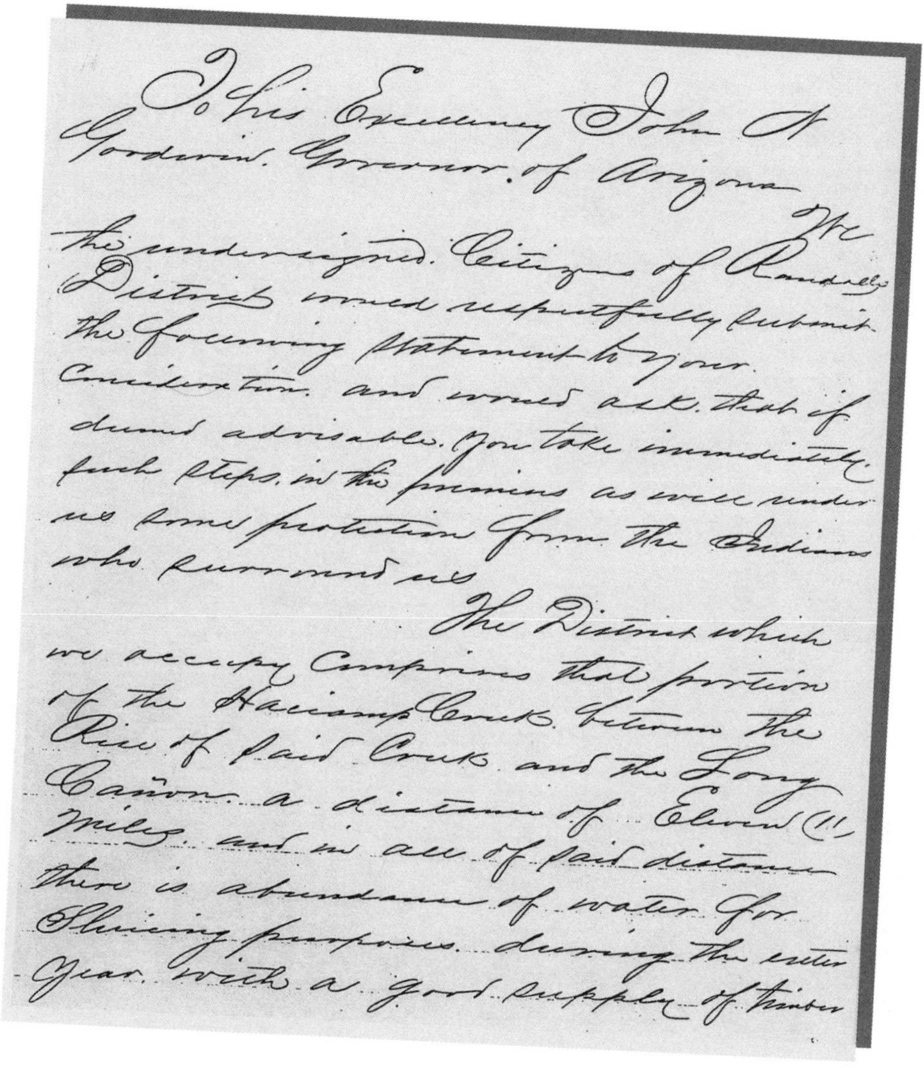
To His Excellency John N Goodwin. Governor of Arizona

We the undersigned. Citizens of Randell's District would respectfully submit the following statement to your Consideration. and would ask. that if deemed advisable. You take immediately such steps. in the premises as will render us some protection from the Indians who surround us

The District which we occupy Comprises that portion of the Hassiamp Creek. between the Rise of said Creek. and the Long Cañon. a distance of Eleven (11) Miles. and in all of said distance there is abundance of water for Sluicing purposes. during the entire Year. with a good supply of timber

Petition to governor of Arizona regarding Indian raids, signed by Jacob Waltz.

skirting the banks of the Stream
and a luxurious growth of grass
over the extensive Valley and surround-
ing Foot Hills. There have recently
been very rich deposits of Gold discovered in
almost all the Bars of said Creeks
and extensive preparations are now
being made to carry on active mining
opperations. besides many Claims
which are already opened. and
paying largely. We are in an isolated
District. being twenty miles from
Indian Gulch. the nearest Point at
which we can obtain supplies. and
we are Compelled to Keep the animals
necessary to Pack into the District
all such supplies. This District
combines probably more advantages
than any other portion of the Territory
and will be one of the last points
abandoned. as the mines are Rich
and the facilities for working them. great

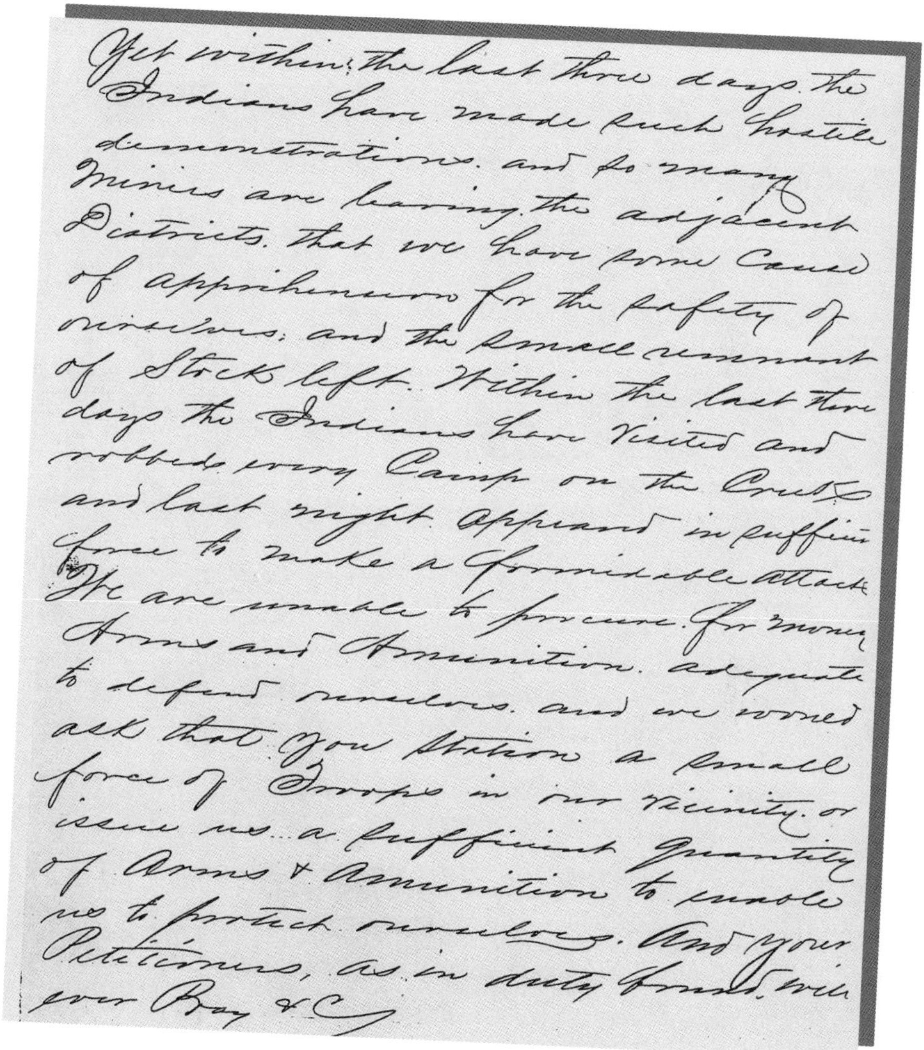

Yet within the last three days the
Indians have made such hostile
demonstrations. and so many
Miners are leaving the adjacent
Districts. that we have some Cause
of apprehension for the safety of
ourselves: and the small remnant
of Stock left. Within the last three
days the Indians have Visited and
robbed every Camp on the Creek
and last night appeared in sufficient
force to make a formidable attack
We are unable to procure. for money
Arms and Ammunition. adequate
to defend ourselves. and we would
ask that you Station a small
force of Troops in our vicinity. or
issue us a sufficient Quantity
of Arms & Ammunition to enable
us to protect ourselves. And your
Petitioners, as in duty bound. will
ever Pray &c

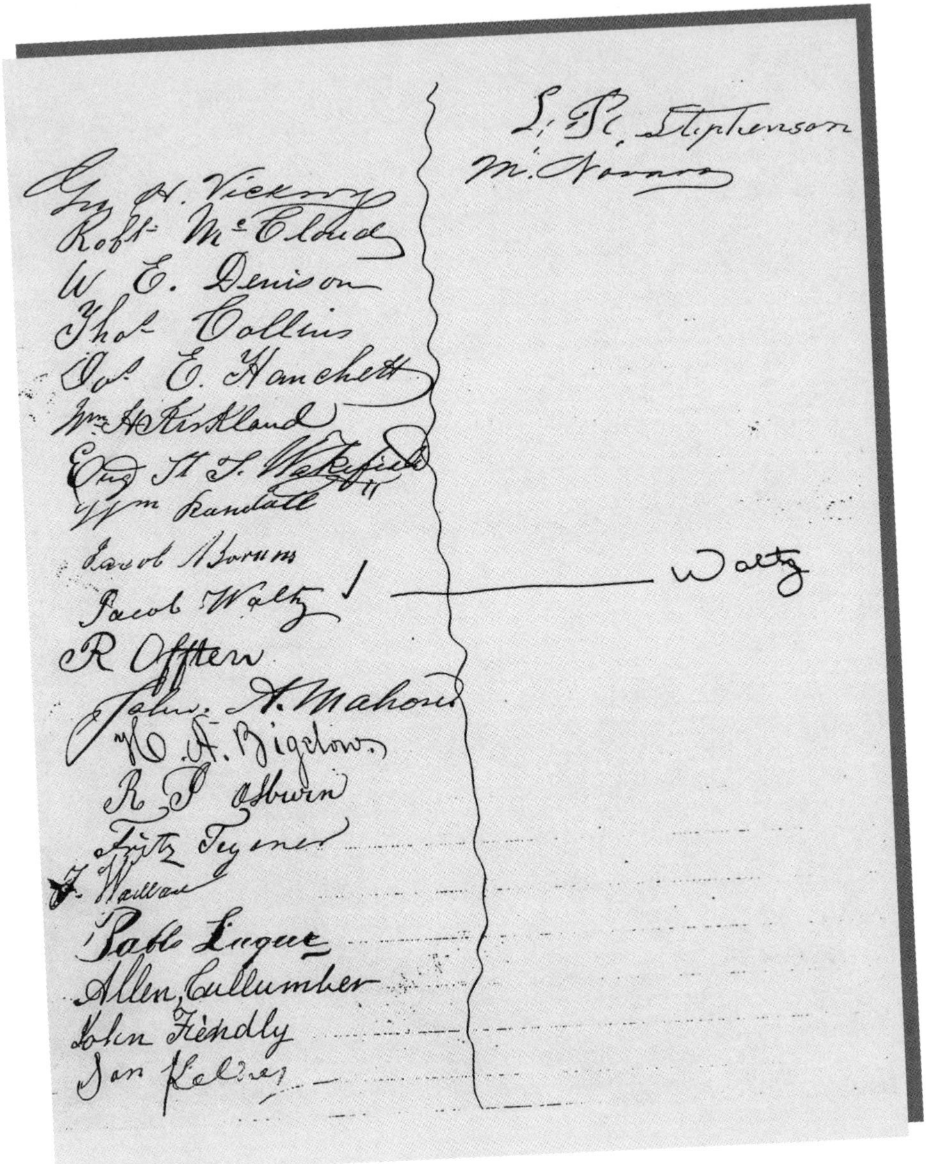
L. P. Stephenson
M. Navarro
Geo H. Vickery
Robt McCloud
W. E. Denison
Thos Collins
Jos E. Hanchett
Wm H Kirkland
Eng H. T. Wakefield
Wm Randall
Jacob Barums
Jacob Waltz — Waltz
R Offten
John A. Mahoney
H. A. Bigelow
R. P. Osburn
Fritz Tegner
J. Wallace
Pablo Luque
Allen Cullumber
John Kendly
San [illegible]

Chapter 4

Mining in Prescott
Part I

Since Jacob Waltz had already mined with A.H. Peeples and George Roberts in Grass Valley, about 20 miles from Sacramento, California, he was rehired in Kernville in 1861. Waltz arrived in the Kernville camp with his 20-year-old nephew Peter Backens. (It is assumed the young man was his sister's son.) Several other Germans arrived there looking for work. Floods in the mines they had previously been working had forced them out.

Waltz was known to the group, and there were others who were also renewing old acquaintances. George Roberts hired Matt Cavaness and other members of his family. Matt Sr. and Matt hauled timber off nearby Greenhorn Mountain. (Later, Matt Jr. would haul freight for the famous Silver King Mine near Superior, Arizona, and would own a well-known cattle and milk ranch [as they were referred to in those days] in the Superstition Mountains that Jacob Waltz referred to as the "Cow House.")

Early in 1862 the Kern River mines in northern California were played out, and the Owens Valley Indians were on the warpath – stealing, killing and raiding, which denied the supply trains a way into the camps. A.H. Peeples, Henry Wickenburg (sometimes spelled Wickenberg), Joe Green and George Roberts decided to explore the Arizona Territory after hearing numerous rumors of minerals near the San Francisco Mountains. Peeples mustered Pauline Weaver, the famous mountain man, into service as their guide.

In March 1863 Peeples, Weaver and Joe Green set out for Fort Yuma, where they outfitted provisions for 30 men. Excitement mounted. Now, they had the expertise of the seasoned Weaver.

It was agreed that a dozen men would lead the expedition, searching for minerals and the best route. The rest of the group would follow as needed. Weaver safely led them through Apache-infested mountains.

Near present-day Stanton, they shot two antelope, which gave the mountain and nearby creek their names. They set up camp on the newly named Antelope Creek, and shortly they located a rich placer deposit. Later, a pack mule strayed to the top of a nearby hill; one of the four went to retrieve him. Bonanza! Up on top was discovered a fabulous deposit of gold nuggets just lying on the ground waiting to be picked up and much more lying just beneath the surface. Within a few days, they had retrieved over $100,000 in nuggets. This strike was one of the richest ever made in Arizona and eventually was called "Rich Hill." (Jack Swilling took credit for the find, but it seems pretty certain the Peeples expedition actually found it.)

Historian Bertie Roberts revealed a quote from A.H. Peeples: "We dug the nuggets out with butcher knives and our bare hands, some were as big as hens' eggs. We filled our canteens and everything we had with gold." Later, one nugget was sold for $900 when gold was selling for $16 a Troy ounce; so, the story was not an exaggeration.

Shortly, two other strikes were made by the follow-up group that included Jacob Waltz. Gideon Roberts discovered some rich benches along a creek that became known as Walnut Creek, and Samuel Groupa discovered fine gold in the sand along a three-mile stretch of what would later be named the Hassayampa River.

Provisions were running low, and Weaver knew they could stock up at the friendly Pima villages near Maricopa Wells. Along with the mountain man, Peeples and several others left to prospect the mountains and river drainages on the way to buy supplies.

Along the way, Henry Wickenburg, a 44-year-old Prussian, made a tremendous, accidental strike. The gold lode became the famous Vulture Mine. (The mine was later sold to unscrupulous folks who cheated Henry out of his fortune. In 1905 at age 85, disillusioned and no longer able to live on his own, he committed suicide. However, two towns grew out of the strikes – Vulture and Wickenburg. The town of Wickenburg survives and is a popular winter resort area.)

For the following six years, the tough band of prospectors worked their rich finds in Prescott in the vast range called the Bradshaw Mountains. It was noted that Jacob Waltz often bought ore from the Vulture to mill out the gold. (Although it was rumored that he worked at the Vulture and that the ore from the Dutchman mine came from there, a complete search of the Vulture's records contradicts that statement.) Waltz staked out claims at Turkey Creek and Lynx Creek, which were recorded in his name. There are no records of the successes of those sites; however, one would guess Jacob had done pretty well thus far. In 1868 it is recorded that he homesteaded farmland in Phoenix along with many of his partisans.

We must also take note of another name that is to become prominent in the Waltz saga. There was a German man who was known to be a friend of Waltz by the name of Weiser, Weiss or Weisner, whose first name was also Jacob. The reader already knows the pair sailed to America together. And, it must be noted that Weisner's name crops up in other stories, but in actuality that person is Jacob C. Weiss.

To reiterate an earlier statement: "Most of these folks could not read or write even their own names correctly." We know that Jacob Waltz could write his own name and did so on many documents that occur in this manuscript, but his friend Jacob Weiss is not recorded in his own hand anywhere to our knowledge.

It is believed that Weiss prospected around Prescott from 1864 until 1869. His name appears in several newspaper articles about mining and Indian troubles, always in conjunction with other noted German prospectors of his day. No mention of him is made after 1869. However, he is mentioned in the Weedin letters as the man who wandered from a watering hole to the Pima Reservation where he ultimately died, having taken an Apache arrow.

The name Weiss sounds like Weisner when another German speaks the name, especially a German of Swabian dialect as was spoken in Württemberg. An analogy would be a person with a heavy Boston accent who pronounces an *r* at the end of certain words where there is no *r*. (President Kennedy had this speech trait.) If Jacob Weiss's name were never seen in writing but only heard from being spoken, it would have indeed sounded like Weisner.

It's important to remember that in the Old West, people didn't

"The original records of that (first 'Mineral Meeting') and subsequent meetings, in long hand, now repose in the vaults of Yavapai County's Courthouse. The procedure outlined by California laws was followed, and the first part of the records reads as follows:

"Mineral Meeting on the Oolkilsipava River"
"May 10th, 1863".

"Mr. S. Shoup, President, Mr. J.V. Wheelhouse, Secretary
"Names of Original Prospectors as required
by
Resolution Number Seven"

------------------"Capt. J. R. Walker, Sr."----------------

"Jos. R. Walker Jr.
John Dixon
Jacob Linn
Jacob Taylor
Jas. V. Wheelhouse
Jack Swilling
Frank Finney
S. C. Miller
George Blosser
A. C. Benedict
S. Shoup
T. J. Johnson

B. Ellis
A. B. French
Chas. Taylor
H. B. Cummings
Wm. Williams
G. Gillalan
Jackson McCracken
Rodney McKinnon
Felix Cholet
M. Lewis
Jas. Chase
George Coulter"

"Preamble, Laws & Resolutions adopted and passed by the "Walker" prospecting & Mining Company for their mutual guidance & protection at a meeting of said company on the Oolkilsipava River, May 10th, 1863".

From *The History of Mining in Yavapai County, Prescott, Arizona* by H.R. Wood.

ask questions. Some "new arrivals" were wanted by the law; others were trying to disappear for sundry personal reasons; still others wanted to start a new life and never did become overt in their new communities. Many people went unrecorded and were killed by Indians or died from natural causes. There were few doctors and no hospitals. Death was common. It was normal to keep one's secrets and even to give an assumed name. It wasn't healthy to ask questions of anyone; therefore, it was not uncommon for people to disappear and never be heard from again.

The late Senator Carl Hayden, who avidly researched early pioneers, was known to have said he believed one in five pioneers was never recorded anywhere other than some offhand reference in an early newspaper. "Many were on the run and shied away from documentation, many died an untimely death due to Indian raids and some never appeared in established communities." This matter is extremely relevant as it concerns Waltz's partner.

Waltz is purported to have told friends that he and his partner Jacob Weiss were prospecting in Mexico when they happened into a gaming tent and saw a Mexican rancher (Peralta) being cheated by the men running the games. They are supposed to have stepped in when Peralta was knifed. After the rescue, they grabbed his money and took the wounded man to his hacienda. He gratefully offered them hospitality while he recuperated. Waltz said later that Peralta had told them of a mine in the Arizona Territory owned by his father through a Spanish land grant.

The Mexican was afraid to go back because the Apaches had killed his father during a mining expedition. He wanted the Germans to join him and "ride shotgun" in exchange for half of whatever gold was mined. They went. The band included 18 peons, Peralta and the two Jacobs. The trip netted $60,000 in ore; however, upon their return Peralta is supposed to have offered them a map and a document that gave them the rights to the mine if they would give him their half of the profits. Apparently Peralta had run up huge gambling debts.

Waltz spoke of Weiss as if they were brothers. He showed true remorse over their decision to accept the offer and return to the mine. Their intention was to collect the rich ore stored in three caches.

While at the mine, their mules were untethered and trampled their food stores, except for the meat that was strung in a tree.

They really couldn't do without flour. Waltz asked Weiss to come with him to the mill, but Weiss refused. While Waltz was gone to get supplies, his horse threw a shoe, and he was forced to wait for the Indian shoer. That matter took an extra day. Meanwhile, the Apaches attacked their camp in the Superstitions, and Weiss was shot in the upper arm and shoulder with an arrow. His horse was brought down also.

When Waltz returned, he saw the camp in disarray and assumed the Indians had captured and possibly tortured his partner. Waltz mourned his friend for the rest of his life. It was 1871.

Following are other facts that should not go unnoticed. In an 1863 mining claim notation of the early Walker Mining District records, Jacob Weiss purchased 300 feet of the Grose Lode for $50. The Grose Lode Mine was one of Jacob Waltz's claims. (Incidentally, eventually both men sold out, but the Grose Lode became a successful venture and was mined for many years after it was sold to new owners.)

In the meantime Henry Wickenburg sold his mine to a man from Virginia known as Doc Jones. He had a medical degree but rarely practiced. Instead he opted to become a cattle rancher in a canyon near Mormon Flat. The place was known as Agua Escondido (hidden water). Waltz was a friend of the doctor and his son Johnny Jones. Because of their friendship, a story developed that was given some credence.

Fifteen years after Waltz and nephew Peter Backens became part of the Peeples expedition, Backens disappeared and was never seen or heard from again. The date was around 1876. By that time Doc Jones kept a store and office on Washington Street in Phoenix, where sundry friends gathered regularly. The rumor mill claimed that earlier Jacob Waltz had found a mine, and his nephew Peter demanded they file a claim on it. Jacob, it is said, refused. The pair are reputed to have quarreled bitterly and wound up fighting. It was claimed that Jacob killed the nephew and buried his body in the soft sand along one of the bluffs at the spring. A Phoenix man – Richard Holmes – who heard the story went to the named canyon to search for the body. He said he found it right where the rumor said it would be. He brought the skull back to town and gave it to Doc Jones, who put the artifact on the shelf behind his desk.

Historian Bertie Roberts was then living in Phoenix on a farm

RECORDED AT REQUEST OF M MORSMER IN BOOK OF A OF RECORDS WALKER DISTRICT ARIZONA TERRITORY. JANUARY 5TH, A D 1864 AT 6 O'CLOCK, & 20 MINUTES A M.

V H SMITH
RECORDER //

WILLIAM GROSE
TO
J. C. WEISS

THIS INDENTURE MADE THE TWENTY FOURTH DAY OF NOVEMBER IN THE YEAR OF OUR LORD ONE THOUSAND EIGHT HUNDRED & SIXTY THREE BETWEEN WILLIAM GROSE, PARTY OF THE FIRST PART AND J. C. WEISS PARTY OF THE SECOND PART AT PRESENT BOTH RESIDENTS OF THE MINING DISTRICT OF LA PAZ, TERRITORY OF ARIZONA. WITNESSETH - THAT THE SAID PARTY OF THE FIRST FOR & IN CONSIDERATION OF THE SUM OF FIFTY DOLLARS LAWFUL MONEY OF THE UNITED STATES OF AMERICA TO HIM IN HAND PAID BY THE SAID PARTY OF THE SECOND PART AT OR BEFORE THE ENSEALING AND DELIVERY OF THESE PRESENTS THE RECEIPT WHEREOF IS HEREBY ACKNOWLEDGED, HATH REMISED RELEASED & QUIT CLAIMED AND BY THESE PRESENTS DOTH REMISE, RELEASE & QUIT CLAIM UNTO THE SAID PARTY OF THE SECOND PART AND TO HIS HEIRS & ASSIGNS FOREVER. ALL OF THAT UNDIVIDED FEET OF MINING GROUND SITUATED IN THE WALKER DISTRICT TERRITORY OF ARIZONA OR NEW MEXICO, BEING THREE HUNDRED FEET IN THE LODE OF GOLD & SILVER KNOWN AS THE "GROSE LODE" STANDING IN THE NAME OF THE PARTY FIRST PART IN THE RECORD OF SAID LODE IN THE RECORDER'S OFFICE OF SAID WALKER'S DISTRICT TO WHICH REFERENCE IS HEREBY HAD. TOGETHER WITH ALL THE DIPS SPURS & ANGLES AND ALSO ALL THE METALS ORES GOLD AND SILVER BEARING QUARTZ ROCK AND EARTH THEREIN AND ALL THE RIGHTS, PRIVILEGES, AND FRANCHISES, THERETO INCIDENT APPENDANT APPURTENANT OR THEREWITH USUALLY HAD & ENJOYED AND ALSO ALL AND SINGULAR THE TENEMENTS HEREDITAMENTS AND APPURTENANCES THEREUNTO BELONGING AND THE RENTS ISSUE AND PROFITS THEREOF: AND ALSO ALL THE ESTATE RIGHT TITLE INTEREST POSSESSION CLAIM AND DEMAND // WHATSOEVER OF THE SAID PARTY OF THE FIRST PART OF, IN, OR TO, THE PREMISES AND EVERY PART & PARCEL THEREOF, AND ALSO ALL THE ESTATE RIGHT TITLE INTEREST PROPERTY POSSESSION CLAIM & DEMAND WHATSOEVER AS WELL IN LAW AS IN EQUITY, OF THE SAID PARTY OF THE FIRST PART OF, IN, OR TO THE ABOVE DESCRIBED PREMISES AND EVERY PART & PARCEL THEREOF, WITH THE APPURTENANCES TO HAVE & TO HOLD ALL & SINGULAR THE ABOVE MENTIONED AND DESCRIBED PREMISES TOGETHER WITH THE APPURTENANCES UNTO THE SAID PARTY OF THE SECOND PART HIS HEIRS & ASSIGNS FOREVER. IN WITNESS WHEREOF SAID PARTY OF THE FIRST PART HAS HEREUNTO SET HIS HAND & SEAL THE DAY & YEAR FIRST ABOVE WRITTEN.

SIGNED SEALED & DELIVERED
IN PRESENCE OF
H HIMLEY

WILLIAM GROSE (SS)

RECORDED AT THE REQUEST OF M MORSMAN - JANUARY 5TH 1864 & DULY RECORDED AT 6.30 O'CLOCK A M.

V. N. SMITH
RECORDER //

Purchase of Grose Lode claim by Weiss.

owned by her father. Their farmland was across the road from Jacob Waltz's. It was not uncommon for all the people, who were by then old friends, to go to town regularly and during the trip stop by Doc Jones's office for a visit. Bertie wrote in her memoirs that she could not verify the story but would swear the skull was in Jones's office for many years.

It is widely speculated that Waltz and his partner located the mine between 1867 and the early 1870s. And we know that Jacob Weiss died in 1871 in Dr. John D. Walker's home. Waltz filed on his homestead in 1868. He was then living in Phoenix. It has also been reported that once the German was established on the farm, he lived in very meager surroundings. However, barley seed, which Jacob grew, was expensive to buy, and it was noted that within a few years he built a really nice adobe, had chickens, ducks, a cow and a beautiful vegetable garden. That is when he began to supply Julia (Helena) Thomas with eggs and milk for her bakery.

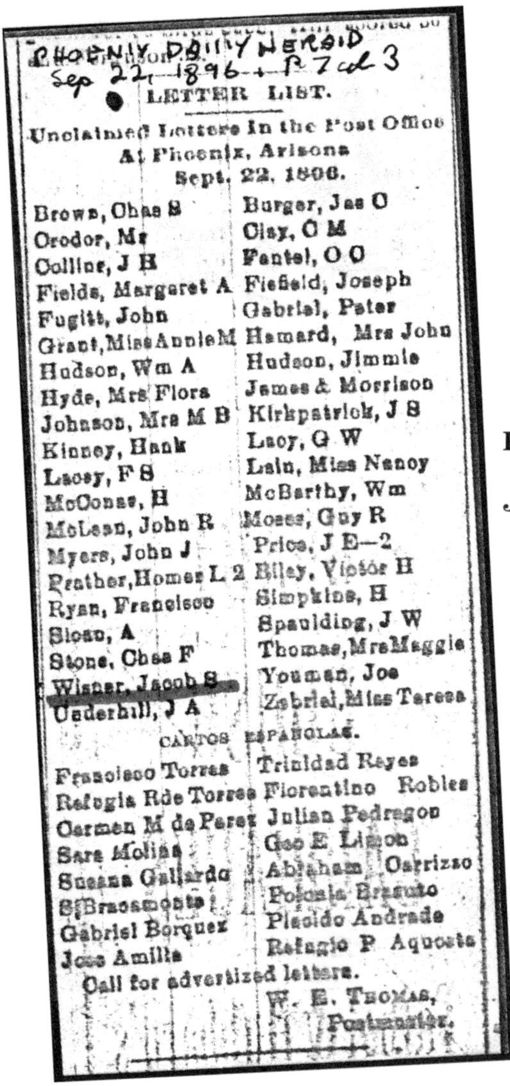

Phoenix Daily Herald
Sep 22, 1896 P7 col 3

LETTER LIST.

Unclaimed Letters in the Post Office
At Phoenix, Arizona
Sept. 22, 1896.

Brown, Chas S	Burger, Jas C
Orodor, Mr	Clay, C M
Collins, J H	Fantel, O C
Fields, Margaret A	Fiefield, Joseph
Fugitt, John	Gabriel, Peter
Grant, Miss Annie M	Hamard, Mrs John
Hudson, Wm A	Hudson, Jimmie
Hyde, Mrs Flora	James & Morrison
Johnson, Mrs M B	Kirkpatrick, J S
Kinney, Hank	Lacy, G W
Lacey, F S	Lain, Miss Nancy
McConas, H	McBarthy, Wm
McLean, John R	Moses, Guy R
Myers, John J	Price, J E—2
Prather, Homer L 2	Riley, Victor H
Ryan, Francisco	Simpkins, H
Sloan, A	Spaulding, J W
Stone, Chas F	Thomas, Mrs Maggie
Wisner, Jacob S	Youman, Joe
Underhill, J A	Zabriel, Miss Teresa

CARTOS ESPAÑOLAS.

Francisco Torres	Trinidad Reyes
Refugia R de Torres	Florentino Robles
Carmen M de Perez	Julian Pedregon
Sara Molina	Geo E Limon
Susana Gallardo	Abraham Carrizso
S Bracamonte	Polonia Brasuto
Gabriel Borquez	Placido Andrade
Jose Amilla	Refugio P Acosta

Call for advertized letters.

W. E. Thomas,
Postmaster.

Dead Letter List published in the* Phoenix Daily Herald *(1896) proved Jacob Weiss did live.

Walker Mining District
Part II

Much has been said thus far regarding the travels, battles and mining efforts of the pioneers who came into the Arizona Territory early on, but we would be remiss if we didn't discuss the legal controversies that were ongoing involving the rights of those men. In fact, the following information thoroughly explains the legal harangue that erupted between newcomers, residents and the early miners.

Readers should be cognizant of the fact that Prescott was in its infancy, plagued with all the problems that implies. There were no women in town; there were still regular Indian raids; food was scarce; and gold was the prime motivator. Clothing could not be purchased, which was the main reason the Indians stripped their victims. Also, settlers believed the army would protect them and therefore welcomed any military presence. Once the rumors of gold surfaced, soldiers were no different than anyone else who came – they went prospecting.

The fact of the location of mining sites and the legal matters that might protect these claims were beginning to cause a ruckus. Rumors spread to other localities, and news items surfaced in distant cities. The problem was not a small matter to those who had come so far and fought so hard to get there.

Mexicans had been mining in the area for many years; their land grants had been given to them by the King of Spain. It didn't take the newcomers long to remedy that situation. They refused to allow Mexicans to own mines. Rather than try to explain this complicated occurrence, we will include a dissertation on this subject done by Harwood Hinton. It is titled "Frontier Speculation: A Study of the Walker Mining Districts."

Before we go on, we might ask, what does this have to do with our subject? Jacob Waltz was a part of all this. He was living in the mountains and by all accounts filing claims and holding gold. No one knows for certain when he actually left Prescott or when he went into the Superstitions or, indeed, when he found his mine. The problems that enveloped Prescott and the miners included Jacob. It may have driven him out of the area and into another where Mexicans, and earlier the Spaniards, had mined successfully for a very long time.

By this time Waltz was seasoned. He had been at it for years and knew the lay of the land and what to look for. He was obviously not afraid of the savages, at least not enough to stop prospecting. They had all made a good haul at Rich Hill, and there was really no place to buy anything. We can safely assume Jacob still had a good stash. One historian was quoted as saying Jacob was never seen entering a bank; therefore, we can conclude he was stashing his metal in a hiding place or carrying it with him.

Frontier Speculation: A Study of the Walker Mining Districts

HARWOOD HINTON

[Harwood Hinton is a candidate for the doctorate in history at the University of Wisconsin.]

ON NOVEMBER 4, 1863, a commentator in the San Francisco *Evening Bulletin* blasted the practice of prospectors converting mining districts and their codes into vehicles for speculation. Throughout the 1860's, in different areas of the West, newspaper reporters, individual capitalists, and federal mining investigators echoed this sentiment. Charles D. Poston, Superintendent of Indian Affairs in Arizona, visited the mining camps near Prescott and left convinced that prospectors erected districts and enacted codes only to protect and enhance the value of claims, which they had no intention of developing. Late-comers to the diggings, he added, invariably left in disgust. J. Ross Browne, in a mining survey in 1865, called attention to the frequency and ease with which claim-seekers manipulated their bylaws. Neither prospector nor capitalist, he concluded, could expect protection or encouragement when "on a few days' notice, a corporal's guard assembles, and, on simple motion, radically changes the whole system." From 1863 to 1865, the Walker mining districts in central Arizona illustrated the extent and significance of these practices on one part of the mining frontier.[1]

Late in May, 1863, twenty-five dusty prospectors, led by the experienced guide, Joseph Reddeford Walker, reached the Pima villages on the Gila with news of a gold discovery in the mountains of northern Arizona. For years Catholic priests, trappers, and railroad surveyors had reported that placer gold lay along the southern slopes of these highlands. In the spring of 1861, Walker piloted an unsuccessful expedition from California into the region; several months later they disbanded on the Rio Grande, Walker and a majority of the men going north to the mines near Denver, Colorado. In the early winter of 1862, he left Colorado with a reorganized band of gold-seekers and swung through New Mexico

[1] San Francisco *Evening Herald*, Aug. 4, 1864; Report upon the Mineral Resources of the States and Territories West of the Rocky Mountains, 39 Cong., 2 sess., *House Ex. Doc.* 29, p. 228. Cited hereafter as Mineral Resources. For activities in the Walker districts, see Daniel E. Conner, *Joseph Reddeford Walker and the Arizona Adventure* edited by Donald J. Berthrong and Odessa Davenport (Norman, 1956), *passim;* and the same manuscript, with slightly different editing, in Thomas E. Farish, *History of Arizona* (Phoenix, 1915), II, 143–149, 243–244. Conner joined Walker in Colorado in the fall of 1862.

to the Gila. Crossing central Arizona they reached the Pimas in mid-April. Here, Walker announced to a friend on the Colorado his plans to explore to the north, and that he would return to the villages within a month. Then, with a heavily laden pack train, he and his band turned down the Gila to the Hassayampa, whose sandy bed pointed north across a sterile desert. A week later the expedition was in the mountains. On the southern edge of an extensive plateau, they located the headwaters of the Hassayampa and established a camp. A few days later, on the Ookilsipava, an eastern tributary, they discovered pay dirt.[2]

Before returning to the villages, the Walker party re-enacted a scene common to the mining camps of the West. On May 10, by a voice vote, they constituted themselves a sovereign body, elected a temporary chairman, and proceeded to erect a placer mining district and to draft a proper code.[3] They included in their "Pioneer District" all the gulches and ravines entering the Ookilsipava, from its head to a "tree below the falls, at the foot of the mountains," and decreed that "no claims be taken for persons outside of original prospectors until they have definitely settled."

Many of those present had helped organize mining districts in California and Colorado; all knew the rudiments of a placer code. In their domain, the sovereigns continued, a mining claim would be one hundred yards square—one hundred yards along and fifty yards back on each side of the stream. An individual could stake two claims—one by right of discovery and one by pre-emption—and could secure title to each merely by registering it and paying a filing fee of two dollars and fifty cents. All altercations, laws, elections, and other affairs of the district would be settled by a majority vote at duly announced meetings. Two annually elected officers—a president and a recorder—would administer the business of the district.

In several of the bylaws, the Walker party expanded, encouraged, and guaranteed their position in the new district. First, they required that their names be entered on the minutes as the "original prospectors." Next, they banned Mexicans from the diggings for six months, assumed

[2] *Bulletin*, Nov. 10, 1860, July 12, 1861; Report of the Secretary of the Interior, 1863, 38 Cong., 1 sess., *House Ex. Doc.* 1, p. 25. Firsthand accounts of the expedition in 1861 are in *History of Arizona Territory* (San Francisco, 1884), 207, 301; and Orick Jackson, *The White Conquest of Arizona* (Los Angeles, 1908), 12–15. For the 1862 venture, see Conner, *Walker*, 3–98; Farish, *History of Arizona*, II, 244–245; *Bulletin*, May 18, June 17, 1863; and *History of Arizona Territory*, 207, 301–302.

[3] See "Journal of the Pioneer and Walker Districts, 1863–1865." The original is in the Yavapai County Court House, Prescott, Arizona; a microfilm copy of the journal is available at the State Historical Society of Wisconsin, Madison. Subsequently cited as "Journal." For the first meeting, see "Journal," 2–5; and Conner, *Walker*, 100.

the privileges of relocating claims on unclaimed ground anywhere in the district, and declared themselves exempt from any filing fee. Finally, they resolved that each member of the party would draw two claims by lottery immediately upon his return from the Gila.

The next day, the prospectors, anxious to report their gold strike, cached the remaining supplies and headed south. They knew that their return to the settlements would signal a wild rush to the diggings, and by forming a placer district and a code they hoped to regulate the ensuing boom. Being prospectors, not miners, and speculators, not laborers, they hoped to protect, enhance, and sell their claims before the placer interest ended. If they could uncover gold-bearing quartz, they would locate claims and hope for a buyer, or would merge their interests with others, wait for the appreciation that actual mining would bring, then sell. To protect their claim values over an extended period, they had to limit further prospecting. Only by controlling the voting at subsequent meetings could they do this. Their control, they knew, could be easily destroyed. Members of the party, because of claim sales, scarcity of supplies, or Indian menace, could scatter to other districts; new groups could enter the diggings, buy up and locate claims, and vitally affect the voting balance. In every instance, the Walker position was precarious. However, knowing that they would locate the first claims in the new district, they hurried down the Hassayampa to promote their discoveries.[4]

Upon reaching the Pima villages, the explorers found that Walker's announcement had been widely circulated. Bands of Americans, Mexicans, and Chinese had eagerly gathered in the neighborhood. The prospectors spent two weeks on the Gila, spreading the particulars of their El Dorado.[5]

One prospector, Albert C. Benedict, in writing to Santa Fe, set in motion forces which vitally affected the entire mining scene in central Arizona. Brigadier General James H. Carleton, commanding the Department of New Mexico, read Benedict's letter, and sprang into action. For over five months, he had been writing to various officials in Washington about the minerals said to exist north of the Gila. Now, in June, he excitedly informed his superiors of the Walker strike, and assured them he would immediately send an expedition to confirm its location and magnitude.[6]

---

[4] Conner, *Walker*, 103–104.

[5] Reports of the Walker strike appeared in the *Bulletin*, July 2, 9, 31, 1863.

[6] Brig. Gen. James H. Carleton, Santa Fe, to Maj. Gen. Henry W. Halleck, Washington, June 14, 1863, Report on the Condition of the Indian Tribes, 39 Cong., 2 sess., *Senate Report*

Early in June the Walker party returned north, passing en route numerous bands of prospectors. Within a month more than one hundred men would be tediously panning streams and scratching for bits of gold in countless gulches in the sierras. From settlements on the Rio Grande, the Colorado, and the Gila, they streamed to the placers. "All the loose men," wrote a traveler at the Pima villages, "have left this vicinity and Tucson for the Diggings. No flies or mosquitoes or any other vile thing" hindered the exodus. A majority of the prospectors stopped in new districts below the Walker discovery, but many continued north to the Ookilsipava.[7]

Soon after reaching their camp in the mountains, the Walker party assembled with others to alter both district and code.[8] Anticipating other placer deposits nearby, they expanded the boundaries of their domain, east, west, and north, practically doubling its area. Next, they amended the mining regulations, thereby initiating a practice which persisted throughout the history of the diggings. An individual could hold a claim one year without working it only if he had marked it with stakes or rocks and properly recorded it within ten days of location. He could hire Mexicans to occupy his holding if he registered each laborer at "four bits" per head and informed the recorder of all dismissals.

The assembly adopted several amendments to regulate situations which could arise in the enlarged district. In no instance, they asserted, could an individual register claims for nonresidents of the district. This amendment struck at the basis of absentee ownership. Though relaxing the barrier against Mexicans, the lawmakers manifested no tolerance toward the Chinese. They banned the race completely from the diggings. Then, in anticipation of gold-bearing quartz discoveries, they set the length of a quartz claim at 200 feet along the outcropping. In conclusion, the group closed the district to all but members of the Walker party for two days.

Three days later, when the district recorder could register claims for any prospector, few claimants appeared. The reasons for this situation, which continued through the remainder of the month, were several. The Walker party staked fifty claims along the Ookilsipava, practically

156, p. 113. See also Carleton to Brig. Gen. Lorenzo Thomas, Washington, Feb. 1, 1863. *War of the Rebellion: A Compilation of the Official Records of the Union and Confederate Armies* (Washington, 1880–1901). Ser. I, Vol. XV, 669–670. Cited hereafter as *Official Records*. Carleton to Montgomery Blair, Washington, May 2, 1863. *Official Records*, Ser. I, Vol. L, Pt. II, 419.

[7] *Bulletin*, July 2, 1863. The best description of the Walker and adjoining districts appeared in the same newspaper on August 24.

[8] "Journal," 6–7.

monopolizing the only known goldbearing stream in the diggings. Elsewhere in the district, prospecting was sporadic. Men repeatedly located and abandoned unworked sites in their rush to investigate new gold strikes. Perhaps the most important reason for the paucity of claim registrations was the searing summer heat, which gradually dried up the mountain streams, thereby forcing the prospectors out of the placers. As June ended, the diminishing mining activity shifted more and more to the northern reaches of the diggings, particularly to Lynx Creek.[9]

On July 12, the men remaining in the placer district held their last summer meeting at the recorder's office, which had been moved to Lynx Creek. As at previous assembles, they amended their code to favor the interests of the Walker party. They declared that only "original prospectors" could hold more than one pre-emption, granted them ten days in which to relocate their initial claims anywhere in the district, and placed the group in control of committees to measure and stake new claims and to certify who were Mexicans. These acts indicated that the Walker influence over local mining legislation remained strong. The prospecting privilege, the control over committees to validate claims and to select the labor force, and the restrictions on claim holding all served to protect and enhance their interests. The fame of the Arizona placers, however, had traveled far. At that moment, numerous groups who would substantially affect activities in the Walker domain were hurrying to the mines.[10]

On August 19, the surveyor general of New Mexico, John A. Clark, with a cavalry escort, rode into the diggings. General Carleton had instructed the party to estimate the magnitude and potential of the new gold strike. Should the region prove rich, the military planned to build a fort nearby to protect the inhabitants. To Walker, Carleton sent his congratulations, reassuring him that helping others to a fortune "afford[s] me not quite as much happiness as finding one myself . . . but nearly as much." For twelve days Clark's party remained in the neighborhood, probing, examining, and assaying the scattered claims. Then, apparently satisfied, they loaded their pack train and descended the mountain to other districts.[11]

Clark's visit signaled an abrupt shift in prospecting in the Walker district. Since the Ookilsipava strike four months before, the local resi-

[9] *Ibid.*, 240–257. Prospectors rushed from the Ookilsipava to the Big Bug, then to Lynx Creek. In every instance, the Walker party located the first claims.

[10] *Ibid.*, 7–8.

[11] Report of the Secretary of the Interior, 1863, 38 Cong., 1 sess., *House Ex. Doc.* 1, pp. 25–27; Carleton to Walker, Walker's Diggings, June 22, 1863, Indian Tribes, 39 Cong., 2 sess., *Senate Report* 156, p. 117; *Bulletin*, Oct. 5, 1863.

dents had expended their energies in the placers. One day after his arrival, however, Clark, with eight others—five were prominent nonresidents of the district—filed the first quartz claim. Styling themselves the "Pioneer Silver and Gold Mining Company," they appropriated 1,800 feet along an outcropping thought to be gold-bearing. The same day, another group, which included Walker and two members of his party, recorded 1,600 feet for the "Maricopa Silver and Gold Mining Company." Even more extensive in area claimed and in the number of claimants involved was the "Agua Frio Mining Company," which, several weeks later, located two separate 4,000-foot claims. Among its community of twenty were the names of Clark, Carleton, and Benedict. A rapid increase in absentee ownership had accompanied the sudden interest in quartz, but Lynx Creek residents made no move to arrest the development. Their inaction was due to the nature of the mining organizations which had appeared in the diggings.[12]

On Lynx Creek, as elsewhere on the mining frontier, companies formed to open quartz mines experienced little restraint. But, as a mining investigator explained, the men who organized these early ventures generally had other ends in mind. In each new diggings, he said, one or several prospectors, whether authorized or not, would include the names of personal friends, notable businessmen, or territorial officials when constituting a new company. These individuals, upon being informed of their good fortune, either bought out their known, or unknown partners, sold them their interests for a small sum, or simply forgot the matter. Whatever the result, prospectors knew that influential names on the district records enhanced the over-all speculative values of their diggings. In the quartz companies organized on Lynx Creek during the early fall of 1863, there were many who had engaged in an agent-client capacity in other districts. Here, as elsewhere, prospectors scrambled for mining claims, not for themselves but for sale to others.[13]

The rising interest in quartz quickly dampened all enthusiasm in the placers. The owners of placer claims, believing the sudden appearance of quartz companies would unsettle claim values, began selling or transferring their holdings. Through October and early November their

[12] "Journal," 261–262. For identification of members of the companies, see Sacramento *Union*, Dec. 5, 1863; Hubert H. Bancroft, *History of California* (San Francisco, 1886–1890), VI–VII (scattered references); *Bulletin*, July 12, 1861; and *Official Records*, Ser. I, Vol. XLI, Pt. II, 111, 114, 169. Most of the nonresidents in the Pioneer Silver and Gold Mining Company lived in California; a majority of the stockholders in the Agua Frio Company resided in New Mexico.

[13] Mineral Resources, 39 Cong., 2 sess., *House Ex. Doc.* 29, p. 230. Browne comments at length on the organization of quartz companies. For an example of authorized agents at work, see Sacramento *Union*, Dec. 5, 1863.

transactions continued, as falling temperatures and chilling winds drove many south to warmer climes. The placer boom in the Pioneer District was over.[14]

On November 24, Walker prospectors and others on Lynx Creek braved the mountain cold to meet in the recorder's office and discuss their changing circumstances.[15] The discussion shifted from the interest in quartz to the possibility of an army post—and perhaps the territorial capital—being located nearby. To protect their claims and to regulate more closely any subsequent quartz boom, they decided to erect a quartz district. The placer officials immediately resigned, and the meeting became, to all intents and purposes, a constitutional convention. During the day, they redesignated the Lynx Creek area the "Walker Quartz Mining District" and hammered out a fairly detailed set of quartz regulations.

The convention prefaced its quartz code with a promise and a polemic. By pledging their full assistance to all "traders and persons, whether citizens or not," in the "prosecution of right business," they strove to counter the "untruthful reports" then spreading concerning opportunities in their diggings. They roundly denounced the perpetrators of these "falsehoods," many of whom had returned to their "shinwarming firesides," as a set unworthy the name of "pioneer." To these introductory remarks they appended an extended list of bylaws.

In framing the quartz code the law makers drew heavily on their placer experience. From their placer code they extracted directly the provisions dealing with the duties of district officers; the principles of settling disputes, laws, and elections; the restrictions on Mexicans and Chinese; and the specifications for a quartz claim. They veered away from the placers, however, in their instructions for claim validation. To secure title to a quartz site, an individual or a group, within sixty days of locating on a mineral vein, must present to the recorder an ore specimen, a copy of the posted claim notice, and pay a filing fee, assessed at fifty cents per name. To hold a claim, the interested party must occupy and work it three days per quarter for one year, or twelve days continually within three months of location. The legislators, however, postponed for four months the effective date of the work stipulation. In conclusion, they elected one set of officers to function over both the

[14] "Journal," 50–55. The first sales and transfers in the district occurred in October. For weather conditions, see *Bulletin*, Dec. 9, 1863.

[15] "Journal," 25–31. A commentator in the *Bulletin*, Sept. 14, 1863, said the reports from the Walker diggings were exaggerated; on Oct. 30, he stated that many prospectors were returning from Arizona in disgust.

placer and quartz districts, restricted their tenure to six months, and adjourned.

The Walker influence became blurred in this meeting. Several members of the party served on the committee to draft quartz bylaws; the final code contained many of their placer regulations. In no way were their holdings and their opportunities limited. But, for the first time, the Walker party received no special privileges. Everyone in the diggings, however, stood to benefit by the formation of the new district. Few in number, with the weather conditions precluding any immediate accretion, they were in a favorable position to locate the richer claims. This opportunity, however, proved short-lived.

During January, 1864, the denizens of Lynx Creek expected a resurgence of mining activity. The preceding month, soldiers from General Carleton's command had begun constructing Fort Whipple in the Chino Valley, twenty miles to the northwest. As the military had previously furloughed men to prospect, the Lynx Creek residents viewed the recent arrivals with a jaundiced eye. On the tenth, at a "called" meeting, they sharply curtailed further placer prospecting. They restricted pre-emption claims to one per person in the district and ordered each individual with a newly acquired claim to occupy it immediately or be prepared to show title. These amendments, they added, were not binding on the Walker party. The assembly, however, in moving to protect their claim values, had acted belatedly. Prospectors had already abandoned the gravel beds in their quest for ore veins. The arrival of each new group at the diggings intensified this activity.[16]

Late in January, John N. Goodwin, governor of the newly created Territory of Arizona, and a retinue of officials and soldiers reached Lynx Creek. More than a month before, at Santa Fe, they had heard and read glowing reports of the Walker discoveries, and at the urgings of General Carleton had quickly altered their intended line of march. Instead of proceeding to Tucson, designated the temporary capital, they had headed west along the thirty-fifth parallel. After a brief stay at Fort Whipple, they turned south to tour the mines.[17]

The governor's party found the Lynx Creek area bleak and uninviting. Jonathan Richmond, one of the group, described the diggings as

[16] General Order No. 27, Oct. 23, 1863, Department of New Mexico, *Official Records*, Ser. I, Vol. L, Pt. II, 654, provided for the establishment of Fort Whipple as a three company post. For activities in the diggings, see "Journal," 7–9, 56–58, 270–274.

[17] Congress created the Territory of Arizona, Feb. 24, 1863. Goodwin described his trip west in Territory of Arizona, *Council and House Journals*, first Legislative Assembly, 1864, p. 15. Cited hereafter as *ACHJ*. See also New York *Tribune*, June 26, 1865; and letters of Jonathan Richmond in Farish, *History of Arizona*, III, 53–60.

practically deserted. Abandoned claims were everywhere. The few residents, he noted, had located quartz lodes, and were "holding on for capitalists to come in with means and machinery." For ten days the officials searched among unclaimed rock outcroppings in the vicinity for gold-streaked quartz. Their efforts were unrewarded, but they were hardly bored. For nearly two months the community had been the prey of Apache raiders. At the governor's departure, early in February, bands of Indians were penetrating even deeper into the district.[18]

By March the Lynx Creek inhabitants found the Indian menace unbearable and took firm steps to protect their lives and property. On the sixth, an assembly of "citizens" suspended all work requirements and declared valid all recorded but unoccupied claims. Eight days later, they petitioned the governor for soldiers to brush aside the Indians who could cut off the "various trains of provisions and clothing" en route to the diggings. Goodwin had already requested the War Department to grant him permission to raise a ranger force. Before the reply came, however, Walker and others led punitive expeditions northeast into the hostile country. After they dispersed the Indian concentrations, life at the diggings returned to normal.[19]

During the late spring, the Walker quartz district sparkled with activity. Prospectors recorded four times as many claims in May as they had in the preceding month. Sales and transfers were few. The quieting of the Indian menace, the ease in securing claims—merely by registration—and a flood of new gold-seekers combined to produce the briskness. The mounting enthusiasm, however, caused growing unrest in some quarters. For weeks the Walker group had seen soldiers from Fort Whipple swarm into the diggings to prospect and to record claims of unauthorized size. On May 22, at a routine meeting to elect new officers, the disgruntled element spoke out. Without forewarning, George Coulter, who had been with Walker since 1861, suddenly suggested that all persons "in the service of the United States" be prohibited from filing claims in the district. This would alleviate the increasing disregard for local regulations and stifle the growing atmosphere of competition—conditions which could unsettle property values. After a short discussion, the assembly adopted the proposal.[20]

---

[18] Richmond's comments are in Farish, *History of Arizona*, III, 218–230. For Indian troubles, see Maj. Edward B. Willis, Fort Whipple, to Carleton, Dec. 21, 1863, *Official Records*, Ser. I, Vol. XXVI, Pt. I, 869–870; and *Bulletin*, April 23, 1864.

[19] "Journal," 9, 11–12. Goodwin's request is reprinted in Farish, *History of Arizona*, V, 93. For expeditions against the Indians, see Goodwin, Tucson, to Carleton, April 4, 1864, *Official Records*, Ser. I, Vol. XXXIV, Pt. III, 202.

[20] "Journal," 67–127, 13.

The Walker edict was untimely. It vitally affected a group that had acquired extensive holdings in the diggings—the territorial party, composed of civil and military officials and their friends. As property owners with grounds for complaint, they reacted immediately. At their request, the district president called a meeting of the Lynx Creek community on May 29. The meeting was short and decisive. By a quick voice vote, those attending quashed the onerous edict and threw the diggings open to "all citizens of the Territory." By this action they swept from the district the Walker influence.[21]

In June the quartz boom ended. The recorder entered half as many claims as he had in May. Sales and transfers tripled. In July the slump became even more pronounced. The enervating summer heat, with its blighting effect on the water supply, caused many to leave. Others viewed unregulated prospecting as a portent of claim depreciation, and sold out. Also, a new influence had come to the fore in the district. Since their arrival in Arizona, territorial officials, by location and purchase, had acquired the more promising quartz veins in the Lynx Creek vicinity. They had indicated further interest in the area by locating the territorial capital at Prescott, ten miles to the west. Throughout the summer and fall, the group expanded its holdings. On June 20, the Lynx Creek assembly struck off the southern portion of the old placer district; here, several weeks later, Governor Goodwin located the first quartz claim. Some of the officials invested heavily in partially opened mines. Charles D. Poston, Indian superintendent, paid $20,000.00 for a one half interest in thirty-eight mining shafts on July 12; twelve days later he invested $10,000.00 more.[22]

As the territorial officials extended their ownership over the Walker diggings, the residents ceased prospecting. In August, with the last placer transaction, the Pioneer District became moribund. The quartz fever had subsided also, though an assembly on October 2 tried to rekindle interest by liberalizing the code. Any person, they stated, could secure a claim by sinking a shaft two feet in depth, four in length, and three in width, and by paying one dollar and fifty cents. The resolutions appeared attractive, particularly to mining companies that could afford the necessary drilling. All in Arizona, however, had momentarily turned their attention to Prescott.[23]

[21] *Ibid.*, 14.

[22] *Ibid.*, 18–337 (particularly 101–170, 281, 32). Goodwin's claim was in the Turkey Creek District, which was formed in the southern portion of the Pioneer District. Hubert H. Bancroft, *History of Arizona and New Mexico, 1530–1888* (San Francisco, 1889), 587 f. For the beginnings of Prescott, see *Bulletin*, August 4, 1864.

[23] "Journal," 163–167. The Pioneer District expired on Aug. 13, 1864. *Ibid.*, 285. For the particulars of the October meeting, see *ibid.*, 15.

ARIZONA TERRITORIAL DEED RECORDS

1864 - 1869
YAVAPAI COUNTY, PRESCOTT ARIZONA

Grantor	Grantee	Date	Bk/Pg.	Deed Recorded
George R. Roberts	*H.W. Ward*	*16 Mar 1867*	*3/649*	*180 acres land, Prescott*
Joseph Ingram & Thomas Cargile	*Geo. Roberts*	*14 Oct 1867*	*3/822*	*Ingram/Cargile Ranch*
John Roberts	*Joseph Tuttle*	*20 Apr 1866*	*3/95*	*300 ft. Goodwin Lode*
John Roberts	*John Baldwin*	*21 Apr 1866*	*3/99*	*50 ft. Oriental Lode*
John Roberts	*Robert McLand*	*8 Dec 1866*	*3/523*	*300 ft. Mammoth Lode*
John Roberts	*P. McAteer*	*10 Dec 1866*	*3/524*	*Ranch, Kirkland Valley*
John Roberts	*Lon C. Gray*	*2 Aug 1867*	*3/727*	*200 ft. Big Bug Lode*
John Roberts	*Wm. Lyttaker*	*13 Feb 1869*	*3/1092*	*162 ft. Mary Ann Lode*
John Roberts	*C. Schroder*	*5 Apr 1869*	*3/1152*	*Lot 31 Blk. 13 Prescott*
Wm. Roberts	*A.H.Peeples*	*2 May 1865*	*1/288*	*400 ft. Gold Reef Lode*
Chas. Roberts	*T. Marsh*	*2 Jan 1865*	*1/181*	*50 ft. Eureka Lode*
Chas. Roberts	*A.W. Adams*	*6 May 1865*	*1/309*	*100 ft. Eureka Lode*
Chas. Roberts	*Geo. Coulter*	*6 May 1865*	*1/310*	*100 ft. Eureka Lode*
Chas. Roberts	*Geo. Coulter*	*12 May 1865*	*1/311*	*400 ft. Eastern Lode*
A.H.Peeples	*Louis Landsberger*	*4 May 1865*	*1/350*	*150 ft. Vulture Lode*
A.H. Peeples	*Wm.O. Davis*	*16 Jun 1865*	*3/925*	*100 ft. Vulture Lode*
A.H. Peeples	*R.P.D. Brooks*	*19 Oct 1865*	*3/1016*	*P.Burns Ranch Walnut Gr.*
A.H. Peeples	*Hannah Meier*	*6 Feb 1866*	*2/516*	*40 ft. Vulture Lode*
J.M. Robbins	*A.H. Peeples*	*25 Aug 1866*	*3/370*	*Blacksmith shop Wickenberg*
P.W. Smith	*A.H. Peeples*	*7 Dec 1864*	*1/23*	*300 ft. Vulture Lode*
J.W. Swilling	*A.H. Peeples*	*17 Apr 1865*	*1/285*	*150 ft. Vulture Lode*
Thomas Worden	*A.H. Peeples*	*24 Mar 1869*	*3/1137*	*Lot 13 Blk. 2 Wickenberg*
Joseph Green	*Robert W. Groom*	*9 Apr 1865*	*1/399*	*50 ft. Vulture Lode*
Joseph Green	*H. Mannassee*	*9 Apr. 1865*	*1/304*	*40 ft. Vulture Lode*
Joseph Green	*Issac Goldburg*	*12 Apr 1865*	*1/296*	*10 ft. Vulture Lode*
Joseph Green	*J.W. Swilling*	*17 Jan 1865*	*1/290*	*150 ft. Vulture Lode*
Robert W. Groom	*Joseph Green*	*19 Apr 1865*	*2/109*	*50 ft. Bully Bueno Lode*
Joseph Lennon	*Joseph Green*	*30 Nov 1865*	*2/381*	*100 ft. Clinton Lode*

On Friday, September 29, the first territorial legislature of Arizona convened at Prescott. The next day, Governor Goodwin, in his message to the members, outlined a program which stressed the need for territorial regulations of quartz mining. For this industry to develop, he explained, the great confusion among district laws must be rectified. Placer codes could continue, for they were ephemeral, but local quartz laws must be superseded by a territorial code. Capitalists, he concluded, would not invest in the mines unless they could obtain clear and uniform titles. The following week, the legislators began discussing a detailed code which William T. Howell, a territorial judge, had prepared the preceding spring. On November 10, when the session ended, they had enacted, and the governor had approved, the major portion of Howell's proposals. By this act, the territorial party imposed its control over Arizona's richest natural asset.[24]

Local approval of the Howell Code came four days later. An assembly on Lynx Creek adopted a section of the code which dealt with sites for ore crushing mills, affirming their action to be "subject to the territorial law." The higher law had been ratified. Lynx Creek residents met twice more to elect officers. At the last meeting, on February 26, 1865, they closed the books of the Walker Quartz Mining District.[25] Local sovereignty on Lynx Creek had ended.

From the beginning, activity in the Walker districts had centered upon the location and sale of mining claims. The Walker party, in particular, had made every effort to control the district voting so as to protect and enhance the value of their holdings. In this endeavor they were eminently successful for nearly a year. Then, new prospectors and a new governmental organization swept their influence from the diggings. During their brief span of activity, however, the Walker party had reenacted a drama which gave credence to the charge that prospectors erected mining districts and enacted codes primarily for speculative ends.

---

[24] *ACHJ*, first Legislative Assembly, 1864, pp. 14–41; *ACHJ*, second Legislative Assembly, 1865, p. 35. The Howell Code appeared in Mining Laws of the Territory of Arizona, 38 Cong., 2 sess., *House Misc. Doc.* 14, pp.1–12.

[25] "Journal," 16–17. The first quartz mill in the Lynx Creek vicinity began operation in 1866. Conner, *Walker*, 284. Before the legislature reconvened the following November, Arizona prospectors raised the cry of monopoly. "The mining law passed last session," asserted an observer at Prescott, "does not suit a majority of the miners." The Howell Code, they avowed, favored the capitalist, not the individual prospector. *Bulletin*, Dec. 7, 1865. Other references to discontent appear in Conner, *Walker*, 268–270. The prospectors clearly implied that they had enjoyed greater opportunity when making their own laws.

Chapter 5

The Man, His Life and His Mine

By 1867 a few of the Prescott pioneers had moved to Phoenix and filed the first water claims ever recorded in the Salt River Valley. Six months later Civil War veteran Jack Swilling and Matt Cavaness, a former pioneer and prospector, convinced their more affluent friends from Prescott to invest in a new venture. The old Hohokam canals were lying fallow and filled with debris but were apparently a good place to grow hay and vegetables, both desperately needed by the army. Cavaness had already been active doing just that at Smith Station, which was a very lucrative venture. Mining in Prescott was winding down. About 40 men left to dig ditches at the new business, Jacob Waltz being one of them. In fact, most of the group who were with Peeples in Kernville and had come to the Arizona Territory in 1863 were now moving to the Salt River Valley.

The Pinal County courthouse in Florence, Arizona, was the place where homesteads could be filed. Jacob filed on the northeast quarter of section 16, Tract 1 N, Range 3 E in 1868. A large irrigation canal cut diagonally across Waltz's land from northeast to southwest. Waltz built his adobe house on the west bank of the ditch about 1,000 feet south of what is now Buckeye Road. It would be his permanent home. Since leaving Germany he had never been permanently settled, and ironically this place would be the reason for his ultimate demise. Jacob would live on that bank for 23 years until the great flood of 1891 caused the canal to overflow and wash away his home. Jacob, who was an old man by that time, was trapped in the water and took cold. Then, unable to fight off the resulting pneumonia, he died.

A Dutch immigrant and builder, Henry Le Vandermark, arranged for the reconstruction of this canal, and eventually it became known as the Dutch Ditch. It would be many years later when local rancher, businessman and prospector Jim Bark would coin the name "The Dutchman" when referring to Jacob Waltz.

Historian Bertie Roberts, as a young girl, lived on the farm adjacent to Jacob's for 13 years. She remembered this quiet man. In her notes she said, "He kept mostly to himself. Traveling to town, tugging at his mule, a couple of times a week he would trade eggs and vegetables for baked goods from Julia Thomas' bakery."

Waltz spoke German and English and was very interested in civic matters but never became embroiled in the resulting skirmishes. It was not uncommon for the old prospector to stop at the Roberts farm for cold buttermilk and a visit with Bertie's father. Afterward he would continue on into town.

Bertie's brothers were available to help out on the Waltz farm or run errands. They often located stray stock or cut hay and were paid with gold concentrate. Jacob was known to have sold gold broken from quartz to Goldman's store. There was no question that Jacob was a curiosity; people gossiped about him. He had friends and neighbors with whom he visited, was cordial to most folks but was definitely reclusive. Naturally, since he wasn't approachable, people thought him odd or at least secretive – having a great secret he would never divulge.

Jacob was observed going away twice a year, walking alongside his mule, headed toward the Superstitions. He would be gone for days without explanation. People assumed he had a mine somewhere in those mountains or had discovered a cache. He always paid his bills with gold and was never seen going into a bank. One explanation given for his seclusion was his desire to keep secret the location of the gold. If Jacob had indeed killed Peralta's peons to secure the mine and there were Peraltas living and working in Phoenix, he kept his own council to protect his caches or because of guilt.

Bertie Roberts gave the following quotes in her manuscript: "A German friend of Waltz (named Magdeberg) lived with Andy Starrar and his son on an adjacent farm. He was in poor health, to [*sic*] ill to help with the farm, he just lived on the place. It was his habit to walk up to our house with Waltz on Sunday afternoons. I believe his name was Magberg [*sic*]. Waltz and Magberg

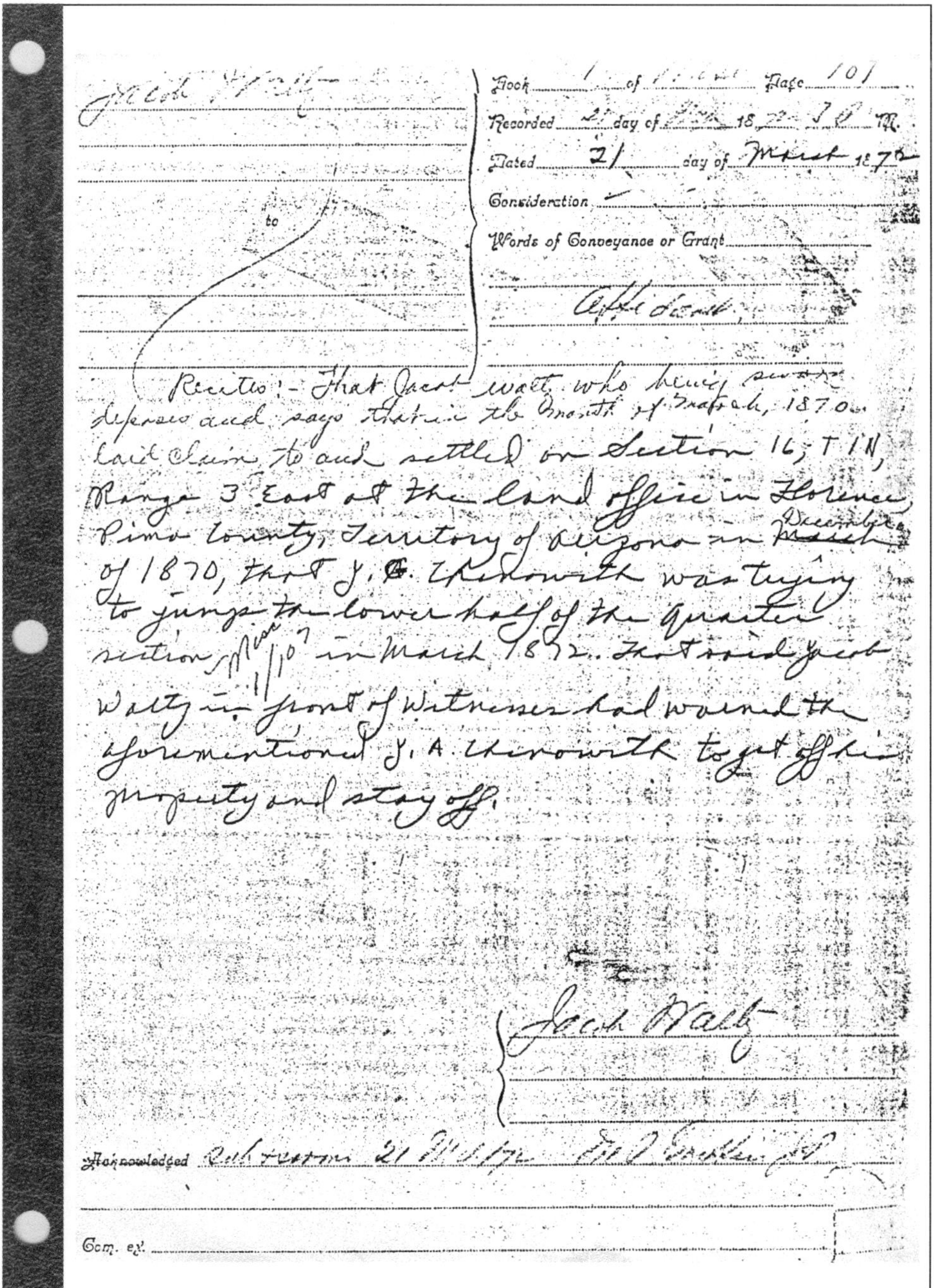

Jacob Waltz [illegible]

to

Book 1 of [illegible] Page 101

Recorded 21 day of [illegible] 18 72 [illegible] M.

Dated 21 day of March 1872

Consideration —

Words of Conveyance or Grant

Affidavit

Recites:- That Jacob Waltz who being sworn deposes and says that in the month of March, 1870 laid claim to and settled on Section 16, T 1 N, Range 3 East at the land office in Florence, Pima County, Territory of Arizona in ~~March~~ December of 1870, that J. A. Chenowith was trying to jump the lower half of the quarter section, Nov 1/10 in March 1872. That said Jacob Waltz in front of witnesses had warned the aforementioned J. A. Chenowith to get off his property and stay off.

Jacob Waltz

Acknowledged [illegible] 21 March 1872 [illegible] J.P.

Com. ex.

Affidavit of Jacob Waltz against claim jumping on his property (circa 1870).

(Magdeberg) would sit on the porch with father and talk politics."

When Magdeberg died, he willed all his possessions to Waltz. A copy of that will is included in this chapter. The bequeathment amounted to approximately $5,000, a considerable sum in the 1800s. Most farmers would borrow at the bank at planting time but not Jacob. He obviously had no need of a loan.

A copy of a transmitting draft on an accountable warrant of the Treasury of the United States dated August 24, 1887, to the National Bank of Lawrence, Kansas, worth $7,000 is also included in this chapter. This money was from Jacob Waltz to

Transcript of the :

Last Will and Testament of Ferdinand Magdeberg Sept 8, 1877
Maricopa County Arizona Territory.

Beginning on page 62 of the Maricopa County Probate Court
Records book A. Wm. A. Hancock Probate Judge.

I Ferdinand Magdeberg of the town of Phoenix, County of Maricopa, Territory of Arizona, being in ill bodily health and of sound and disposing mind and memory and calling in mind the frailty and uncertainty of human life and being desirous of settling my worldly affairs and directing how the property with which I am blessed shall be disposed of after my decease while I still have strength and capacity so to do. Do make this my last will and testament. --------------

Page 63 beginning on line 1.

I give divise and bequeath to my old friend **Jacob Walsh** *(Waltz) of the same town and county, all property real and personal. My three mares, Sorrell mare Neusis, my mule Francsis, three mares ______ and colt. And My three shares of water rights in the Griffin Irrigation Ditch. Also the amount due to me on a promisory note by H. (Henry) Glossen , also the amount due me on a promisory note by H. (Heinrich) Hurssey and lastly, I do hereby nominate and appoint my trusty friend Henry Glossen to be the executor of this my last will and testament. In witness whereas I have herefore set my hand and seal this 8th day of September 1877 One Thousand eight hundred and seventy-seven.*

Ferdinand Magdeberg

Seal

A copy of Ferdinand Magdeberg's Last Will and Testament.

his sister, who lived in Lawrence. The rumor that Jacob had sent a large amount of money to his sister was true. The problem was no one knew his sister didn't live in Germany but in Kansas. The fact that the warrant was purchased with 50 pounds of gold is positive proof that Jacob had a mine.

Remember, it was 1863 when the German came across the desert to Arizona. He didn't send any money to his sister until 1887. That and all the other proof should put an end to the speculation that Jacob was high-grading ore from the Vulture or that

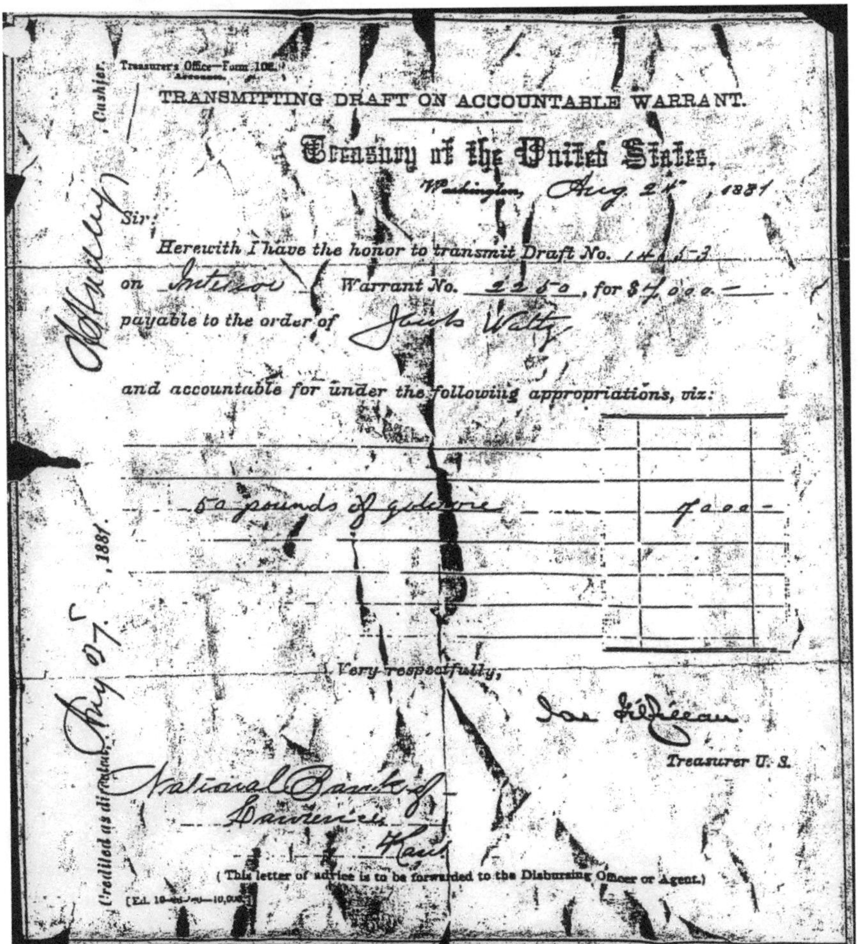

Treasurer's Office—Form 102.

TRANSMITTING DRAFT ON ACCOUNTABLE WARRANT.

Treasury of the United States,

Washington, Aug 24, 1887

Sir:

Herewith I have the honor to transmit Draft No. on Interior Warrant No. 2250, for $7,000 payable to the order of Jacob Waltz and accountable for under the following appropriations, viz:

50 pounds of gold ore 7000

Very respectfully,

Treasurer U. S.

Credited as directed Aug 27, 1887, Cashier.

National Bank of Lawrence Kans

(This letter of advice is to be forwarded to the Disbursing Officer or Agent.)

A transmitting draft on an accountable warrant of the Treasury of the United States dated August 24, 1887, to the National Bank of Lawrence worth $7,000. This is proof that Jacob Waltz sent money to his sister in Kansas.

he had saved his gold from the Peeples expedition. There is also the large payment he made on Julia's behalf and the gold brought to him by Reiney from the farm. Later 50 pounds of gold would be taken from under his deathbed.

Jacob became close friends with Julia Thomas and her adopted son Rheinhart (Reiney) Petrasch. The Petrasch family, husband and wife, sailed to the United States from Hanover, an industrial city in Germany, on the ship *Cimbria*. After their exit from Hamburg, they landed in New York on April 10, 1871.

Andrew Starrar, neighbor of Jacob Waltz, provided a home for Magdeberg.

Julia (Corn) Thomas (1865-1914)

Rheinhart Petrasch (Julia's adopted son) who committed suicide in Globe.

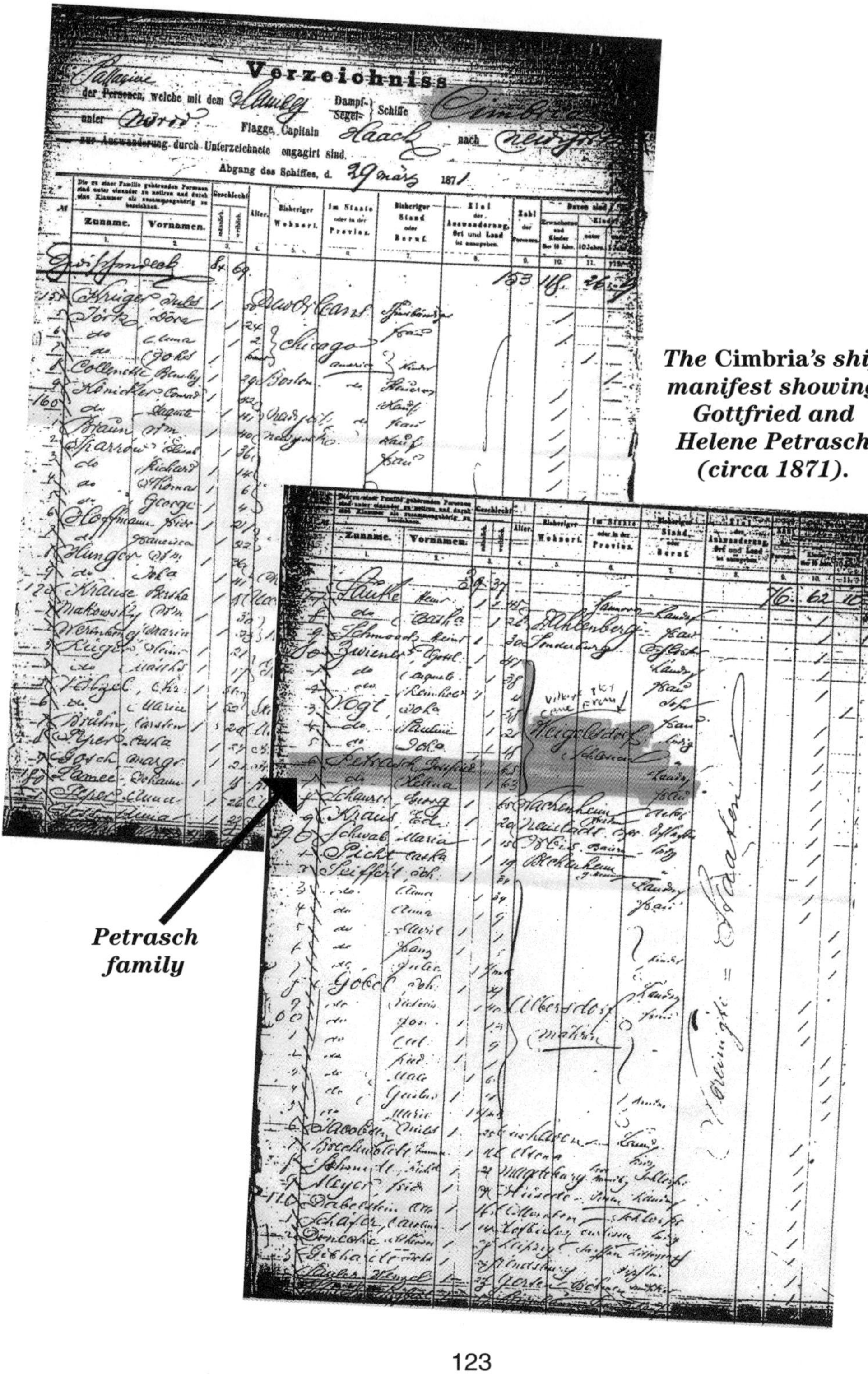

The* Cimbria*'s ship manifest showing Gottfried and Helene Petrasch (circa 1871).

R	AGE	SEX	OCCUPATION	PRV IVL DES
	9	M	CHILD	SRZZZZUSA
R.	8	M	CHILD	SRZZZZUSA
A	7	F	CHILD	SRZZZZUSA
N	6	M	CHILD	SRZZZZUSA
	5	M	CHILD	SRZZZZUSA
	4	F	CHILD	SRZZZZUSA
	3	M	CHILD	SRZZZZUSA
NE	.11	F	INFANT	SRZZZZUSA
JOH.	27	M	FARMER	SRZZZZUSA
	22	F	W	SRZZZZUSA
AVER	46	M	FARMER	SRZZZZUSA
COB	31	M	FARMER	SRZZZZUSA
	33	F	W	SRZZZZUSA
AD	8	M	CHILD	SRZZZZUSA
	7	M	CHILD	SRZZZZUSA
	3	M	CHILD	SRZZZZUSA
, GEORG	27	M	FARMER	SRZZZZUSA
SAVETTE	28	F	SGL	SRZZZZUSA
	25	M	FARMER	SRZZZZUSA
HLE, CARL	29	M	FARMER	SRZZZZUSA
EGER, LUDWIG	29	M	FARMER	SRZZZZUSA
JOHS.	29	M	FARMER	SRZZZZUSA
NICOLAUS	35	M	FARMER	SRZZZZUSA
H.	24	M	FARMER	SRZZZZUSA
SAM.	22	M	FARMER	SRZZZZUSA
I, F.	20	M	FARMER	SRZZZZUSA
JACOB	74	M	FARMER	SRZZZZUSA
	17	F	SGL	GRZZZZUSA
HR.	14	M	LABR	SRZZZZUSA
	12	M	LABR	SRZZZZUSA
IN, HEINR.	26	M	JNR	GRZZZZUSA
R.	19	M	JNR	GRZZZZUSA
CHAEL	19	M	LABR	GRZZZZUSA
. MARIE	27	F	WO	GRZZZZUSA
	8	F	CHILD	GRZZZZUSA
JACOB	19	M	LABR	GRZZZZUSA
I, JOH.	16	M	LABR	SRZZZZUSA
, JOHS.	35	M	LABR	SRZZZZUSA
	33	F	W	SRZZZZUSA
NNA	55	F	M	SRZZZZUSA
	7	M	CHILD	SRZZZZUSA
	6	F	CHILD	SRZZZZUSA
IA	3	F	CHILD	SRZZZZUSA
SE	1	F	CHILD	SRZZZZUSA
IA	.10	F	INFANT	SRZZZZUSA
JOHS.	27	M	LABR	SRZZZZUSA
. ANDREAS	25	M	LABR	GRZZZZUSA
. C.	40	M	LABR	GRZZZZUSA
I, LEOPOLD	38	M	UNKNOWN	GRZZZZUSA
IY, MARIE	25	F	SGL	GRZZZZUSA
. AROS.	8	M	CHILD	GRZZZZUSA
DORA	29	F	SGL	GRZZZZUSA
I, JACOB	16	M	CL	GRZZZZUSA
I, SARAH	20	F	SGL	GRZZZZUSA
, AUG.	28	M	PNTR	GRZZZZUSA
LFRED	20	M	MCHT	GRZZZZUSA
JACOB	22	M	MCHT	GRZZZZUSA
DI, JOH.	28	M	LABR	SRZZZZUSA
ETER	30	M	LABR	SRZZZZUSA
A.	29	F	SGL	SRZZZZUSA
A.	27	F	SGL	SRZZZZUSA
N, DORA	35	F	WO	SRZZZZUSA
	3	M	CHILD	SRZZZZUSA
Z. JOHS.	31	M	MCHT	GRZZZZUSA
UDW.	28	M	MLNR	GRZZZZUSA
R, CARL	20	M	MCHT	GRZZZZUSA
MORITZ	17	M	MCHT	GRZZZZUSA
. J.	26	M	MCHT	GRZZZZUSA

PASSENGER	AGE	SEX	OCCUPATION	PRV IVL DES

SHIP: ROANOKE

FROM: LA GUAIRA AND PUERTO CABELLO
TO: PHILADELPHIA
ARRIVED: 06 APRIL 1871

PASSENGER	AGE	SEX	OCCUPATION	PRV IVL DES
SCHACHT, HENRIETTE	22	F	SVNT	GRZZZZUSA
BESTPHALEN, LOUISE	21	F	SVNT	GRZZZZUSA
MEGO, OSCAR	33	M	UNKNOWN	GRZZZZUSA
GRANCKO, ERNST	31	M	UNKNOWN	GRZZZZUSA

SHIP: CINBRIA

FROM: HAMBURG
TO: NEW YORK
ARRIVED 10 APRIL 1871

PASSENGER	AGE	SEX	OCCUPATION	PRV IVL DES
SCHINDLER, JOS.	30	M	TLR	GRZZZZUSA
CLAUSSEN, MARIE	17	F	SGL	GRZZZZUSA
H.	16	M	UNKNOWN	GRZZZZUSA
SCHWARZ, LISETTE	20	F	SGL	GRZZZZUSA
GROENKE, JULIANNE	24	F	SGI	GRZZZZUSA
BIBER, AUG.	51	M	LABR	GRZZZZUSA
ELEONORE	45	F	W	GRZZZZUSA
ED.	9	M	CHILD	GRZZZZUSA
ROSALIE	7	F	CHILD	GRZZZZUSA
PORSCHONKE, ELISE	21	F	SGL	GRZZZZUSA
ANDERSAN, FERD.	30	M	TLR	GRZZZZUSA
PETERS, CLAUS	48	M	FARMER	GRZZZZUSA
ANNA	45	F	W	GRZZZZUSA
JOH.	17	M	S	GRZZZZUSA
KLAHN, TH.	30	M	TCHR	GRZZZZUSA
KOSCHMIFDER, HERRM.	19	M	MCHT	GRZZZZUSA
MICHAEL, AUGUST	17	M	MCHT	GRZZZZUSA
WIESE, NICLS.	36	M	SEMN	GRZZZZUSA
AHRENS, WM.	31	M	SHMK	GRZZZZUSA
BOLLOW, CHRIST.	33	M	SHMK	GRZZZZUSA
EMMA	33	F	W	GRZZZZUSA
C.	4	M	CHILD	GRZZZZUSA
BERTHA	3	F	CHILD	GRZZZZUSA
EMMA	.06	F	INFANT	GRZZZZUSA
KNUTH, MARIE	21	F	SGL	GRZZZZUSA
BODENDICK, WM.	26	M	FARMER	GRZZZZUSA
ECKHARDT, JOS.	29	M	CGRMKR	GRZZZZUSA
BOLLOW, MARIE	19	F	SGL	GRZZZZUSA
NEIDHARDT, EMILIE	17	F	SGL	GRZZZZUSA
FLUEGEL, MATHA.	24	F	SGL	GRZZZZUSA
BERGSCHICKER, SOPHIE	20	F	SGL	GRZZZZUSA
MISSELBECK, CHR.	18	M	FARMER	GRZZZZUSA
KLENK, WILH.	16	M	FARMER	GRZZZZUSA
HAFENRICHTER, ADAM	29	M	BKR	GRZZZZUSA
KNEPPLE, WILHE.	17	F	SGL	GRZZZZUSA
L.	15	M	UNKNOWN	GRZZZZUSA
SCHAFER, HEINR.	34	M	STCTR	GRZZZZUSA
SOPHIE	69	F	W	GRZZZZUSA
BOLTER, JOHS.	31	M	UNKNOWN	GRZZZZUSA
HUNSINGER, PHE.	50	F	WO	GRZZZZUSA
JUNG KATHA.	21	F	SGL	GRZZZZUSA
ZINSMEISTER, JOHS.	47	M	FARMER	GRZZZZUSA
ELISE	33	F	W	GRZZZZUSA
J.	10	M	CH	GRZZZZUSA
H.	9	M	CHILD	GRZZZZUSA
ELISE	5	F	CHILD	GRZZZZUSA
M.	1	M	CHILD	GRZZZZUSA
F.	.01	M	INFANT	GRZZZZUSA
SCHUCK, GEORG	36	M	FARMER	GRZZZZUSA
KRENER, MARG.	19	F	SGL	GRZZZZUSA
GREIS, GEORG	32	M	FARMER	GRZZZZUSA
HILBERT, MAGDA.	16	F	SGL	GRZZZZUSA

The ship* Cimbria *manifest (1871).

PASSENGER	AGE	SEX	OCCUPATION	PRV VIL DES
BENGEL, ELISE	17	F	SGL	GRZZZZUSA
ZACHMANN, JOHS.	35	M	FARMER	GRZZZZUSA
ROSSLEIN, MICH.	42	M	FARMER	GRZZZZUSA
KURZMANN, A.	39	M	FARMER	GRZZZZUSA
WAIKER, BARB.	57	F	WO	GRZZZZUSA
BUSSMANN, CATH.	27	F	SGL	GRZZZZUSA
STEIERWALD, MARIE	18	F	SGL	GRZZZZUSA
GERST, WENDELINA	55	F	WO	GRZZZZUSA
PRIMUS	15	M	S	GRZZZZUSA
KRAUSE, THEODOR	27	M	LABR	GRZZZZUSA
FRANZ	2	M	CHILD	GRZZZZUSA
DALL. MARIE	20	F	SGL	GRZZZZUSA
BRAUN. JOS.	31	M	LABR	GRZZZZUSA
MARIANNE	28	F	W	GRZZZZUSA
THERESE	5	F	CHILD	GRZZZZUSA
R.	3	M	CHILD	GRZZZZUSA
LOUISE	.11	F	INFANT	GRZZZZUSA
FUCHS, ELISE	21	F	SGL	GRZZZZUSA
PETZINGER, CAROLE.	24	F	SGL	GRZZZZUSA
DIETH. JACOB	36	M	LABR	GRZZZZUSA
MOHR, LENA	29	F	SGL	GRZZZZUSA
UHLEMANN, RICHD.	25	M	HTR	GRZZZZUSA
KEHRER, IGNATZ	20	M	LABR	GRZZZZUSA
SANKE, HERM.	45	M	FARMER	GRZZZZUSA
CATH.	26	F	W	GRZZZZUSA
SCHMOOCK, HEINR.	30	M	LKSH	GRZZZZUSA
ZWIENER, GOTTL.	47	M	FARMER	GRZZZZUSA
ANGE.	38	F	W	GRZZZZUSA
R.	4	M	CHILD	GRZZZZUSA
VOGT, JOHA.	58	F	WO	GRZZZZUSA
PAULI	21	F	CH	GRZZZZUSA
JOHA	18	F	CH	GRZZZZUSA
PETRASCH GOTTFR.	65	M	FARMER	GRZZZZUSA
HELENE	35	F	W	GRZZZZUSA
SCHAURER, GEORG	65	M	LABR	GRZZZZUSA
KRAUS, ED.	20	M	UNKNOWN	GRZZZZUSA
SCHWAB, MARIE	15	F	SGL	GRZZZZUSA
PICHT, CATH.	19	F	SGL	GRZZZZUSA
JACOBSEN, NICLS.	25	M	FARMER	GRZZZZUSA
BRECKWOLDT, EMMA	16	F	SGL	GRZZZZUSA
SOBEL. JOH.	39	M	FARMER	GRZZZZUSA
VICTORIA	40	F	W	GRZZZZUSA
JOS.	13	M	CH	GRZZZZUSA
ABEL.	9	F	CHILD	GRZZZZUSA
R.	7	M	CHILD	GRZZZZUSA
MALE.	6	F	CHILD	GRZZZZUSA
GUSTAV	4	M	CHILD	GRZZZZUSA
MARIE	.09	F	INFANT	GRZZZZUSA
DABELSTEIN, OTTO	16	M	LABR	GRZZZZUSA
SCHAFER, CAROLINE	14	F	SGL	GRZZZZUSA
DENECKE, THEODOR	28	M	LITGR	GRZZZZUSA
GIBHARDT, JOHS.	22	M	JNR	GRZZZZUSA
PAULER, WENZEL	23	M	MUSN	GRZZZZUSA
STRICK, SIEGBERT	17	M	MCHT	GRZZZZUSA
LUDWIG. MATH.	29	F	SGL	GRZZZZUSA
ZORNIG, WM.	16	M	CL	GRZZZZUSA
WALTER, EMIL	28	M	MCHT	GRZZZZUSA
KELMEREIT, HERM.	27	M	BKR	GRZZZZUSA
LOUISF	26	F	W	GRZZZZUSA
WICKSTROEM, NILSON	43	M	SHMK	GRZZZZUSA
WETZ, MARIF	20	F	SGL	GRZZZZUSA
SCHAACK, MARIE	30	F	WO	GRZZZZUSA
GLUCK, JACOB	16	M	CL	GRZZZZUSA
KLAWIKOWSKA, MARIANNE	18	F	SGL	GRZZZZUSA
STERNBERG. ANDREAS	47	M	SEMN	GRZZZZUSA
HENRINGER, G.	61	M	PNTR	GRZZZZUSA
SPIRA, JULS.	41	M	MCHT	GRZZZZUSA
THIESSEN, JURGEN	31	M	MCHT	GRZZZZUSA
CATH.	29	F	W	GRZZZZUSA
H.	4	M	CHILD	GRZZZZUSA
C.	.11	M	INFANT	GRZZZZUSA
TAESCHMANN, MARIE	23	F	SGL	GRZZZZUSA
RASMUSSEN, ANDR.	22	M	FARMER	GRZZZZUSA
GLUCHAN, JOH.	19	M	FARMER	GRZZZZUSA
JESPERSEN, RASMUS	19	M	FARMER	GRZZZZUSA

PASSENGER	AGE	SEX	OCCUPATION	PRV VIL DES
RADEBECK, THOM.	50	M	FARMER	GRZZZZU
GRUMM, HANS	50	M	FARMER	GRZZZZU
BRODERSEN, JOHE.	21	F	WO	GRZZZZU
HANNY	.09	F	INFANT	GRZZZZU
FRITZEN, JUSINA	24	M	SGL	GRZZZZU
SEIFFERT, JOH.	34	M	FARMER	GRZZZZU
ANNA	34	F	W	GRZZZZU
ANNA	9	F	CHILD	GRZZZZU
DAVID	7	M	CHILD	GRZZZZU
FRANZ	5	M	CHILD	GRZZZZU
JULIE	.09	F	INFANT	GRZZZZU
SCHMIDT, RICHARD	22	M	LKSH	GRZZZZU
MEYER, FRIEDR.	29	M	FARMER	GRZZZZU
HANGER, WM.	36	M	MCHT	GRZZZZU
JOHA.	41	F	W	GRZZZZU
KRAUSE, BERTHA	18	F	SGL	GRZZZZU
MAKOWSKY, WM.	30	M	UNKNOWN	GRZZZZU
WERENBERG. MARIE	23	F	SGL	GRZZZZU
RUEGER, HEINR.	21	M	SHMK	GRZZZZU
MATH.	17	M	FARMER	GRZZZZU
VOELZEL, CHR.	59	M	FARMER	GRZZZZU
MARIA	52	F	W	GRZZZZU
BRUHN, CASSTEN	20	M	FARMER	GRZZZZU
PEPER, CATH.	27	F	SGL	GRZZZZU
GOSCH, MARG.	21	F	SGL	GRZZZZU
LAMEE, JOHANN	18	M	FARMER	GRZZZZU
PEPER, ANNA	26	F	SGL	GRZZZZU
JESS, ANNA	23	F	SGL	GRZZZZU
KUEHL, HEINR.	47	M	FARMER	GRZZZZU
ANNA	36	F	W	GRZZZZU
ANNA	4	F	CHILD	GRZZZZU
STEINEL, OSCAR	29	M	UPHST	GRZZZZU
WALDEMAR	25	M	UPHST	GRZZZZU
BUELL. HANS	30	M	FARMER	GRZZZZU
BRAMER, CARL	38	M	BKR	GRZZZZU
SZIBLEWSKI, JOH.	37	M	CPTR	GRZZZZU
BRASCH, WILHE.	28	F	SGL	GRZZZZU
REICH, HINR.	27	M	PNTR	GRZZZZU
ROST, JOH.	28	M	TLR	GRZZZZU
MARIF	22	F	W	GRZZZZU
STRUBACH, GG.	24	M	TLR	GRZZZZU
BOEHNER, JOH.	19	M	FARMER	GRZZZZU
REDER, ANDR.	23	M	FARMER	GRZZZZU
WIGGERS, HANS	17	M	FARMER	GRZZZZU
DAASCH, FRIEDR.	29	M	PNTR	GRZZZZU
BUETTNER, ANNA	17	F	SGL	GRZZZZU
SCHLAGHEIT, HERRM.	28	M	FARMER	GRZZZZU
ELISE	28	F	W	GRZZZZU
STOCK, JOHA.	45	F	WO	GRZZZZU
WM.	17	M	CH	GRZZZZU
LUDWIG	13	M	CH	GRZZZZU
FR.	5	M	CHILD	GRZZZZU
HUBER, CHRIST.	29	F	SGL	GRZZZZU
MARG.	24	F	WO	GRZZZZU
HAEVE, PETER	4	M	CHILD	GRZZZZU
HENSEN, HANNI	.01	F	INFANT	GRZZZZU
BRODERSEN, JOH.	25	M	FARMER	GRZZZZU
PEPER, DETLEF	34	M	FARMER	GRZZZZU
SIERTS, HEINR.	28	M	FARMER	GRZZZZU
STUENE, HELLMUTH	66	M	FARMER	GRZZZZU
MARIA	64	F	W	GRZZZZU
ERNST	46	M	FARMER	GRZZZZU
LISETHE	43	F	W	GRZZZZU
ED.	19	M	CH	GRZZZZU
FERD.	18	M	CH	GRZZZZU
ALBE.	17	M	CH	GRZZZZU
ELISE	15	F	CH	GRZZZZU
META	14	F	CH	GRZZZZU
JOH.	9	M	CHILD	GRZZZZU
OTTO	8	M	CHILD	GRZZZZU
SANDHER, CARL	29	M	FARMER	GRZZZZU
FRANZISCA	28	F	W	GRZZZZU
JUS.	6	M	CHILD	GRZZZZU
WILFERT, FRANZISCA	22	F	SGL	GRZZZZU
FELDMANN, BERTHA	20	F	SGL	GRZZZZU

Petrasch family

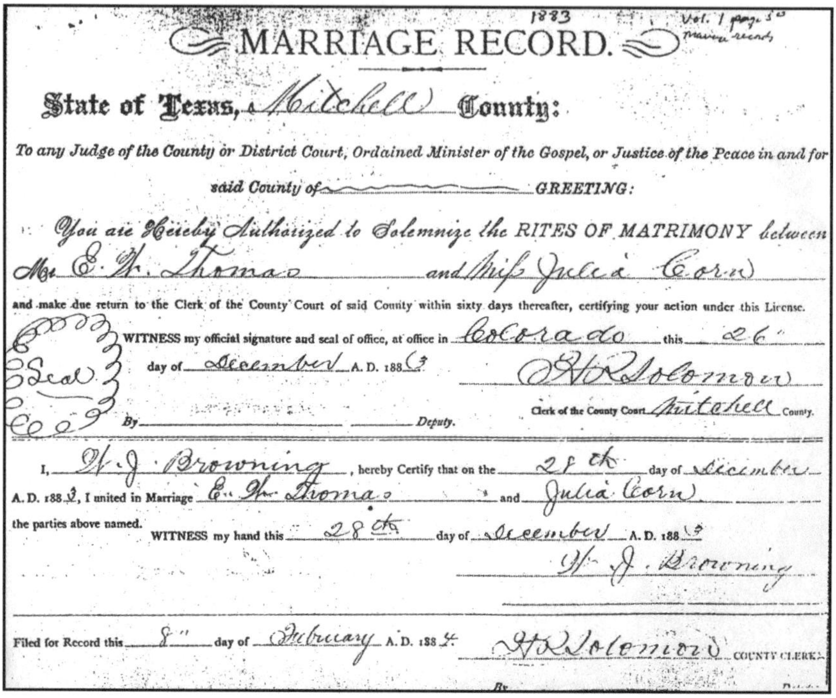

1883 Vol. 1 page 50 marriage records

MARRIAGE RECORD.

State of Texas, Mitchell County:

To any Judge of the County or District Court, Ordained Minister of the Gospel, or Justice of the Peace in and for said County of ——————— *GREETING:*

You are Hereby Authorized to Solemnize the RITES OF MATRIMONY *between* Mr. E. W. Thomas and Miss Julia Corn

and make due return to the Clerk of the County Court of said County within sixty days thereafter, certifying your action under this License.

(Seal) WITNESS my official signature and seal of office, at office in Colorado this 26 day of December A. D. 1883

H. R. Solomon
Clerk of the County Court Mitchell County.

By ———————— Deputy.

I, W. J. Browning, hereby Certify that on the 28th day of December A. D. 1883, I united in Marriage E. W. Thomas and Julia Corn the parties above named.

WITNESS my hand this 28th day of December A. D. 1883

W. J. Browning

Filed for Record this 8th day of February A. D. 1884 H. R. Solomon COUNTY CLERK.

Julia's marriage certificate to E.W. Thomas.

Julia's bakery/ice cream parlor at the corner of Central and Washington Streets in Phoenix, Arizona (circa 1884).

The docket listed them as Gottfreid and Helene. Ultimately, the family arrived in Colorado where the father and sons went to work as hard-rock miners. The boys' mother left to go back East, leaving the men to raise their new son Rheinhart. Julia, a mulatto woman who also spoke German, took care of the boy and decided to adopt him. (Although the adoption was widely rumored, there have never been any documents to prove it.) She, her husband and the boy came to the Arizona Territory and opened a bakery in Phoenix.

Julia and her husband ran the E.W. Thomas Bakery on Washington Street across from the Plaza between 1888 and 1889. Since the business was doing well, the pair invested in an expensive Arctic soda fountain that was to be shipped from Chicago. Unbeknown to Julia, her husband had taken up with another woman and soon after ran off with her, taking all their savings in the bargain. Julia was suddenly faced with the shipment due soon and the flour and sugar bill, both totaling $2,000. She was hard working but was mulatto, and no bank would have considered her for a loan. It was a dark day.

Interior of E.W. Thomas Bakery.
Left to right: Joseph Gilmore (standing); James Lee (first man seated); Jacob Waltz; unknown (third man seated); Emil Thomas, owner of the parlor, (behind the counter on left); and Rheinhart Petrasch (behind the counter).
Photo, circa 1887-1889, courtesy of Paul Pettit.

As fate would have it, Jacob Waltz showed up in the store bringing milk and eggs. He found Julia collapsed in tears over the flour table. In an attempt to console the woman, he asked her to confide her problem. There is no doubt Julia thought the confidence futile; after all the German was just a poor farmer no matter what the gossips said. After she finished, Jacob told her to wipe her eyes; he would help her. She might have been disbelieving, but Jacob was a man of his word.

According to the Roberts manuscript, he returned the next day with a bundle of cloth. Wrapped inside was 35 pounds of gold concentrate. In Ely's book Jacob is supposed to have said, "Send Reiney to me tonight and he can help me." Reiney is said to have been totally enamored of the gold when it was exposed in the lamp light after they returned to Julia's store from the farm. Either way, Jacob was in possession of a great deal of rich ore. Jacob said he would take the ore to the Wells Fargo office and purchase a draft that was then sent to the soda fountain company. The debt was satisfied. He paid the local debt with concentrate. Within a few days Julia was a successful business woman again. Now, the friendship was cemented.

During these years Phoenix consisted of a few dirt streets. (A grid of the town's structure is included.) Once Jacob paid the flour and sugar bill with gold concentrate, the speculation had been terminated. The old man did indeed have a mine – a rich mine. It is also evident the man had put himself in peril. From that moment on he would be prey to any unscrupulous citizens who would make it their business to trail him into the mountains. It is obvious that is the reason he wouldn't allow Peter Backens to file a claim on the mine.

Julia became a sort of caretaker to the old German. Of course, Waltz was a friend of Rheinhart's biological father. Waltz also may have decided he was getting old and might need someone to look after him if he became ill. As things turned out, he was right.

There is always speculation about the extent of Jacob's wealth. He lived modestly in the adobe he had built, living off his chickens and ducks and the vegetable garden he planted. But, it is incumbent upon the reader to note that Phoenix was in its infancy, and there was little to buy. No one dressed or lived lavishly. Jacob's needs were sparse. He had been poor all his life, and perhaps, it never occurred to him to live otherwise.

Homesteads on Salt River, Phoenix (1872)

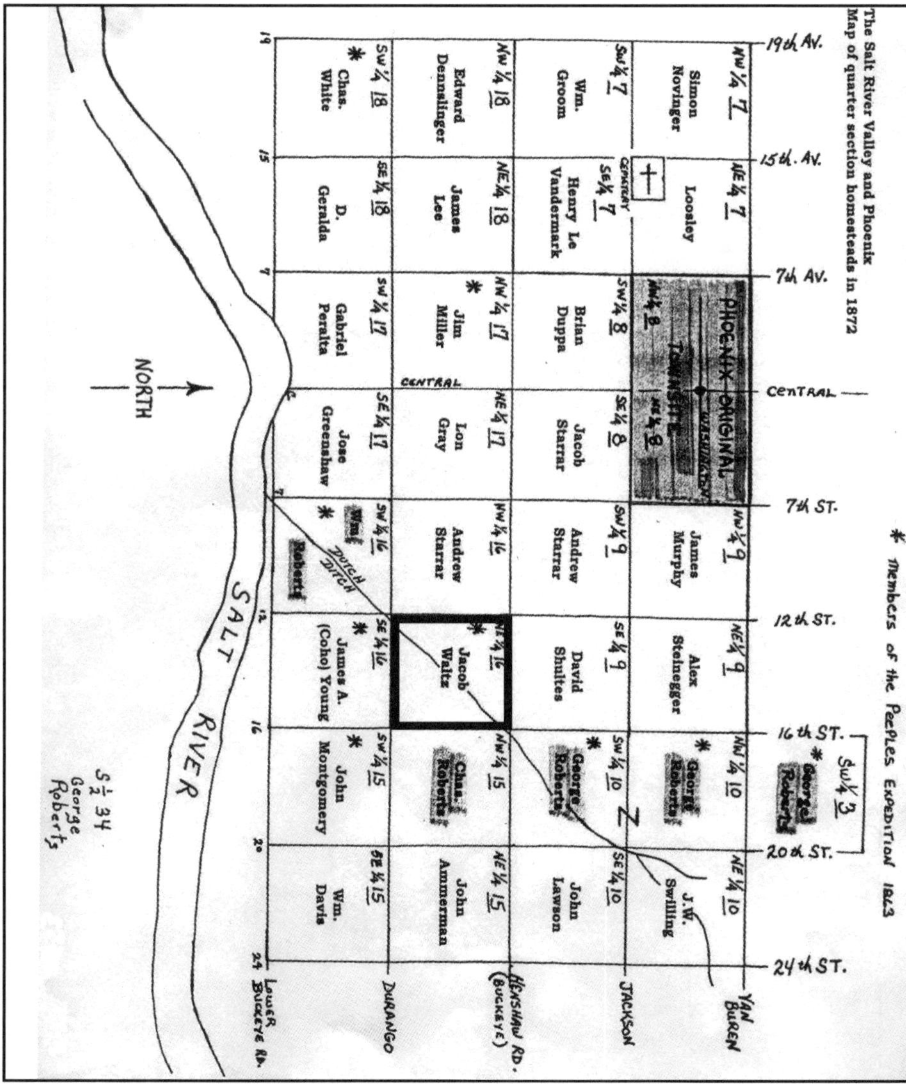

Julia Thomas's bakery on Washington Street (1891)

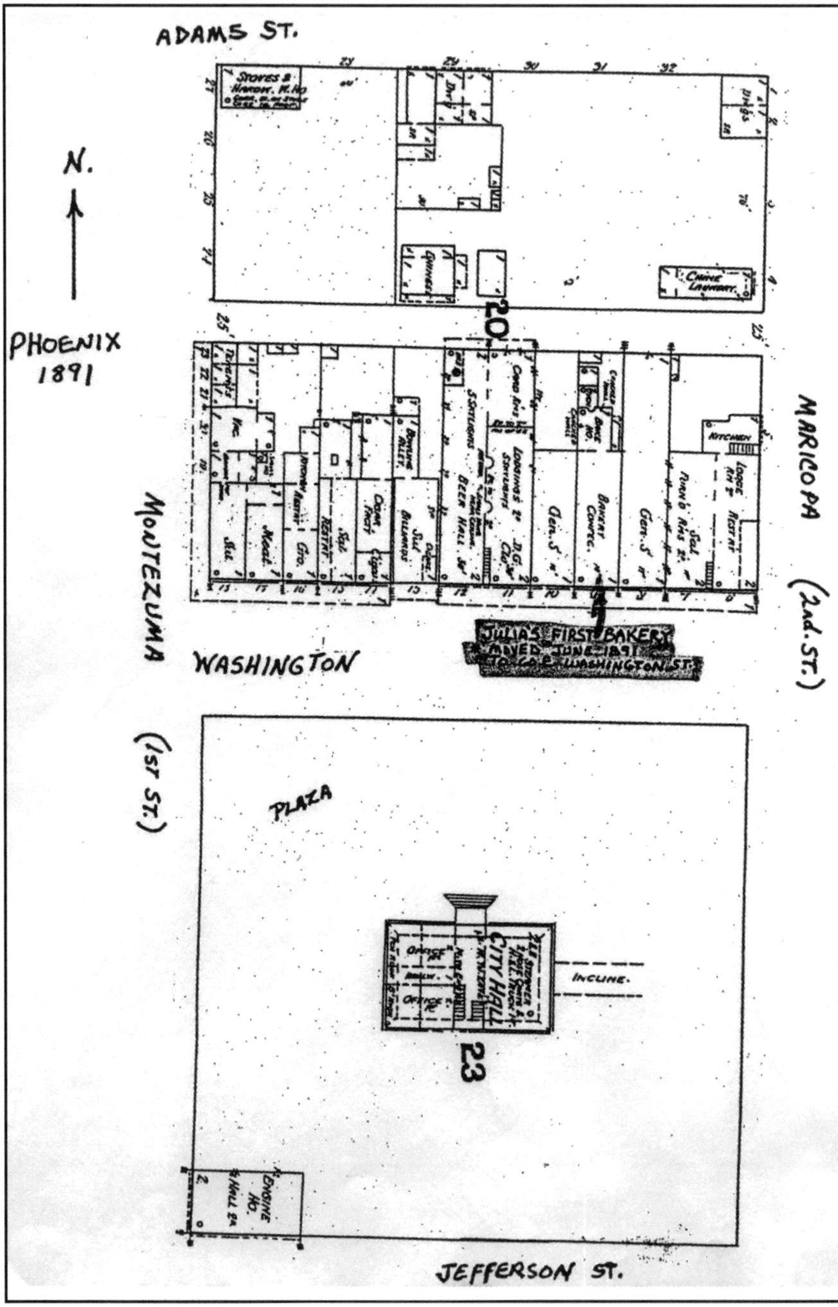

Phoenix homes of Julia Thomas and Dick Holmes (1891-1892)

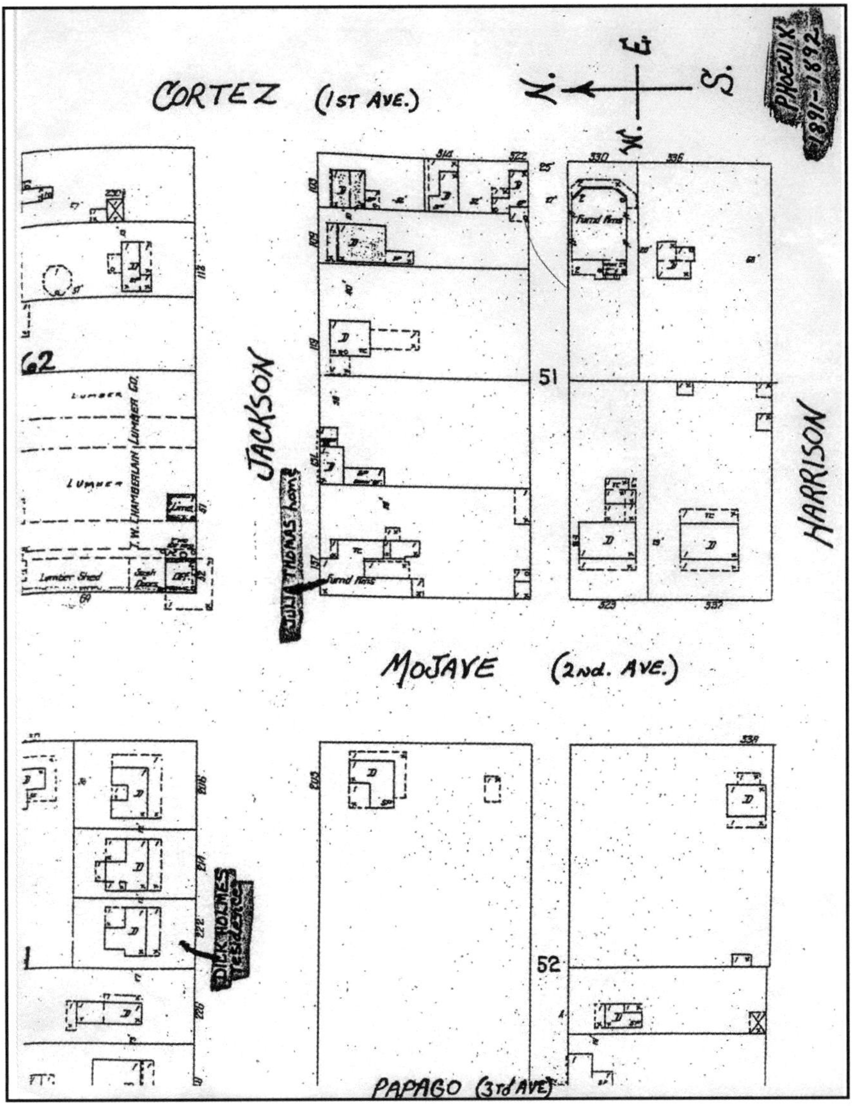

Plot map of Phoenix (1891)

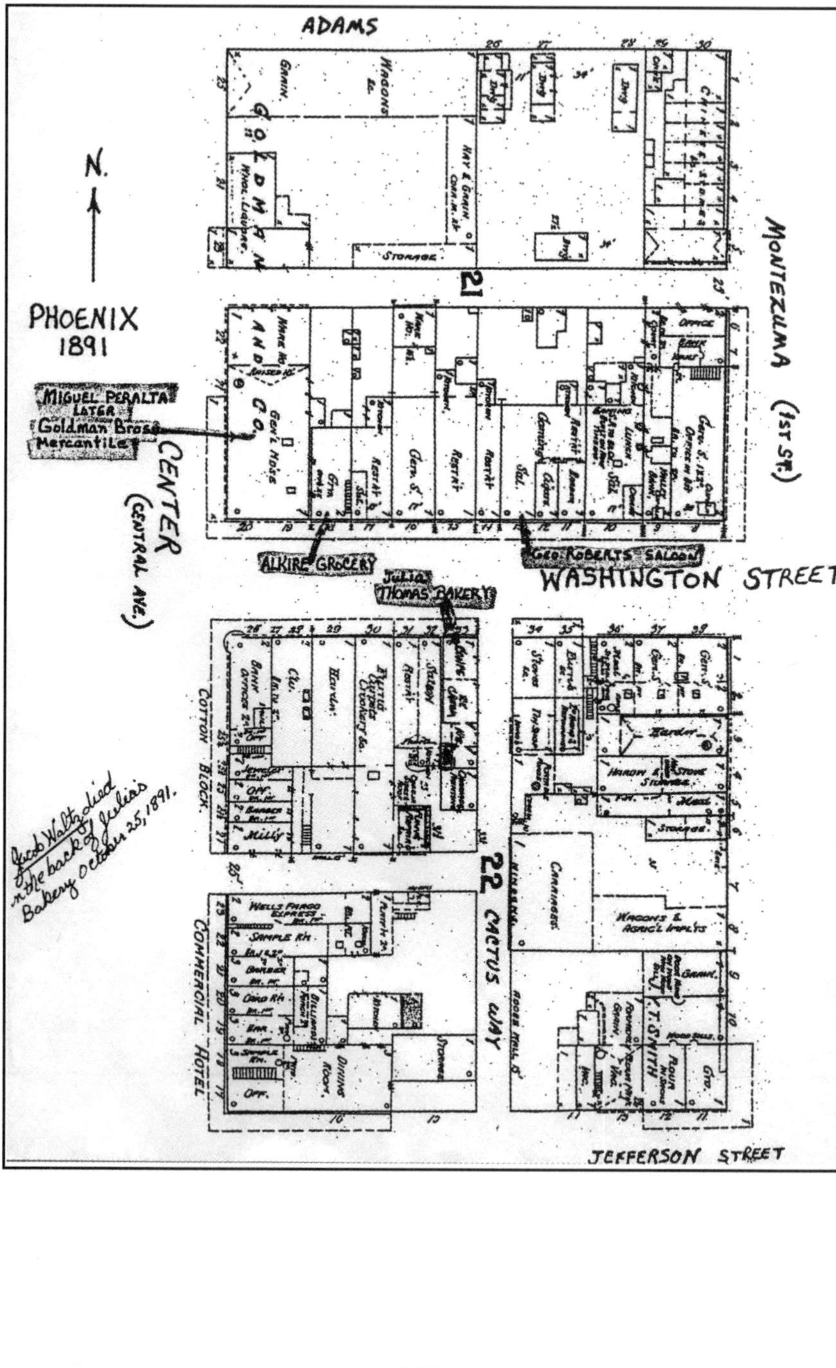

Now he had proven he was a generous man; Julia's debt was enormous for that time and place. Jacob never even questioned the amount.

Ultimately, as the friendship deepened, Waltz opened up to the pair. Reiney Petrasch told Bertie Roberts the following story in 1933. This story is essentially the same as the one Sims Ely tells in his book *The Lost Dutchman Mine*, which he said Julia told him.

It was Christmas day 1890. Julia had prepared a holiday meal to be taken to Jacob's farm. After they finished dinner, they all went outside to sit in the sun. Reiney said, "Waltz began to talk about the mine and preparations to go into the mountains and bring back more of his gold. He looked over to the East and pointed and said, 'My mine is over there, in those Salt River Mountains.' (The Superstition Mountains were originally known by this name, and Waltz always used it.) You could see the mountains off in the distance. He told us the country was very brushy and told of the items we would need to take with us.

"When I told Waltz that I knew where I could borrow some mining tools, he just waved his hand and told me we wouldn't need them. He said he hid his tools in a hole not far from the mine. From the way Waltz spoke to us I got the notion that we weren't going to mine any gold, rather we would just go in and pick some up that had already been dug. Waltz told us he had three holes (caches) near the mine . . . one large and two smaller ones."

Rheinhart continued by saying that during the next few months Waltz gave them directions and continued to talk about their visit to the mine. He also talked about how he often took circuitous routes to the mine to throw off the trackers who wanted to steal it from him. Julia and Rheinhart had never been in the mountains, and there were no discernible landmarks. These two were completely confused as to the location, but they didn't concern themselves because they knew Jacob was going to take them.

In 1933 when Reiney was talking to Bertie Roberts, he said a very profound thing: "I never thought about Waltz dying before we would get to go to the mine."

In February 1891 the worst flood in Arizona's recorded history came down the Salt River, overflowing its banks and devastating

This is an exact copy of the drawing made by Jacob Waltz and given to Rheinhart Petrasch, probably a month before his death on Oct. 25, 1891. It was made on a leaf torn from a notebook that was used for a collection of foreign stamps, some of them German.

the farms and homes around the Dutch Ditch for two miles. Since adobes are mud based, they melted. Jacob took refuge on some high ground surrounded by flood water. Rheinhart Petrasch and Albert Schafer waded through freezing water and carried the old man to safety. He had taken a chill, was unable to walk on his own and was taken immediately to the doctor and then to Julia's home.

She ran a small boarding house on the south side of Jackson Street at the intersection of Mojave (2nd Avenue). The boarding house had three rental rooms that were already occupied by Rheinhart Petrasch, Albert Schafer and two friends of the Petrasch family, Joe Gilmore and Simon Dandridge. Rheinhart said later Waltz was put out on a sun porch but later, wishing his privacy, insisted he be taken to the storeroom in the back of Julia's new bakery. Jacob's reclusive nature certainly would verify that request. Besides, Jacob was used to living alone and probably preferred it.

In 1920 the Arizona pioneers had their annual picnic in Phoenix. Bertie Roberts was always present at these parties. Until her death it was her custom to interview people, take photographs and document comments they made regarding the events that had passed. Rheinhart Petrasch was there. Bertie asked him to discuss what happened after he took Jacob to Julia's home during the flood. The information that follows is from Bertie's notes:

> Rhinehart said that a month or so before the old man's passing he arranged with Reiney to build him a coffin so he could have a proper burial when his time came. The boy also sold Waltz a grave site in the city cemetery which belonged to his family. Reiney's mother had left Denver when he was born and settled in the East. She was not coming back. Waltz paid $30.00 in gold for his final resting place in what is now known as the Pioneer Cemetery in the Petrasch family plot.

Waltz spent months recuperating in the storeroom. During much of that time he talked of taking Julia and the boy out to the mountains to obtain the ore in the caches. He told them mining was too arduous for them. Again, he tried to give directions. They either didn't listen or were confused, but whatever happened was later regretted. That fact probably, more than any other, ruined their lives.

Jacob's pneumonia was taking its toll. During the early morning hours of October 25, 1891, he began to fade. Julia, realizing he had never been this bad, ran out of the store crying, "Come quick! Jacob Waltz is dying," to Gideon Roberts and Dick Holmes, who were standing outside. She told them she was going to get the doctor.

As the story goes, they went inside and stayed with the old man, who proceeded to give them his gold and directions to the mine. That is the account Julia gave to Jim Bark and later to Sims Ely for his book. She also did not go to the funeral, which would seem rather odd to most, after the care she had given the old man. However, the men who stayed with Jacob while she was gone took the 50 pounds of gold from under the bed and said it was a gift. They also said Jacob had given them directions to the mine. This particular matter caused a feud that has lasted for six generations between Petrasch believers and Holmes believers.

Richard "Dick" Holmes, Jr. and his wife, Ida Roberts Holmes (circa 1892).

George "Brownie" Holmes in the Superstition Mountains.

It seems relevant to discuss a few questions at this point. According to Jim Bark's notes, Dick Holmes and Gideon Roberts were waiting outside the store when Julia emerged looking for the doctor. They agreed to stay with the dying German. First, does it seem realistic for this man to give his gold and directions to the richest mine in the country to two men who admittedly had been following him for sometime? He is reputed to have told them,"Follow me again and I'll kill you." No one doubted Jacob's threat. Also, Julia and the boy were his closest friends. They had cared for him, protected him, brought him to their home and been privy to his directions to the mine. Would it make any sense to have given his precious gold and the mine to two people other than his caregivers? These people were like family.

There are other factors that are important to note. Jacob had a heavy accent. He was ill with a respiratory disease that obviously made speaking difficult. How accurate could his information have been or how audible? Of course, he could have been delirious and babbled on without comprehension, which they took as fact. It is fairly certain that a man in delirium would answer if questioned, but who knows what he might say or if it would be coherent?

Dick Holmes never found the mine and spent a lifetime looking, as did his heirs. Holmes had the gold after Jacob's death. He sold a portion of it to Goldman Brothers Mercantile Exchange for $4,800. The assay was done by Joe Porterie of Phoenix. Holmes later had jewelry, a matchbox and a stickpin made up at Goldman's. Those pieces are still in existence. Dick Holmes left them to his son Brownie, who gifted them to a close friend before he died. Gideon Roberts never had any gold; he also didn't live too much longer than Waltz.

The Roberts family had their own version of the Waltz demise: As Jacob lay dying, a crowd of men gathered on the boardwalk and street outside Julia's bakery. The old man was wavering in and out of consciousness for several hours; he had pneumonia and would soon die. Inside the front room of the bakery, Julia was putting on a strange show for everyone. She began chanting a strange incantation and throwing powder into the fireplace, which caused a brilliant flash of colored fire and smoke that hung eerily in the night air.

Julia and Reiney realized Jacob's death would incite curiosity with the townspeople, so they decided to make some money out of it. Earlier they passed the word around town that Jacob was delirious. They said he was talking about his mine and gold. Men flocked to the site. Julia began to charge people to go in and spend a few minutes with the sick man. Frank and George Alkire, well-known grocers, told Bertie Roberts that earlier in the evening Julia was charging two bits (25¢) to spend a few minutes with Waltz. Later, as the crowd increased, she raised the price to $1.00. All the residents who were around later recalled the incident in the same manner.

Waltz's body was taken to Julia's home to be prepared for

burial, a common practice in that era. Several friends came by that day to pay their respects only to learn that Jacob Waltz had already been taken to the cemetery and buried.

Julia did not go to the funeral. Everyone speculated as to the reason. If she believed Jacob had given the gold to Dick Holmes after all she had been to the old man, that would have been reason enough for anyone. But, it is best to try to sort out the truth from other sources.

Rheinhart Petrasch relates the story of retrieving some of Waltz's property after the flood in 1891. The account, which follows, was written down by Rheinhart and told to Bertie Roberts and Christine (Cavaness) Morell at the Pioneer Days Celebration held in Phoenix Riverside Park in 1933.

> A short time after Jacob Waltz came to stay with Julia Thomas in Phoenix he paid me to go back out to his farm about 2 miles southeast of town and fetch back something he'd buried under the hearthstones of the fireplace in his house. Old Waltz's house had been ruined and most of it swept away by the flood of February 91. Albert Schafer, Sheriff Montgomery and I had gone out to see the flood damage and find old Waltz who hadn't made his usual stops in town for a couple days. When we got out near Waltz's place we could see his house was most all down and the little barn was completely swept down the ditch bank about a quarter mile or more. Jacob was sitting on a wall of the house clinging to it and just a shivering, soaked through to the skin by the cold water. He couldn't talk when we called out to him so Schafer and me waded out through the muddy, freezing water which was about waist deep for fifty rods or more to get to him. We had to carry Waltz out to Henshaw road which was partly underwater itself at the time. Schafer and I had walked out from town and Sheriff Montgomery had his horse and we had no wagon to get Waltz back to town in. We started back for town carrying Waltz and we hadn't gone far when a neighbor come along in a wagon and told us to put Jacob in the back and we'd take him on up to the Roberts farm which was just up the road a piece. Old Waltz looked real bad and Sheriff Montgomery told us that we better get him on into town and a doctor soon or the old man would die so _____ and Schafer and I hove out for Phoenix with Waltz a laying in the back of

the wagon. Jno Cloud was a doctor and had a little office at the head of Wall Street just below Washington in Phoenix back then so we made a line for there. We got Waltz inside the Doc's place and we give him some lognum and made him drink some hot Yerba mansa an this brought old Waltz around some. We warmed him up by wraping him in a load of blankets. Waltz come around and then slept for about a day and a half after that. When he recovered his senses we took him over to Julia's house on Jackson Street and put him up in the sun porch off the kitchen. Old Waltz raised a fit the whole time he was there and Julia moved her bakery business from across the plaza and moved in next to Archer's on the south side of Washington Street. Julia put Waltz up in a little storage room in back of that bakery. All's he had in there was a pallet for a bed and a chair and some shelfs [*sic*] but he quit fittin and fussin all the time and finally seemed satisfied.

Well sir, for a time after Waltz got his senses back he was beside himself and kept a trying to get out and walk but his legs was too weak to go more than a few steps. Julia asked him what was troubling him and he just said he had to get back out to the farm and see to his place. We told him the farm was gone and all his stock save for the ducks was gone drowned and nothing left of the house and barn. He fretted something terrible and finally he say to me to take a cart and shovel out there and dig up the hearthstones in his fireplace and I would find two tins buried there. I was to fetch him back them tins and I weren't to open them or he wouldn't pay me the three dollars he promised me for completing the job. I say's to Waltz, sure I go and I figure he wouldn't know anyhow if I looked inside them tins so I didn't ask him any questions about what it was I was fetching for him.

I figured it was money as nobody would bury anything that weren't valuable in a fireplace like that. Well sir, the laugh was on me cause when I got out to his place and pryed up the hearthstones and dug down about a foot an found them two tins the old man had sealed the lids back on them with some lead. I couldn't open them and look inside without bustin them seals.

The old man had outsmarted me and he knew it. The cans was heavy for sure and weighed 30 – 40 pounds apiece I reckon. They was like dead weight and one of them shook some and sounded like rocks or gravel rollin around inside. I figured right off they was full of gold but I wanted my three dollars so I left them seals where they was and never looked inside to see what was in em. I fetched them back to town and give them to Waltz, and he was mighty relieved when he seen them and his fingers worked all the way around both seals to tell if I'd messed with em any. Finally he was satisfied and he told me to come back the next morning and he would settle with me. The next day he give me a piece of gold in quartz rock that looked to have most an ounce of gold in it. Waltz told me to break the gold out of the rock best I could and sell it at Ganz's store across from the cotton block there. Waltz said for me to keep three dollars and bring him back the difference. I wasn't to tell nobody where I got the gold or nothing about the errand I run for him. Well sir, I busted up that rock and got me six dollars for it. I went back to Waltz and give him back six bits (75¢) and he had a coniption over it. He rounded me real good over that but I stuck to my story. I guess Waltz figured it was worth the difference if I didn't tell nobody about the matter so he let it pass but he never forgot it and he never forgive me for it neither. I felt bad later on for pinching an old man like that but I was young at the time and times were hard.

After my father come back to Phoenix I told him and my brother about the fireplace and the tins and we went on out to Waltz's farm and dug around some more but we never found nothing more. I know that was gold in them tin's, Waltz kept em in his room with him but Julia was the only person that went in there an she said they must have been hid cause she never seen them. Waltz didn't go too far after he settled in Julia's storeroom, just up and down Washington and up to Steinegger's was the furthest I reckon. I seen a couple times when he went visitin he had a lock on his door and he always carried that key with him in his coat-pocket.

We must realize if the Waltz, Weiss, Thomas and Petrasch stories are true, the Lost Dutchman Mine is really rich – rich enough to have created the unending search into the Superstition Mountains of Arizona.

THE OBITUARY OF JACOB WALTZ

by

Gregory E. Davis

For a number of years there has been some debate as to the the date of the demise of the fabled "Lost Dutchman". Jacob Waltz did not die in the Phoenix flood of 1891 as some writers would have us believe, but left this earth some eight months later in October. He died at 6 A.M. Sunday, October 25, 1891 at the residence of Julia Thomas. He was buried at 10 o'clock Monday morning in the old city cemetery.

The articles below are copies of the obituary notices from the Phoenix Daily Herald and the Arizona Daily Gazette pertaining to Jacob Waltz.

Phoenix Daily Herald

Jacob Waltz, aged 81 years, died at 6 a.m. Sunday, October 25, 1891, and was buried at 10 o'clock this morning, from the residence of Mrs. J. F. Thomas, who had kindly nursed him through his last sickness. Deceased was a native of Germany and spent the last thirty years of his life in Arizona, mining part of the time, ranching and raising chickens. [His] honest, industrious, amiable charac[ter] [le]d Mrs. Thomas to care for him [dur]ing his final days on earth, and he [...] with a blessing for [...]

Oct. 26, 1891, p. 3 col 1

Phoenix Daily Herald

Jacob Waltz, aged 81 years, died at 6 a.m. Sunday, October 25, 1891, and was buried at 10 o'clock this morning, from the residence of Mrs. J. F. Thomas, who had kindly nursed him through his last sickness. Deceased was a native of Germany and spent the last thirty years of his life in Arizona, mining part of the time, ranching and raising chickens. [His] honest, industrious, amiable charac[ter] [le]d Mrs. Thomas to care for him [dur]ing his final days on earth, and he [...] with a blessing for [...]

Oct. 26, 1891, p. 3 col 1

Chapter 6

The Mine and the Maps

We have now followed Jacob Waltz from birth to the grave. The story is filled with intrigue regardless of one's attitude regarding treasure and the seeking of it. One might think we have come to the end of this tale, but that is hardly true. In fact, it is just the beginning. For the next 100 years or more there was an ongoing feud, a desperate search for Jacob's gold and the destruction of many people who searched for the mine and some who actually found it. However, what followed is extremely important to the untold masses of Dutchman hunters.

In order for the reader to evaluate the notes of people who have been or are intrinsically involved in solving this mystery, it is necessary to include relevant notes from specific people. What follows are the notes written by Brownie Holmes that reflect the information he was given by his father regarding Jacob's deathbed story. These notes are, of course, just hearsay. We will also offer copies of Tom Weedin's letters that contain an actual description of his recollection of the route to the mine.

THE TRUE STORY OF THE LOST DUTCHMAN OF THE SUPERSTITIONS AS TOLD TO MY FATHER, DICK HOLMES, BY JACOB WOLZ ON HIS DEATH BED.

The old man was very sick and frequently, during his conversation, was compelled to rest for a moment. But in spite of his weaked condition he related the following: "Dick, I know that you've wanted to locate this mine—you tried to follow me once, and I might have killed you. I'm glad that I didn't because I like you—and there is no one I'd rather see find it than you. Years ago I worked in the Vulture at Wickenburg. . . I made a little stake and wanted to take a trip to Picket Post (Picket Post is located near where Superior now is). After I got there I didn't want to stay so I started back to the Vulture by the way of Fort McDowell over the Military Trail. This trail as you know is used by the soldiers at McDowell. This was in 1877. I had three mules with me and a pretty fair camping outfit—that I had bought a few months before. First night out from Picket Post—I guess I was about fifteen miles from there—I made camp. Next morning about dawn—just as I was about to start fire, to make some coffee—three Apaches came running towards my camp. I knew the ornery devils were after me. . .We played hide-and-seek among the rocks for quite a spell—and I knew they'd get me sooner or later—so I slipped out of the arroyo and got away from them. They looked for me for a while—I guess—and then made off with my outfit. That's all they wanted anyway. Well, there I was afoot with no water and no breakfast. I had gotten off the trail considerably and was pretty much lost. I know general directions to the Fort but it was a long way off. I had to keep close watch for other Indians because that whole country was full of them and they were pretty damn mean too. I didn't go very far that first day because the country was rough and in some places I'd have to walk a half mile to get around a rough canyon. That night I found a cave in the side of a hill. It was full of bats and every once in awhile I heard a noise a lot like a rattler. But I would rather sleep with the bats and the rattlers too than take a chance on them damn Apaches. Early next morning I started out again. I had not had any grub since the Indians ran me out of my camp and my feet were getting sore. I had just crossed a wash and got up the other side, when I saw human footprints. So I followed them. They looked like they were fairly fresh—and they pointed in the general direction of Fort McDowell." (Here the old man stopped and rested and felt of his side and emitted a suppressed groan as a sudden severe pain had pierced his congested lungs. After a short interval he continued.) "I had only followed the trail for about half a mile when I came upon a camp. I could tell by the size of the outfit there was more than one person belonging to it. There wasn't a soul about—but there was grub and water. It wasn't Indian grub either and after I tasted some of it I know it was white man's grub and I ate plenty. I didn't drink much water though just took a sip now and then till I quenched my thirst. It was just getting about sundown when I woke up—I had fallen asleep. I heard some voices and I jumped up and grabbed my rifle. Pretty soon three Mexicans came into camp. They looked friendly enough but they acted surprised to find me there. One of them who spoke fairly good English asked me what I was doing there—and I told them about the affair with the Apaches. They all laughed and said I could stay there with them until they were ready to go to Fort McDowell—which would be in a few days. They said I could go with them if I wanted to stay until they were ready or if I wanted to go ahead they'd fix me some grub. I was about in the notion of taking the grub and going on my way when I saw one of them unload a sack of ore from a burro. I asked them if they would let me see the ore. They were simple Mexicans and not at all suspicious so they let me take a look. Dick, that sack was full of gold ore. I asked them where the mine was located and they said I could go with them the next morning. I was still dead-tired when I went to bed but I couldn't sleep a wink thinking of that gold. *Next morning I went with them to the place about a quarter of a mile up the canyon from the camp.* They showed me the shaft and asked if I'd like to go down and take a look. The hole was only about four feet across and about twelve feet deep. I walked down the ladder and then I got to the bottom." (Here the old man paused again; it seemed to father that he was reliving the scenes of that exciting day; his feeble old hands shook as he continued.) "There was the gold, Dick, wire gold in the rocks as large as peas. Several pounds of ore was lying on the bottom. On the bottom of the shaft I could see where they had been following the vein for just a few feet. It was about this thick." (Here he indicated the approximate thickness of the vein of rich ore with his hands—he held them about two feet apart.) "As the Mexicans worked, they talked freely. They told me they had only been working the mine a few months and that it had been originally operated by one Pedro Peralta, a Spanish Don. He had left Spain many years ago and had taken up residence near Mexico City. This Don Peralta made only three trips from Mexico to the mine; his party would consist of peon laborers, soldiers as escorts, their women and children and many animals, including cattle and sheep." (Here father interrupted the old man by admonishing him that he should rest for a moment. He wasn't particularly interested in the activities of the Peraltas and he realized the old man was sinking fast. What he wanted was the secret of the mine before it was too late. The old man waved aside his protestation and continued with his story.) "On these trips every precaution was taken to insure protection from the Apaches, who were always hostile towards the Mexicans, resenting their coming for gold and always attempting to raid the party to capture the animals and whatever else they could get. As means of protection, the Mexicans would herd their stock by day and place them in corrals at night under heavy guard. About 1858, after a winter of successful mining, the Peralta party broke camp and started his party on the long trip to Mexico. They were attacked by a large band of Apaches who were not as well equipped as the Mexicans—yet their number was greater and the Peralta party was handicapped because they had to defend their stock and women and children. Peralta tried to fight his way into the open country where he would be in a better position to defend the party. The Apaches guessed he would do this and were prepared for it. They tried to make their way to the west and then go over the slope of the mountain but they were completely surrounded by the Indians. All were massacred except one twelve-year-old boy who managed to make his escape during the battle. The lad found his way to a little Mexican settlement on the Gila River. The village people helped the boy make his way back to his people in Mexico. The lad never forgot the rich mine and in 1877 he returned with his two cousins and began working the property. The Mexican identified himself as the twelve-year-old boy and the other two Mexicans were the Cousins who had come to help with the mine. They explained to me that the early Spaniards put rocks in the forks of trees leading from the desert to the mines. Some of these rocks are so embedded now that they can only be removed by chopping them out. They took me into their confidence as though I was one of them. Their simple stories and absolute frankness—together with their extreme generosity impressed me not in the least. I had the gold lust in my blood, Dick—I made up my mind to get sole possession of that mine, at any cost. In a few days I saw my chance—they both had their backs to me. I grabbed my gun and shot them both. I had watched them closely and when one of them left and went to put some tools away as they were getting ready to break camp and go to Fort McDowell and had left the two alone with me. I saw my chance and took it. I covered the bodies and waited for the third man to return to camp. I didn't have to wait long. It seemed only a few moments until he came down the wash. I waited until he was fairly close to me before I shot him. (A pause and a noise coming from his throat that sounded very much like a whimper, the old man was extremely nervous now.) I buried their bodies close to their camp then took their outfit and went on to Fort McDowell. I stayed there two days, then came to Phoenix. I made another trip to the mine sometime later. I worked it for several weeks and took out several thousand dollars worth of gold. Before I made plans for my next trip I wrote to my sister in Germany and asked her to send her son over to help me. I sent her several thousand dollars and told her there was a lot more to be had. Soon afterwards the nephew arrived in Phoenix. I took him out to the mine. He hadn't been there long until he insisted on me recording the property and taking out all the gold at one time, rather than make secret trips every year. *I was not a citizen of the United States nor had I declared my intentions to become one*—so for that reason I couldn't locate and record the mine. That is according to law as if I had been a citizen. I was afraid there might be other Mexicans that knew of the mine and that they would come to investigate. I knew from the start that there would be trouble between my nephew and me because he was very quarrelsome and pretty much the type to tell other people what to do. It got so bad that we were quarreling all the time and two or three times I thought he would strike me. Somehow I got to hating him as if he was a bitter enemy instead of a nephew. We left the mine one morning and started for Phoenix, by the way of Fort McDowell. That night we camped at Agua Escondido. We had a terrible argument that night, the worst we'd had before. The next morning when we got up he said, 'I'm going to record the mine myself, whether you want to or not.' The dying old man looked up at father when he got to this point of the narrative—tears were gathering in his fast-dimming eyes and he said feebly, "I shot him in the forehead between the eyes—I then took a piece of chain and, putting it around his neck, dragged him under a shelving rock and where the dirt was soft I dug—a grave—and buried him." (The old man was softly crying now—it was evident that the slaying of his nephew unnerved him far more than the other murders he had committed.) "I returned to Phoenix and when the people asked me where my nephew was, I told them he'd gone back to Germany. I took care of my sister, though, Dick—I guess I sent her over two hundred thousand dollars all together. When you locate the mine I want you to give a portion of it to her. Will you? (Father nodded his head.) After the confession of this, his fourth murder, the old man remained silent for several minutes—with his eyes closed—after which he rallied and indicated that there were more killings of which he must confess before he would give directions to the mine.)

Author's note: Permit me to digress from this story for a moment to mention facts which should precede Waltz's further confession. Fort McDowell was one of the main United States Cavalry posts in the Southwest. It was located near the Verde River about 35 or so miles northeast of Phoenix. There were several other posts in the Territory at that time. One of them was at Picket Post and another at San Carlos. Part of the duties of the soldiers was to escort pack trains through the mountains as well as carry mail between the various posts. The mail carriers on one route consisted of two men, who made the trip from Fort McDowell to Picket Post once a week. After a series of trips back and forth, the two soldiers began to appear at the Silver King with a considerable quantity of gold in their possession. They boasted of having found a place where there was plenty more. There came a time when they did not arrive at their whereabouts and their bodies were found a few days later within a mile east of Weaver's Needle in the Superstitions. Circumstances surrounding their condition indicated that they had been murdered and horribly mutilated by the Apaches.)

"I had been gone from the mine for several months, Dick," continued the old German, "and on the next trip took with me some lumber to build a rocker as I was wasting some of the finer gold. I had to abandon some of this lumber in a rough brushy canyon and on returning to the mine I found it had been disturbed. That gold had done something to me—somehow I felt I was the sole owner of all the Superstition country. It made me mad to see anyone going in the direction of this mountain. I made up my mind to hide for two days, and then two soldiers came up to the mine and began working it. They only worked for a few hours and then got on their horses and rode away. I followed them and killed them both. I made it look like Indians had done it. After I killed the two soldiers, I went back to the mine and took out some more ore, then back to Phoenix. When I returned the next winter I saw where someone else had been there. I knew then that I was going to have to do something if I intended to keep the mine. So I started concealing it so no one could find it. I enlarged the shaft two and one half feet all around and left a ledge about six feet below the surface. I then went up the mountain and sawed timbers the right length to fit the ledge. I worked all winter sawing, dragging, and placing those timbers. They're in the shaft now—criss-crossing to a depth of six feet. I left about two feet near the top so I could fill in with dirt and rocks. *No one will ever find it unless he finds the rock house down in a brushy canyon.* It's almost impossible to get there unless you know how. But when you get there and follow my simple directions, you can go straight to the mine. The hideout is well-covered, with a natural growth of brush and trees and is impossible to see until you are almost upon it. Well, Dick, I hid some of my belongings in the hideout and took what gold I had dug up and started out to Phoenix. About three miles from camp I met a prospector with two burros loaded with camping equipment." (Here the old man hesitated again and maintained his customary silence—his eyes closed.) "And I shot him without giving him a chance to explain where he was going or what his business was in the mountains. I drove his burros quite some distance from where I killed him. I unloaded the equipment and set fire to it, then drove the burros away." (Wolz thought in disposing the aged prospector he had eliminated the last man who knew of the mine's location or existence. He did not realize that at the time there was a man at that moment at the Silver King whose name was Deering, and who was gathering supplies to make a return trip to his mine.) "That prospector's death made seven men that I've killed, Dick—seven human lives over gold—and look at me now. My last trips were made by the way of the Salt River. Traveling up to Monroy's Ranch and taking the old Government Trail which crosses the river at that point and then following it into the Superstitions. I worked the mine for two weeks this winter before the rains set in. It was very cold and somehow the weather seemed to affect me more. One morning I had a chill. I knew there was something wrong and somehow I couldn't help but believe that I was on my last trip to the mine. I hid most of the gold I had gotten out and brought the rest with me. There's not much there, but after I am gone you can have it—it's in a box under my bed. Take it—I know there is enough to grubstake you for awhile anyway, Dick. Dick," the old man gasped, "I must tell you," he breathed laboriously and there appeared a peculiar pallor on his cheeks, as if his life's blood was quickly drawn away from his body. "*Go to First Water,*" he whispered, "*then to Second Water—then take the old Government Trail to San Carlos—where the trail turns south—you will see—over the point of the ridge—a rock—standing in the brush—*

it looks like a man. When I first saw it—I thought it was—a man—so one day I rode over to it—and with my pick—carved the eyes, the nose—and mouth on it, then I said—You look like a man so I'll make you a man. *This is where I always leave the trail.*" (His strength failed him and once again he lay with closed eyes.) Father thought he was dead. The peculiar pallor of his face seemed to become more noticeable and then he rallied with renewed strength and continued. "*Go to the left of the trail, follow up the long ridge and you will come to a saddle. In this saddle is a round Indian ruin of rocks. Go through this saddle and on up a low ridge and when you get on the highest point of the ridge you can look north and the Four Peaks are lined up to look like one peak. In the other direction you will see a high needle (pointed rock). In the canyon under you is my hidden camp. You can't get down there because it is too steep, but go to the mouth of the canyon and then back. You can find the rock house with very little difficulty. You won't be able to see it until you are right upon it. After finding the camp then come back out of the canyon.*" (Here he gave a direction to the mine.) "*You will never be able to find the mine until you first find the rock house as the shaft is completely hidden.* A prospector will never be able to find it because there is no ledge in view. In the mine you will find about $75,000 dollars in gold already dug out. *Be careful, Dick, in opening the shaft as I have set a trap that will instanty kill any person who accidentally finds the mine.*" Here father interrupted with: "Wolz, did you dig all the gold out or is there some left?" "There is enough gold to make twenty men millionaires," Wolz replied. And, holding his hands about two feet apart, continued: "*The ore is this wide and runs the mountain side about four hundred feet where it crops out in the bottom of a wash. I dug this cropping away and erased all signs of my digging. . . .*" (Here the old man stopped and rested for a while, and suddenly appeared to becoming weak again.) When he returned to his narrative he talked in a whisper. And the whispers became more faint as he went on. At times he would gasp for breath and clutch his breast faintly or weakly. "*Follow my directions and you will have no difficulty in finding the mine. First find the rock with the face on it—then the rock house—*

you—have—to do that. Don't forget to take the candle—the box of ore—under my bed. When you find the mine—send my sister—some money." He closed his eyes, taking a deep breath as if to continue the conversation. There was a slight stiffening of the giant frame, then a gurgling sound emanated from somewhere in his throat—and Jacob Wolz—sole owner of the fabulous gold mine of the Superstitions, was dead.

Father and Roberts remained at the bedside until the doctor arrived with the much excited housekeeper. It was fortunate, so far as listening to the story of the mine was concerned, that the colored woman had some difficulty in locating the physician. However unfortunate it might have been for poor Mr. Wolz, it is doubtful that anything could have been done anyway to save his life. The doctor gave the corpse a hasty examination and returned to the bedside of a woman who was soon to present the town of Phoenix with a new citizen to replace one who was no more. Father turned from his rather long and thoughtful observance of the corpse of Jacob Wolz to speak to the colored woman, but she had vanished. He then reached under the bed and brought out the candle box of ore—which the Dutchman said he could have. He and Roberts carried the box home, and ground the contents by hand, laying aside some of the specimens of quartz for future study. The gold content of the rich high grade ore was approximately forty-eight hundred dollars, which was a godsend to father at that time as he had left the Government Service as a packer to be near Mother, who was soon to bring another citizen of Phoenix into the world, no less than George Brownie Holmes. Father was at that time running short of funds as a result of unemployment. *I have in my possession at the present time a beautiful pair of cuff links, each of which contains an elaborate setting of polished quartz taken from the Lost Dutchman Gold Mine. In addition, I also have a slabbed match box made by a San Francisco jeweler, a tie pin and a priceless ring, together with a number of pieces of quartz.* Wolz was buried amid very simple ceremonies in the cemetery on West Madison Street in the City of Phoenix. This cemetery is today still in existence, but in a very sad state of repair.

We have already heard the story about being given the rights to the mine that Julia (Helena) told Sims Ely, a Phoenix newsman. Again, this is hearsay. Jacob is supposed to have revealed that story to her and Reiney outside the adobe when they got together for Christmas dinner. These facts relate to the Peraltas in Mexico, the gambling debt and a parchment map. It is speculated that when the Germans returned to the Superstitions, they went to the mine and saw two men working it whom they thought were Indians. They shot and killed them.

It is incumbent upon this writer to include other information from the Ely manuscript, which is obviously not true. We have documented in the first chapter that Jacob Waltz and Jacob Weiss came to New Orleans at ages 28 and 29, respectively, along with two other Germans named Gerhard W. Schmidt and Peter Schmidt. Apparently, they all came from Württemberg, from a settlement named Horb, which is about eight miles from the main town.

Ely says Waltz and Weiss were boyhood friends, and their families lived near each other in St. Louis, Missouri. This is patently untrue. Also, he states they joined the army together to fight in the Civil War, which ended in 1864. We can document Jacob Waltz with the Peeples's Expedition during that time. There are many other discrepancies that really don't seem to be worth repeating.

What is important is the fact that Sims Ely was an old man when he wrote his book in 1953. Jim Bark was already dead. Sims never knew Jacob; he was relying solely on hearsay from Julia. He said he interviewed her over many years including times when she had been drinking. Sims found her broke, living in a shack by the railroad tracks from which she was selling hand-drawn maps to the mine. Sims was quoted as saying, "Hers was a sad lot indeed." The woman died in 1917 of Bright's disease.

The newsman also had a copy of Bark's notes, some of which are included in this chapter. Sims didn't write his book for another 36 years. One might imagine that being a newsman, he could have decided to take license with the truth to enhance the story or possibly had forgotten the truth in 36 years. For many years Sims Ely's book was considered the gospel on the Lost Dutchman Gold Mine, because it was the first book written on the subject. Naturally most people believed he was privy to Jim

Bark's information, and Bark was well respected in Phoenix during his life and afterwards. It makes for very good reading and certainly elicits intrigue, but it is quite a yarn. (The book is a collector's item now worth $100.)

Real Dutchman hunters hang on every word of documents, diaries, notes, myths and invented tales from those who supposedly know. That is why this author decided to include the verbatim notes of the pioneers and those who came after. It seemed extremely important to include all their words because those who search are desperately trying to follow the clues. It makes a great deal of difference if one reads the words of those who were there or the translation of those words by someone who has a literary goal.

Another item that warrants repeating is the matter of Jacob's character. In the Brownie Holmes material, Jacob seemed to be someone who wanted gold so badly he would kill for it. As a matter of fact, he has been called a killer by numerous people who lived during and after his lifetime. However, it was Dick Holmes who went to the ranch near the spring where he said he found the skull of Peter Backens, which he brought back to Doc Jones's store in Phoenix. This is the skull Bertie Roberts said was on the shelf in the store for many years.

There is also the matter of Jacob's words to Dick Holmes when he returned from his mine and accosted McFall (a cowboy sidekick) and Holmes in downtown Phoenix. Jacob knew they were following him. He stated, "Follow me again and I'll kill you." Does that make Jacob a murderer? In fact, there was no law in the Superstitions. If you got into trouble, it was "kill or be killed." And what did Holmes expect to do if indeed he had been able to follow Jacob to the mine? We will never know, but it is a good bet Jacob believed that would have meant his demise. That was the reason he took circuitous routes into the Superstitions, kept a nightly vigilance and a careful watch as he worked. That care kept him alive.

Brownie Holmes relied upon his father's information completely. No one would expect him to do otherwise. Surely, he believed every word. However, some of his father's actions seem acutely suspect as was discussed in an earlier chapter.

What is evident is Jacob's ability to be a good friend. He certainly befriended Julia and Reiney. He sent a large amount of money to his sister in Lawrence, Kansas. He befriended Magdeberg – enough that the old man left Jacob his entire estate. He was reclusive; he was obviously a hard worker, and he was dedicated to the people he traveled and fought with. He certainly cared about his friend Jacob C. Weiss and mourned him even until his death. It is evident this man believed that if he had come back to their mining camp sooner he could have saved his friend from capture and torture. Jacob never used the gold for his own pleasure or luxury. If the Indians were right and the mine was cursed, perhaps Jacob was the only one who was allowed to own it until he decided to give it away – then, he died.

In Bark's notes, which appear later in this chapter, he talks about Julia and Reiney stopping at his ranch overnight when they were hunting for Jacob's mine. Bark said, ". . . the handbag that Helena always kept in her possession disappeared. Who got it, or where it went, I do not wish to say. The handbag contained Jake's parchment [*sic*] map of the mine, and how to get to it, Peralta's note to the two Jakes for sixty thousand dollars, bill of sale to the two Jakes to the mine and Jake Walsh's naturalization certificate." Bark must have seen those documents that almost prove the Peralta story. (In reading Bark's notes, the reader must bear in mind the inability of locals to record or spell names consistently. Julia is either Julia, Helena or Mrs. Thomas; Waltz is Waltzer, Walsh, Woltz or sundry other spellings; and Weiss is Weisner, Weiser, Wiser, etc.)

One would wonder about the veracity of the information given by others. However, a story came to light much later that changes the complexion of this entire matter and seemingly verifies Jacob's story.

By 1911 Arizona was preparing for statehood. Thomas F. Weedin, editor of the *Florence Blade* newspaper, had decided to run for governor. All the locals whose names had become prominent were preening their feathers in anticipation of the first big election for the legislature. Jim Bark was one of them, and he was elected. Arizona became a state the following year.

Sims Ely had achieved notoriety as a newsman in Phoenix. Weedin dropped into Sims's office for a visit – probably to shore up his political clout with the prominent newsman. They greeted each other cordially and had a satisfactory chat about the upcoming race. Before he left, Tom mentioned he had seen Jim Bark earlier that day at a train station. He said he knew the two

Thomas F. Weedin, editor of the* Florence Blade *newspaper, with his wife and daughter (circa 1911).

men were mutually interested in a lost gold mine in the Superstitions. Weedin admitted there was bad blood between Bark and himself because of political differences, but he did have some information about the mine he was certain would interest both men. He said he had been meaning to pass a map on to them for a long time and just hadn't gotten around to it.

Sims's interest peaked. He and Bark had been searching in earnest for years, and this man had just said he had a map. One can only imagine Sims's pulse racing at the thought of it. Bark had been tracking the mystery for 15 or 16 years. Everyone knew he believed the mine was out there.

Weedin discussed his close ties with Dr. John Walker who, because of his friendship with the Pima Indians, had become famous in the West. Initially he had been hired by the army to fight Apaches. During one skirmish, he trapped a large band of them on a precipice and attacked. Rather than be defeated, some of the savages leaped to their deaths; the mountain forever retained the name Apache Leap. Before this time most of the range was untitled. There were few, if any, landmarks except where some tragedy or historic event occurred. (Bob Corbin, when Attorney General of Arizona, was given a territorial copper map of Arizona as a gift from the Department of Public Safety. It shows the minimal named locations in the Superstition Mountains. Other maps included in this chapter show the scarcity of landmarks.)

Walker became ill during his tour of duty and took up residence in a friendly Pima village. As he recovered he ministered to the Pima and the Maricopa tribes. They were passive farmers and ranchers. These Indians had been Christianized since the Spanish conquest. Walker soon became a mentor, healer and friend who gave unselfishly of himself.

One of his nurses during his long illness was a beautiful Indian maiden named Churga. The two fell in love and were married in a typical Indian wedding ceremony. Their union produced a daughter named Juana, who soon became the center of the doctor's life. Since the Indians accepted the doctor's medical assistance, Juana, schooled by her father, became a capable nurse.

To repay Dr. Walker's kindness and because they had no use for precious metals, the tribesmen gave him a valuable silver deposit ultimately known as the Vekol Mine. It was so rich that Walker eventually became one of the richest men in Arizona. He altruistically spent over $500,000 improving living conditions on the reservation.

Meanwhile the Apaches were still marauding in and out of the Superstition Mountains. Their favorite trick was to raid the reservation, stealing livestock and vegetables at harvest time. Many peaceful Indians were killed. The army enlisted Dr. Walker's aid. That battle was at Apache Leap. Walker gained much respect in the territory because he refused to turn the captured Apaches over to the conquering Indians, who wanted to torture their adversaries. Walker took them to Fort McDowell to be tried in a military court. That battle ended the Apache raids outside the mountain range forever.

Six years later a wounded man struggled out of the Superstitions. Two Pima squaws found Jacob Weiss and brought him to Dr. Walker's medical facility at his home. It was 1871.

The German had been shot in the shoulder and arm with an arrow that he broke and pulled from his flesh before collapsing beside a water hole. The Indian women fed the patient, cleaned his wounds and put him to bed at Dr. Walker's. The festering wounds in the upper shoulder and arm looked extreme. Although bleeding had stopped and no major artery or vein had been damaged, Weiss was ghostly pale. The man was suffering from loss of blood, exposure and lack of medical care. He was nonetheless able to choke out that he was the victim of an Apache arrow. He was given drugs all night long, but by morning it was apparent he had pneumonia and wouldn't survive.

Dr. Walker began to extract information from his patient while his mind was still clear. He naturally wanted to know if the German had any relatives in Arizona. Weiss answered, "No. All of my relatives are in Germany, and they are just cousins. My parents are dead."

"Any friends in Arizona or elsewhere in this country?" Dr. Walker asked. "I had a friend," Weiss answered. "My partner. But, the Apaches got him the same day they got me. Why you askin' me these questions, am I goin' to die?" Being kind, the doctor skirted the answer but told the patient he was a very sick man. Weiss seemed accepting and told the doctor he had a feeling he was going to die.

Then he was momentarily silent before uttering, "Did I have a sack with me when the Indians brought me in?" The doctor nodded and, looking toward the door, he said, "It is still over there on the floor." Weiss asked to have it. Walker went to retrieve the cloth bag about the size of a flour sack. Walker opened it to expose a few toilet articles and a parchment-like rawhide. The doctor asked if there should have been something else. Weiss said no, he had left his rifle and revolver and anything else that weighed anything at the water hole. "But this," he went on after picking up the rawhide, "This is a map my partner and I got in Mexico. Shows the route from Sonora to a mine in them mountains to the north."

The following notes on this subject are from Sims Ely's book. The case is well documented in history. What is critical for the reader is that we will include copies of letters from Tom Weedin to Sims that came from a collection of Dutchman material held by Al Reser. A longtime Dutchman hunter, Reser walked miles in the Superstitions – a practice he continued into his nineties and almost until his death in 2000. The letters are so old they will not reproduce clearly and will be copied so the reader can profit from the information given. We are also privy to Weedin's recollections regarding the directions to the mine, which should impress any avid Dutchman hunter.

Weiser now interrupted his story to interpret the rawhide map. Deeply interested, Dr. Walker listened closely, and the details remained clearly in his mind long years afterward.

"My partner and I," Weiser said when he was done with the map, "went to the mine first with the Mexican owner. His name was Don Miguel Peralta. We went to it as partners, all three of us. Don Miguel wanted fightin' men with him, and my partner and I were soldiers in the Confederate Army. And we'd saved Don Miguel's life in a brawl in Mexico. We were to have one half of the gold we took out, Don Miguel the other half.

"We took peons with us. Any number of 'em, all well armed. And we brought away about sixty thousand dollars. Then Don Miguel asked us to let him have most of our share. He needed money bad, and he gave us a paper that said we were to have the mine for ourselves long as we wanted it. He was only doin' what he had a right to—the old don, Don Miguel's father, was killed at the mine a few years ago, and Don Miguel had neither wife nor children. And now, it'll be your mine if I die, for my partner is gone."

The sick man's eyes turned from Walker's face and focused in the distance. He was silent for a long while.

"My partner and I, we went back to the mine alone," he continued at last. "After we'd worked it for a number of weeks, our mule destroyed most of our provisions, and my partner undertook to come out to the Adams mill for supplies. He never got back. I expected him on the fourth day. And then, on the mornin' of the fifth day, when the Apaches jumped me, I knew they'd killed my partner first.

"Apaches got all the animals 'cept one horse. I rode him a few miles, but the goin' was slow, and then their arrows stopped him. I got one in the arm, but I kept on goin'. Runnin', then stoppin' behind rocks to fire my rifle at 'em. They were afraid to come close, and at last they turned back. I made for the first water hole on the desert I knew about."

He was suffering a terrible thirst, he said, by the time he reached the water hole, and the wound pained him unbearably. Sometime before he got to the water he guessed he must have gone delirious, because it was then that he tore out the arrow. He stayed at the water hole for most of the day, and then, in a moment of clarity, he realized that he had lost his canteen and must make it across the desert to the Gila without any more water. This he had done and, at last coming to the river, had followed it down to the Pima villages and Dr. Walker.

Weiser had finished his story with considerable difficulty, and Dr. Walker cautioned him not to talk any more for a while. "I'll give you some medicine now," he said, "so that you can sleep."

All that night Walker sat with his patient, noting the steady progress of the pneumonia in both lungs, helpless to stop it. The next day Weiser became delirious, and the doctor realized that there was nothing to be done except to ease the suffering as much as possible and hope that a strong constitution might prevail. But on the fourth full day after his arrival Weiser died.

The dead man was placed in a coffin which the Indians made at the doctor's request and was decently buried in the already extensive burial ground down the river, which held the remains of other soldiers and overland travelers who had died among the friendly Pimas.

Dr. Walker was left with a rawhide map, a conviction that a gold mine did exist in the Superstitions, and a temptation to prepare an expedition to it. It would have to be an exceedingly rich mine, in his judgment, to justify the Mexican owners' traveling that long distance into Apache country. The main question now would be whether the mine had been worked to a point near exhaustion. Gold mines, he knew, were never likely to continue to a great depth. Yet there were other deterrents. As the victor of Apache Leap, Dr. Walker was well aware that his presence in the Superstitions would be a matter of particular interest to the Apaches lingering there, and this at a time when life was especially sweet and full for him....

Thereupon I closed the office door firmly, we lighted fresh cigars, and Tom Weedin told, while I listened with rapt attention, the dramatic story of Jacob Weiser's fate.

Weedin had had the story directly from the lips of Dr. Walker himself on one of his many visits to the latter's house on the Pima reservation, some ten years after the events themselves.

The doctor's large adobe house, Weedin said, was luxuriously furnished for those times, and it overflowed with books. In one small room Walker kept his supply of medicines, and next to this room was a library, in which there was a large table. During the visit in question, the two

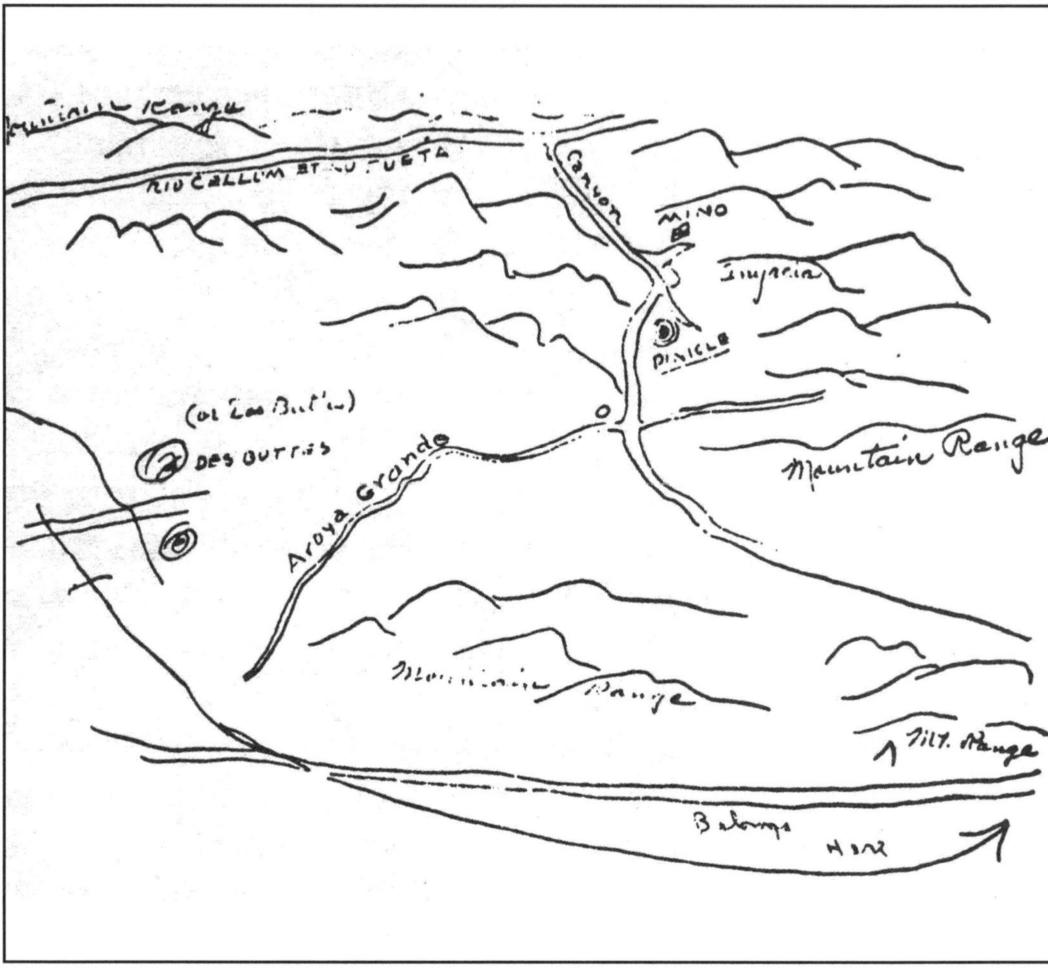

Bark-Ely map drawn by Tom Weedin.
This map was given to Dr. John Walker by a dying white man, Jacob Weiss. Tom Weedin drew it from memory after his wife destroyed his tracing to keep him out of the dangerous Superstition Mountains.

men were sitting at this table when they got to talking about mines.

Suddenly Dr. Walker opened a locked drawer and brought out a flat package, wrapped in paper. As he untied the string, he said, "Here, Tom, is something I've been intending to show you at the first opportunity."

When it had been unfolded and laid on the table, Weedin saw that the article was a piece of rawhide some twelve inches long by eight inches wide, in texture so smooth that it was like parchment.

Both men were now standing, looking intently at the piece of rawhide. "These markings," said Dr. Walker, "and the writing on it are just as the thing was when it was given to me. As you see, it's a map. Note the line running from the Mexican border to Tucson, and then northerly to the Gila River. See, it crosses the river at the Buttes–" the doctor was then pointing to the "Two Buttes" between which the Gila flows some fifteen miles east of Florence– "and here, properly located, is this reservation. From the Buttes, you see, the line runs northwesterly into a range of mountains, after crossing this line here, which the owner of this map described to me as a long draw. Those are the Superstition Mountains, of course. And after the line goes into the mountains–see, there–it runs into a canyon which drains northwest into a river."

The river, as Weedin well knew, was the Salt, once known by the name of the Black River, its parent stream, which forms it in conjunction with the White some fifteen miles east of Fort Apache.

"Into this canyon which runs into the river," Walker went on, "there comes a side canyon from the east. And this mark, right here, near the side canyon, is a gold mine, according to the man who gave me the map. He said it was one of the richest mines in all the world.

"That man's name was Jacob Weiser. He was brought to this house sick and badly wounded. He died here. I've always kept this map very carefully, thinking that someday

I might go and see the mine, for it was a thrilling yarn that Weiser told me. I've never shown this map to anybody before, or repeated what Weiser said to me."

Now Dr. Walker proceeded to recount to Tom Weedin the last days of Jacob Weiser, and when he had finished, Weedin, his eyes shining, urged the doctor to go with him after that mine.

All his conclusions thus far had been against that idea, Dr. Walker explained gravely. "Any mining man who's been as lucky as I have is naturally interested when he hears of another good mine that's easily obtainable. And if I understood that poor man correctly, this mine would now be open to location—that is, unless title has somehow been acquired from the United States. I don't think any valid Mexican grant ever reached that far to the north. But all those things would be investigated if I were considering it in earnest."

"Why should we hesitate?" Weedin argued. "It's your proposition, of course, but I know you wouldn't be telling me about it unless you expected me to share in it after you had recovered all your outlays. You're wealthy and amply able to equip us for the undertaking."

There were many reasons against it, Dr. Walker said, shaking his head, and he mentioned them—chiefly, the thwarted vengeance of the Apaches and his concern for his young daughter.

When Weedin perceived that the doctor was adamant, he asked as a final favor, "May I make a copy of that map?"

"Why, certainly," Dr. Walker said. He then looked for tracing paper, but could find none, whereupon it was agreed that Weedin should take the rawhide map with him to Florence, make an exact tracing of it, and then return the original to Dr. Walker. "But," the latter added, "I'd rather you wouldn't show the map to anyone else, or even talk about it, for the present. Sometime I may be willing to finance you on the proposition. But the way I feel about it, I'll stay out of the thing myself."

When he returned the original map to Dr. Walker and in all subsequent talks with him, Weedin said, the doctor was of the same mind. If anything, he appeared less and less willing to interest himself in the mine. He made no suggestion as to financing Weedin either, and, of course, Tom did not make any such request. It was his feeling that, in refraining from sending him in to find and operate the mine, Dr. Walker was moved by their intimate friendship, the doctor fearing that he might be sending Weedin into dangers which he himself was unwilling to face.

Then, as if excusing his own failure to follow through on the information Walker had given him—much of it was the detail Weiser had revealed about the terrain—Tom pointed out to me that during the years 1879 to 1889, the Apaches were more steadily troublesome in the Superstition Mountains than during any other ten-year period in Arizona history. There were always some hostiles making trouble, and a man could certainly expect Apaches in small numbers to be prowling through the mountains in cool weather, hunting deer and gathering acorns. His wife, Weedin said, was never willing for him to go into the Superstitions.

At the end of his startling narration, he said, "Now, Sims, I've still got that tracing among my papers at Florence. I'll be home again before the end of the campaign, and I'll get that tracing and send it to you. In the one trip I made to the Bark Ranch, years ago, I saw that the landmarks there fitted in with the map. For instance, there's the long draw you'll find on the map. And near there is Minor's Needle, which I decided was the *pichacho* that's mentioned. You're welcome to the map. I hope it leads you and Jim Bark to the mine."

Some two weeks later I received a letter from Weedin saying that he had stopped at home over Sunday and had expected to get the map and mail it to me. To his surprise, however, he couldn't find it anywhere among his papers. Mrs. Weedin had helped him in the search, he wrote, but they had had to give it up for lost. She had suggested that perhaps he had given the map to some prospector and forgotten about it. "But I know that never happened," the letter said. "I'll hunt again, at my first chance, for I feel that the map is somewhere in the house—perhaps tucked away between the leaves of some book."

However, Miss Jennie Weedin, Tom's brilliant daughter for whom, as president of the school board, I found a teaching position in the Phoenix schools, had another idea. After she and I had become warm friends, she told me that it was her belief that her mother had destroyed that tracing of the Weiser map as one means of preventing Tom Weedin from making a search for the mine in dangerous Apache country. She made that statement in confidence, and I have so treated it until now when, both the girl and her parents having died, its revelation can harm no one.

In 1911 Tom Weedin had already grown so heavy that mountaineering could no longer possibly interest him in any practical way, but the loss of the map, when it seemed within our grasp, was, you can understand, a grievous disappointment to both Jim Bark and me. Nor was the original rawhide ever likely to turn up. Dr. Walker had died suddenly in Napa County, California, in the 1890's. He died intestate, and though the local probate court awarded the estate to the beautiful Juanita—for a time she was generally looked upon as the matrimonial catch of the entire Commonwealth of Arizona—the award was later contested successfully by Dr. Walker's brothers and sisters, then living in California. Hurled by this decision from her eminence as a leading heiress of Arizona to the lowly status of a half-breed Indian girl, told that she was illegitimate and without standing in court, even though the marriage between her father and mother had been solemnized according to the tribal ceremonies of the Pimas, Juanita Walker seems to have decided to become an Indian in fact. Thereafter, she associated little with white people. She died in Tucson in 1923, and long before that the rawhide map, which must have been among Dr. Walker's effects, was lost, probably forever.

But what remained was in itself important—the story of Jacob Weiser as it had been revealed by two of Arizona's most honorable citizens. Weiser, it was clear, had actually taken gold from the Lost Dutchman.

There were to be others of equal fortune.

* * *

Copies of Letters Sent to Sims Ely from Tom Weedin

The originals were not in good enough shape to reproduce; in fact, they are barely legible. Every effort to be accurate has been made because the information is very important. Naturally, they are not signed, because the originals were mailed, and these copies were in Tom Weedin's files. They were loaned to us from the Al Reser Collection courtesy of Clay Worst and the Superstition Mountain Historical Society's Acquisitions Director Greg Davis.

Arizona Blade Tribune

Florence, Arizona Feb. 17, 1912

My dear Sims:

I only had about five minutes to make the enclosed sketch, as I looked for the old tracing up to the last minute. I shall continue to look for the tracing. I feel certain, however, that the enclosed sketch follows it closely, except to the location of the pit. I am not certain as to which side of the canyon it was on, hence it would be well to examine every flat on either side of the canyon. The map shows monuments along the trail, after it entered the long draw, but I cannot place them from memory. Wishing you success, I remain

Your friend,
Tom Weedin

This is the first letter. The second will follow. The researcher wrote a note on the copies that the third letter was lost. Also, Sims wrote in his book that Mrs. Weedin had already disposed of the original tracing to keep her husband from going into the Apache-infested Superstitions. This information was learned from Weedin's daughter after Mrs. Weedin's death.

Florence, Arizona, Feb. 27, 1912

Hon. Sims Ely:

For three or four days past my wife and daughter have been assisting me in the search for the tracing and we completed the last box, last night without success. As I made the tracing in the spring of 1881 and thought but little about it after Walker found the Vekol Mine, and was too busy to go with me in search of the mine. It is possible that I gave it to some prospector, but I hardly think so. At any rate, it is thoroughly lost now, as is the mine. Did you find anything that corresponded with my memory sketch?

In answer to your question as to why I assume the "pinnacle" is miner's needle as against the weaver's needle, I will say this, I do so because as I remember the map the pinnacle was on the right side of the trail going in, and the lay of the country as I saw it once when I came down from Rogers District, near Bark's Ranch, it did not seem probable that the trail would swing so far West from the Bark draw, and I feel positive that Bark's draw is the long draw up which the trail led. My recollection of Walker's oral statement of the Dutchman's (Weiser's) description is clearer than my memory of the map. After he repeated it to me many times he always said that the trail led up the first long draw leading into the mountains east of the west end of the Superstitions, on the south side. This must have meant the Bark draw as it is the first long draw leading into the mountains east of the west and on the south side. He never said Miner's Needle or Weaver's Needle, but always referred to it as the "picacho" and the same is true of the maps and he always described the trail as leading up this long draw, on over the mesa and down past the "picacho" on the left side down into a big canyon. I merely assumed that this was La Barge, simply because I did not know of any other canyon in the vicinity that leads down to the Salt River. If there is a mesa leading down from the Bark draw over towards Weaver's Needle and any (bin) canyon leading down to the river and into La Barge as other large canyons leading to the river, it may be safely inferred that the Weaver's Needle was the "Picacho" referred to by the map and Weiser's oral description to Walker. As to the course of the trail out of Bark's draw it is quite possible that the original map was erroneous as the maker undoubtedly had no copies to give him the correct courses. To correct the errors one will have to go into the ground and determine them by observations, and logical deductions as I am sure you are capable of doing. Sometime during the coming summer I will try to make a trip with you over the ground. In the meantime investigate between the Bark draw and the Weaver's Needle and see if the trail could have reasonably led down on the left hand side of said Needle into a big canyon. You know a map made from the eye can be very easily misleading.

Sincerely yours,
Tom F. Weedin

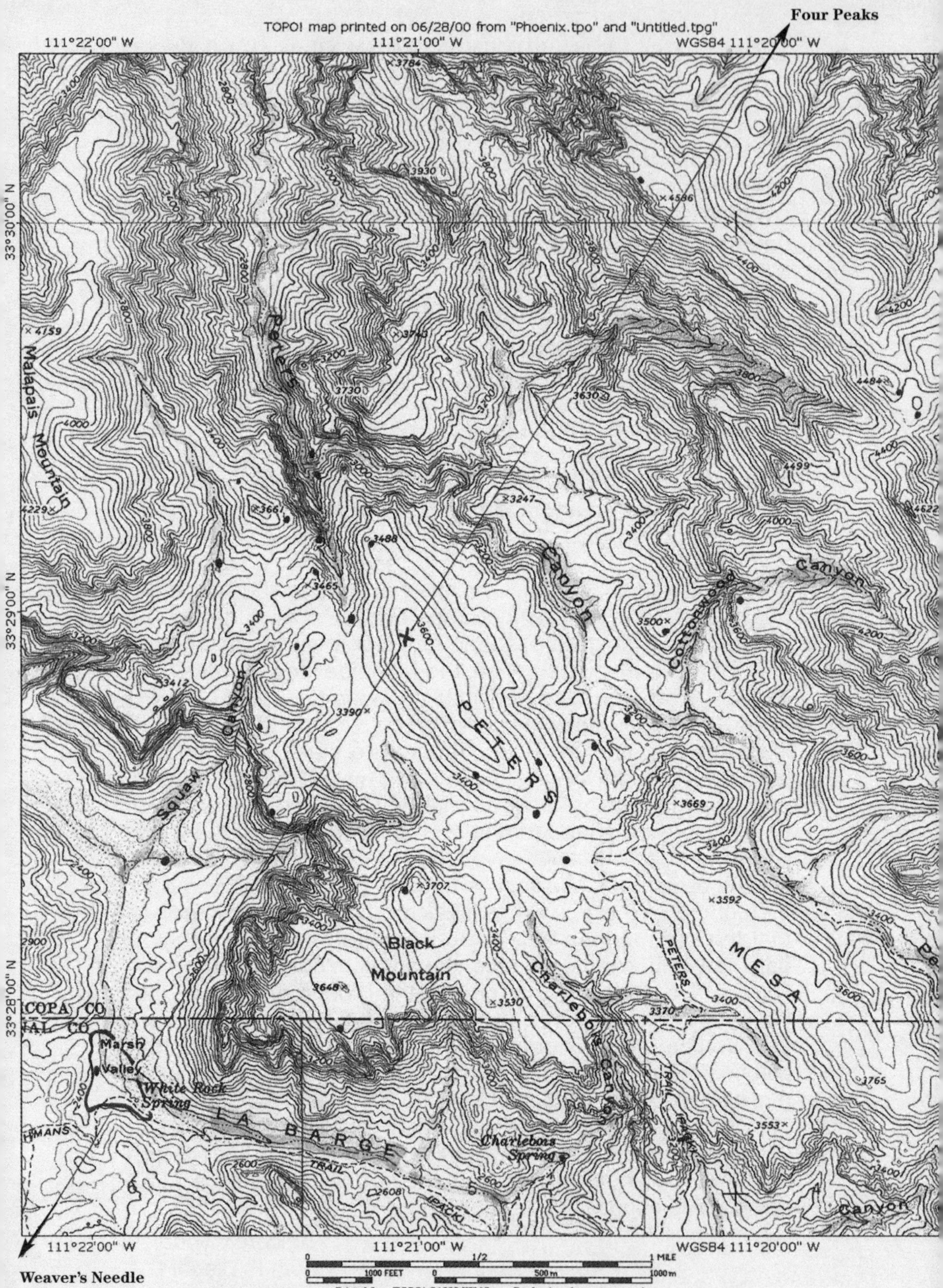

Topographical map of Peter's Mesa, Cottonwood Canyon, Squaw Canyon, La Barge Canyon and Charlebois Spring.

From Jim Bark's Notes – Courtesy of Al Reser

JACOB WALSH

IN 1891, FRANK CRISWELL MY PARTNER, AND MYSELF BOUGHT THE SUPERSTITION CATTLE RANCH FROM THE GEORGE MARLAR ESTATE, KNOWN AS THE ML RANCH, AND TOOK POSSESSION EARLY THAT YEAR. MY PARTNER ATTENDED THE FIRST ROUNDUP, AND AS IT WAS A VERY BAD CATTLE YEAR (ON ACCOUNT OF DROUGHT) TURNED THE MANAGEMENT OVER TO ME, WHICH I RETAINED UNTIL I SOLD MY INTEREST TO HIM ABOUT THE YEAR 1912.

SCARCELY HAD I TAKEN CHARGE UNTIL THERE WERE RUMORS AND VERY INDEFINITE STORIES ABOUT THERE BEING A VERY RICH GOLD MINE OVER THE MOUNTAIN, IN THE MAIN SUPERSTITION RANGE, BUT NOTHING THAT A CATTLE MAN WOULD PAY ANY ATTENTION TO. AS I WAS RIDING THE RANGE ABOUT SIX MILES WEST OF THE HOME RANCH IN AUGUST 1892 OR 1893, I MET SOME CAMPERS NEAR AN OLD WELL BELONGING TO THE RANCH, AND THEY PROVED TO BE A COLORED WOMAN BY THE NAME OF THOMAS AND A YOUNG MAN ABOUT 18 YEARS OLD BY THE NAME OF RHINEY PETRACH, WHOM MR. AND MRS. THOMAS HAD ADOPTED. MR. THOMAS WAS GERMAN, AND ALL THREE SPOKE THE GERMAN LANGUAGE. I HAD A SPEAKING ACQUAINTANCE AT THE TIME WITH ALL THREE AS I HAD MET THEM IN THEIR BAKERY IN PHOENIX. I THINK IT WAS THE ONLY BAKERY AND ICE CREAM PARLOR THERE AT THE TIME. IT WAS GENERALLY KNOWN IN PHOENIX THAT THOMAS HAD RUN AWAY WITH ANOTHER WOMAN, AND THAT MRS. THOMAS NO. 1 AND THE BOY PETRACH WERE STILL RUNNING THE BAKERY. SO I WAS SOMEWHAT SURPRISED TO MEET THEM CAMPED NEAR THE SUPERSTITION MOUNTAIN, AS IT WAS VERY HOT AND AN EXTREMELY DESOLATE COUNTRY TO CAMP IN. I ASKED THEM WHY THEY WERE CAMPED THERE, AND SHE REPLIED THAT THEY WERE OUT ON A VACATION. I INVITED THEM TO COME UP TO THE HOUSE AS THEY AT LEAST WOULD HAVE SHADE. SHE REPLIED THAT THEY WERE GOING BACK IN A SHORT TIME, SO I RODE ON UP THE CANYON, FOLLOWING THEIR BUGGY TRACKS, AND THEY HAD GONE UP THE CANYON TOWARD THE SUPERSTITION MOUNTAINS, AS FAR AS IT WAS POSSIBLE TO GO WITH A HORSE AND BUGGY, AND MUCH FARTHER THAN I THOUGHT WAS POSSIBLE. AT THAT TIME, I DID NOT KNOW THAT THEY HAD CLOSED SHOP IN PHOENIX FOR GOOD, BUT SUCH PROVED TO BE THE CASE. AFTERWARDS, MRS. THOMAS AND THE BOY WERE FREQUENTLY SEEN BY THE COWBOYS, CAMPED FIRST

2 - JACOB WALSH

AT ONE WATER HOLE, THEN ANOTHER, OVER IN THE MAIN SUPERSTITION RANGE, BUT ALWAYS WITH SADDLE HORSES AND PACK ANIMALS. THEY CAME BY THE RANCH, ONE DAY, AND THERE WERE FOUR IN THE PARTY. THE OTHER TWO PROVED TO BE THE BROTHER, HERMAN, AND THE FATHER OF THE TWO BOYS OR MEN, KNOWN AS OLD MAN PETRACH, AND THAT THEY HAD JUST ARRIVED FROM MONTANA. SHORTLY AFTER THAT, THE PARTY CAME BY THE RANCH AND STOPPED OVERNIGHT. THERE APPEARED TO BE MORE OR LESS DISSENSION AMONGST THEM, AND RHINEY TOLD ME THAT THEY HAD BEEN LOOKING FOR A GOLD MINE BUT COULD FIND NO TRACE OF IT: THAT HE HAD WRITTEN HIS FATHER AND BROTHER IN REGARD TO IT AND THEY HAD CONCLUDED THAT RHINEY AND MRS. THOMAS KNEW NOTHING OF MINING OR THE MOUNTAINS, AND AS THE MEN WERE BOTH PROSPECTORS, THEY DECIDED THAT THEY HAD BETTER GO TO PHOENIX, AND THEN GO UP IN THE SUPERSTITION MOUNTAINS AND FIND IT. RHINEY SAID THAT THEY HAD MADE A THOROUGH SEARCH AND THAT HE WAS GOING TO QUIT, AS HE AND MRS. THOMAS HAD BLOWN IN ALL THE MONEY THEY HAD AND THEY WERE ALL GETTING PRETTY GROUCHY. AND THAT IF I WOULD SAY THAT I WOULD DO WHAT WAS RIGHT WITH THEM IN CASE I FOUND THE MINE BY THE DESCRIPTION AND INFORMATION, HE WOULD TELL ME THE STORY AND ANSWER ANY QUESTIONS THAT HE COULD: THAT AS I WAS GOING TO BE RIDING THE MOUNTAINS, I MIGHT UNRAVEL THE STORY AND POSSIBLY FIND THE MINE: THAT HE KNEW IT WAS THERE, AND HE BELIEVED IT TO BE THE RICHEST GOLD MINE IN THE WORLD. I TOLD HIM ALL RIGHT, AND TO GO AHEAD WITH HIS STORY. THIS WAS IN THE EVENING.

HE SAID THAT IN THE LATE 80's THERE WAS A RAYMOND EXCURSION GOING TO THE PACIFIC COAST OVER THE SOUTHERN PACIFIC RAILROAD AND THEY STOPPED OVER AT OLD MARICOPA AND TOOK THE STAGE FOR A VISIT TO PHOENIX, AND AMONG THEM WAS A SODA FOUNTAIN MANUCATURER WHO SOLD TO THOMAS AN EBONY AND MARBLE FOUNDAIN FOR TWENTY-FIVE HUNDRED DOLLARS, FIVE HUNDRED DOLLARS DOWN. WHEN THE FIRST PAYMENT CAME DUE, THOMAS, INSTEAD OF MAKING THE PAYMENT, SKIPPED OUT WITH A WHITE WOMAN, AND AS FAR AS RHINEY KNEW, HAD NEVER BEEN HEARD FROM. THERE WAS AN OLD GERMAN WHO HAD FORTY ACRES OF SCHOOL LAND AND SOME CHICKENS, AND WHO HAD OFTEN SOLD HIS EGGS TO THE BAKERY: HE WOULD FREQUENTLY HAVE A CHAT IN GERMAN WITH MR. AND MRS. THOMAS

3 - JACOB WALSH

AND THE BOY, SO WHEN HELENA (FIRST NAME OF MRS. THOMAS) REALIZED THAT THOMAS HAD ACTUALLY SKIPPED OUT WITH ANOTHER WOMAN, SHE WENT CRYING UP TO OLD JAKE'S ADOBE HOUSE JUST EAST OF PHOENIX, AND TOLD HIM THAT SHE DID NOT KNOW WHAT SHE WOULD DO, AS SHE OWED A WHOLE LOT OF MONEY, AND THAT THOMAS HAD TAKEN ALL THERE WAS BELONGING TO THEM WITH HIM WHEN HE SKIPPED: AND THAT THEY OWED TWO THOUSAND DOLLARS ON THE FOUNTAIN AND QUITE A FLOUR BILL, AND SEVERAL OTHERS, AND THAT SHE HAD NO MONEY TO PAY THEM WITH. THIS WAS EARLY IN 1891. OLD JAKE STEPPED TO THE DOOR OF HIS ONE-ROOM ADOBE, PLACED ONE FOOT IN FRONT OF THE OTHER, HANDED HELENA AN OLD PICK AND TOLD HER TO DIG THERE: SHE DID SO, UNEARTHING AN OYSTER CAN IN WHICH WAS ABOUT EIGHT HUNDRED DOLLARS IN GOLD NUGGETS. OLD JAKE SAID, "TAKE IT HELENA, THIS WILL DO UNTIL I CAN COME DOWN TO THE BAKERY AND WE WILL GO OVER THE BILLS TOGETHER," HELENA REPLYING, "NO, JAKE, I DON'T KNOW WHEN I CAN REPAY YOU." HE SAID, "NOW, NOW, DON'T WORRY AS I HAVE FIFTY MORE OF THESE BURIED AROUND." SO SHE TOOK IT, AND PAID SOME OF HER SMALLER BILLS: AND THAT WHEN SHE AND JAKE WENT THROUGH THE AFFAIRS, THEY FOUND THAT THERE WERE SOME SEVENTEEN HUNDRED DOLLARS OF BILLS, BESIDES THE TWO THOUSAND ON THE FOUNTAIN. THEY BOTH SAID THAT JAKE SQUARED ALL OF THE BAKERY ACCOUNTS. AFTER THOMAS SKIPPED, OLD JAKE WAS QUITE A FREQUENTER AT THE BAKERY AND WAS FAMILIAR WITH THE EARNINGS. ONE EVENING SITTING OUT ON THE PORCH OF THE HOUSE, HE TOLD THEM ABOUT HAVING A RICH GOLD MINE IN THE SUPERSTITION MOUNTAINS, AND THAT THERE WAS NO NEED OF THEIR WORKING SO HARD FOR JUST A LIVING, AND THAT IF THEY WOULD LISTEN HE WOULD TELL THEM WHERE IT WAS AND HOW HE BECAME THE OWNER.

THE DUTCH JAKE STORY:

HIS PARTNER JACOB WISER AND HIMSELF WERE PROSPECTING IN THE SIXTIES IN THE STATE OF SONORA, MEXICO, AND ONE EVENING THEY CAME TO A RANCH AND ASKED A MEXICAN WHOM THEY SAW IF THEY AND THEIR BURROS COULD CAMP THERE FOR THE NIGHT. HE SAID HE DID NOT KNOW, BUT THAT HE WOULD GO AND SEE THE DON: HE RETURNED AND TOLD THEM THAT THE DON WANTED TO KNOW IF THEY WERE AMERICANS. HE TOLD THE DON THAT THEY

4 - JACOB WALSH

DID NOT TALK LIKE AMERICANS, BUT WERE WHITE MEN. "WELL, TELL THEM THEY CAN CAMP AND FOR BOTH OF THEM TO COME UP AND SEE ME THIS EVENING AT SEVEN," WHICH THEY DID. THE DON ASKED THEM WHAT THEY WERE DOING, AND THEY SAID, "PROSPECTING." HE ASKED THEM IF THEY HAD FOUND ANYTHING, AND THEY TOLD HIM THEY HAD NOT. HE WANTED TO KNOW IF THEY WERE AMERICANS, AND THEY SAID THEY WERE NATURALIZED CITIZENS. THE DON SAID THAT HE WAS GETTING AN EXPIDITION TOGETHER TO GO UP INTO THE UNITED STATES TO A MINE THAT HAD BEEN IN THE POSSESSION OF THEIR FAMILY FOR SEVERAL GENERATIONS. A GRANT OF SOLE RIGHT TO MIGUEL PERALTA (WHO WAS HIS GRANDFATHER) AND HIS HEIRS, TO A MINE IN CERTAIN TERRITORY, DESCRIBING THE CORNERS, THREE OF WHICH WERE NATURAL, AND THE OTHER WAS A MONUMENT OF STONE LAID IN MORTAR OR MUD (WHICH IS THE SAME IN MEXICAN) TEN FEET SQUARE AND EIGHT FEET HIGH, AND BUILT LIKE A PYRAMID, THE THREE NATURAL CORNERS ARE EASILY FOUND, BUT THE FOURTH CORNER HE HAD BEEN INFORMED BY AN APACHE INDIAN, WAS TORN DOWN BY THEM, AND IT MUST BE SO AS IT WAS NEVER FOUND: THAT HE WANTED SOME AMERICANS IN HIS PARTY, AS HIS TITLE TO THE MINE WAS SIMPLY A SOLE RIGHT TO MINE WITHIN CERTAIN DESCRIBED BOUNDARIES AND NOT A DEED OF CONVEYANCE AND PERPETUITY: THAT IT WAS NOT RECOGNIZED BY THE UNITED STATES: THAT IF THEY WOULD GO WITH HIM, HE, PERALTA, WOULD GIVE THEM A THIRD OF WHAT HE TOOK OUT. THEY AGREED TO GO, AND THEY WENT UP INTO THE SUPERSTITION MOUNTAINS, HAD A BRUSH WITH THE APACHES COMING OUT AND TWO WERE WOUNDED. UPON THEIR ARRIVAL AT THE PERALTA RANCH IN SONORA, THEIR SHARE OF THE SPOILS WAS A LITTLE OVER THIRTY THOUSAND DOLLARS. I FAILED TO MENTION THAT THERE WERE TWENTY-ONE IN THE PARTY (EIGHTEEN PEONS). DON PERALTA PROPOSED TO GIVE THEM A BILL OF SALE TO THE MINE FOR THEIR SHARE OF THE GOLD. THEY ACCEPTED THE BILL OF SALE AND CAME BACK TO THE MINE. AS THEY WERE ABOUT TO CAMP, THEY HEARD SOME ONE BREAKING ROCK UPON THE CANYON. SO THEY TOOK THEIR GUNS, A DOUBLE BARRELLED MUZZLE LOADER SHOT GUN, LOADED WITH BUCKSHOT, THE OTHER A MUZZLE LOADING SHARP'S RIFLE, AND CRAWLED UP TO WITHIN SIGHT OF THE TUNNEL, AND THERE WERE TWO NAKED MEN (ALL BUT A GEE STRING) WHOM THEY THOUGHT WERE

5 - JACOB WALSH

INDIANS. THEY EACH SELECTED HIS MAN AND FIRED. BOTH WERE KILLED, BUT WHEN THEY WENT UP TO THEM, THEY DISCOVERED THAT THEY WERE TWO OF THE PEONS THAT WERE IN THE PARTY THAT CAME UP ON THE PREVIOUS TRIP. THEY BURIED THEM, MINED ABOUT THREE WEEKS, AND WENT BACK TO PERALTA. THEY FOUND THAT PERALTA HAD BEEN TO MEXICO CITY DURING THEIR ABSENCE, HAD BEEN GAMBLING HEAVILY AND WAS IN DEBT. HE ASKED THEM HOW MUCH GOLD THEY HAD, AND THEY TOLD HIM ABOUT SIXTY THOUSAND DOLLARS: HE SAID HE NEEDED MONEY BADLY, AND IF THEY WOULD LET HIM HAVE THE SIXTY THOUSAND THAT HE WOULD GIVE THEM A MORTGAGE ON HIS RANCH, WITH GOOD INTEREST; AFTER TALKING IT OVER, THEY DECIDED THAT THEY DID NOT WANT A MORTGAGE BUT WOULD TAKE HIS NOTE, AS THEY SAID THEY COULD RETURN TO THE MINE AND GET MORE. SO THEY TOOK THE NOTE AND STRUCK OUT AGAIN FOR THE MINE. AFTER THEIR ARRIVAL THERE, THEY CONCLUDED THAT THEY WERE GOING TO RUN OUT OF GRUB. SO JACOB WALSH TOOK HIS BURROS AND WENT DOWN TO ADAMSVILLE (JUST BELOW FLORENCE) WHERE THERE WAS AN ADOBE GRIST MILL AND A KIND OF A GENERAL STORE KEPT FOR THE PIMA INDIANS; IN FACT, A KIND OF A TRADING POST. WALSH SAID HE WAS GONE THREE OR FOUR DAYS. RIGHT THERE, THE FIRST DOUBT ABOUT THE STORY ENTERED MY MIND ABOUT HIS TELLING THE TRUTH, AS I COULD NOT UNDERSTAND HOW A MINER HAVING MADE TWO TRIPS TO THIS SAME DEPOSIT JUST PREVIOUSLY, SHOULD GO IN THERE WITH A SHORTAGE OF GRUB, BUT THE EXPLANATION TO THIS COMES LATER ON IN THE STORY.

UPON HIS RETURN TO THE MINE, HE FOUND HIS PARTNER, JACOB WISER, STRIPPED NAKED AND HIS NEW HICKORY SHIRT HANGING ON A BUSH NEAR BY WITH HIS MASONIC PIN STICKING IN IT. DOUBT NO. 2. I NEVER KNEW AN APACHE INDIAN TO OVERLOOK ANY KIND OF A SHIRT, MUCH LESS A NEW HICKORY SHIRT. THE CAMP IN GENERAL WAS SHOT UP AND DESTROYED; THAT THE HANDLE OF THE FRYING PAN WAS BROKEN OFF AND THREE BULLET HOLES THROUGH THE PAN AND MANY OTHER UTENSILS DESTROYED BY BULLETS.

DOUBT NO. 3. IF THERE WAS ONE THING IN THE WORLD THAT AN APACHE WAS CAREFUL ABOUT, AND SPARING AND JEALOUS OF, IT WAS AMMUNITION, AND FOR THEM TO WASTE IT ON FRYING PANS AND COFFEE POTS WAS ENTIRELY OUT OF THE QUESTION IN MY MIND. HE SAID THAT HE BURIED HIS PARTNER, WALLED UP THE TUNNEL, ROLLED DIRT

6 * JACOB WALSH

AND ROCK OVER IT AND TOOK THE FRYING PAN WITHOUT A HANDLE WITH THE BULLETHOLES IN IT. THAT THERE WERE MOUNTAIN PEAKS JUST WEST AND ABOVE THE MINE, AND THAT HE PLACED THIS FRYING PAN WITH FOUR SMALL ROCKS IN IT ON THE CENTER PEAK, AND THAT IF WE WOULD GO DOWN THE PEAK DUE EAST, WE WOULD FIND THE MINE. THE CENTER OF FOUR WAS ANOTHER STUMBLING BLOCK. I HAVE UNRAVELLED ALL THE OTHER DISCREPANCIES, BUT THIS ONE STILL REMAINS, AND THE ONLY SOLUTION TO IT IS THAT HE MAY HAVE SAID THREE; THAT NO COWBOY COULD EVER FIND IT, WHICH I TAKE TO MEAN THAT ONE CANNOT RIDE A HORSE TO IT, OR WITHIN SIGHT OF IT; AND FOR SOME REASON, THERE IS AN INTERRUPTION IN THE TRAIL GOING TO IT, AS A COWBOY WILL GO ANYWHERE THAT A TRAIL LEADS.

JACOB ALSO SAID THAT NO PROSPECTOR WOULD FIND IT, WHICH I TAKE TO MEAN THAT IT IS NOT A MINERAL FORMATION: IN FACT, HE SAID IT WAS NOT, AND THAT HE HAD NEVER SEEN ANYTHING LIKE IT AND THAT IT WAS VERY DIFFICULT TO FIND: THAT THE TRAIL THAT WENT OVER THE MOUNTAIN TO THE MINE WAS MONUMENTED WITH TWO LITTLE ROCKS PLACED UPON A LARGER ROCK, OR SOME OTHER CONSPICUOUS PLACE.

HE WOULD ALWAYS START THE SUBJECT OF THE MINE WHENEVER HE FOUND EITHER OR BOTH RHINEY AND HELENA IDLE AND UP AT THE HOUSE, AND APPEARED PROVOKED WHEN HE THOUGHT THEY WERE NOT PAYING ATTENTION; HE WOULD CHIDE THEM AND SAY, "WHEN I AM GONE, YOU WILL WISH YOU HAD LISTENED TO ME."

JAKE WAS EIGHTY-SIX YEARS OLD AND EVIDENTLY WAS NOT EXPECTING TO LIVE LONG. IN EARLY MARCH, 1891, HE TOLD THEM THAT HE DID NOT THINK THEY COULD FIND THE WAY TO THE MINE, AND THAT HE HAD BETTER GET A COUPLE OF PONIES AND A COUPLE OF BURROS, A WAGON AND SOME GRUB, AND GO WITH THEM AS FAR AS THE NEW BOARD HOUSE, WHERE THERE WAS A WOMAN AND THREE CHILDREN, AND THAT WAS AS FARAS HE COULD GO. FROM THERE ON, IT WOULD HAVE TO BE ON HORSEBACK, AND HE THOUGHT HE COULD SHOW THEM THE TRAIL OVER THE MOUNTAIN LOOKING FROM THE HOUSE; AND THEY MUST WEAR THEIR OLDEST CLOTHES BECAUSE THE BRUSH WAS BAD. OLD JAKE SAID THAT HE HAD BEEN TO THE MINE BUT ONCE SINCE HE LEFT IT AND CAME TO PHOENIX AND THAT WAS FOURTEEN YEARS AFTER HE WALLED UP THE TUNNEL, AND THAT EVERYTHING WAS JUST

AS HE HAD LEFT IT, AND HE DID NOT DISTURB ANYTHING. THE REASON I AM SO SURE HE WAS REFERRING TO THE BARK AND CRISWELL RANCH IS THAT MATT CAVENESS AND HIS WIFE BUILT THE BIG BOARD HOUSE IN 1877, AND MATT TOOK HIS FOUR DRAFT HORSES AS HE HAD A CONTRACT TO HAUL ORE FROM THE SILVER KING MINE TO PINAL, WHERE THEY WERE BUILDING THE MILL SOME FIVE MILES FROM THE MINE. MRS. CAVENESS AND THE THREE CHILDREN STAYED AT THE RANCH AND RAN A DAIRY. SHE SOLD BUTTER AT PINAL AND THE KING, SOME TWENTY MILES AWAY, AND GOT AS HIGH AS A DOLLAR A POUND. AS THERE WAS NO OTHER WOMAN OR CHILDREN, OR BOARD HOUSE IN THAT ENTIRE COUNTRY, THERE IS NOT MUCH DOUBT AS TO THE NEW BOARD HOUSE HE WAS REFERRING TO. ALSO, THE TRAIN OVER INTO THE SUPERSTITION MOUNTAINS WAS MONUMENTED IN JUST SUCH A MANNER AS JACOB DESCRIBED. AT THIS BOARD HOUSE, THE WAGON ROAD ENDED. MRS. CAVENESS TOLD MANY OF THE CATTLE MEN THAT SHE FREQUENTLY SOLD FLOUR TO THE APACHE INDIANS, AND THAT THEY ALWAYS PAID HER IN GOLD NUGGETS; AND THAT ONE TIME, AN INDIAN CAME TO THE HOUSE ABOUT TEN O'CLOCK IN THE MORNING AND ASKED HER FOR SOME FLOUR, SAYING THAT HE WOULD GO OVER THE BIG MOUNTAIN AND GET THE GOLD. SHE REFUSED THE FLOUR, TELLING HIM TO GET THE GOLD FIRST. HE RODE UP THE TRAIL OVER THE MOUNTAIN AND RETURNED BEFORE SUNDOWN WITH THE GOLD. WILL WHITLOW, A NEIGHBOR CATTLEMAN, WAS PRESENT AT TIMES WHEN SHE WAS SELLING TO THE INDIANS, AND HE SAID THAT SHE SOLD A MILK PAN FULL OF FLOUR FOR ABOUT TEN DOLLARS WORTH OF GOLD. SHE TOLD THE SAME THING TO MANY OTHERS.

JAKE FINALLY GOT HIS OUTFIT TOGETHER, LOADED THE WAGON WITH GRUB AS THEY WERE PLANNING ON STARTING THE NEXT DAY, BUT THAT NIGHT THE GREAT FLOOD CAME DOWN SALT RIVER AND IN THE MORNING, WATER WAS COVERING ALL OF LOWER PHOENIX. HELENA TOLD RHINEY THAT HE HAD BETTER GO AND SEE HOW OLD JAKE WAS SITUATED. WHEN THE BOY GOT IN SIGHT OF JAKE'S HOUSE, WATER WAS ALL AROUND IT. WHEN HE GOT TO THE HOUSE, HE FOUND JAKE STANDING UP ON HIS BED IN ABOUT SIX INCHES OF WATER. THE BOY TOOK HIM ON HIS BACK TO HELENA'S HOUSE, AND FINALLY JAKE WITH ALL HIS EFFECTS MOVED THERE AND STAYED UNTIL HE DIED IN THE SUMMER OF 1891. HE LIVED SEVERAL MONTHS AFTER THE FLOOD, BUT WAS NEVER ABLE TO GO TO THE MOUNTAINS.

8 - JACOB WALSH

HE COULD SIT OUT ON THE PORCH AT TIMES, AND WHEN HE DID, HE WOULD POINT TO THE SUPERSTITION MOUNTAINS AND SAY, "THE MINE LIES RIGHT OVER THERE". AMONG HIS THINGS THAT WERE MOVED OVER TO HELENA'S HOUSE WAS A SOAP BOX, VERY HEAVY, WITH LEATHER HINGES, A HASP AND A LOCK, WHICH HE ALWAYS INSISTED SHOULD NOT BE OPENED, BUT WAS TO REMAIN UNDER HIS COT. EVERYONE THOUGHT IT CONTAINED ORE, BUT NO ONE KNEW. WHEN JAKE WAS PASSING AWAY, HE KEPT MURMURING, "GOD FORGIVE ME: I HAD TO DO IT." AND THEN HE WOULD REPEAT IT, AND THOSE AROUND HIM SUPPOSED HE HAD KILLED HIS PARTNER JACOB WISER. THERE WAS AN OLD PROSPECTOR BY THE NAME OF OLD GERMANY, WHO WOULD DROP IN TO SEE JAKE, AND JAKE WOULD SAY, "TELL HIM TO GO OUT; I DON'T WANT HIM HERE." HE GAVE RHINEY THREE SMALL PIECES OF ORE, AND SAID THEY WERE FROM THE MINE. RHINEY GOT HARD UP, POUNDED TWO OF THE PIECES UP, WASHED THE GOLD OUT AND GOT SOME EIGHT DOLLARS FOR IT. JAKE DIED WITHOUT ANY OF THEM GOING TO THE MOUNTAIN TO LOOK FOR THE MINE. WHEN THEY WENT TO THE FUNERAL, THE SOAP BOX DISAPPEARED, AND HELENA AND RHINEY EACH SUSPECTED THE OTHER WITH MAKING AWAY WITH IT; BUT BOTH WERE INNOCENT, AND IT IS NOT NECESSARY TO THE STORY TO MENTION THE NAME OF THE GUILTY PARTY, SO I WILL PASS IT.

IT WAS SOME TIME BEFORE HELENA AND RHINEY COULD DISPOSE OF THE BAKERY AND GO LOOK FOR THE MINE. THEIR FIRST TRIP WAS THE TIME I MENTIONED IN THE FIRST PART OF THIS STORY. RHINEY TOLD ME THAT THE FIRST TIME HELENA AND HE WENT OVER THE MOUNTAIN TO LOOK FOR THE MINE, SHE GOT LOST ON TOP OF A MOUNTAIN, AND AS IT WAS SUNDOWN, SHE BECAME FRANTIC; THAT THERE WAS QUITE A NUMBER OF THE DRY RATTLE WEED UP THERE AND THAT SHE GOT UP ON A LARGE ROCK AND STOPPED THERE ALL NIGHT. SHE CARRIED A SMALL 32 REVOLVER AND A BOX OF CARTRIDGES. AS DARK WAS OVERTAKING HER, THE RATTLE WEEDS BEGAN TO RATTLE AND SHE MISTOOK THEM FOR RATTLESNAKES. SO SHE STAYED ON THE BOULDER ALL NIGHT AND FIRED ALL HER CARTRIDGES AWAY AT THE SNAKES. THIS PRACTICALLY IS THE STORY I GOT OUT OF RHINEY THAT EVENING, AND BY MORNING, I HAD A LIST OF FORTY-ONE QUESTIONS THAT I ASKED HIM. SOME HE ANSWERED "YES" OR "NO", AND SOME HE DID NOT KNOW. I KEPT THE QUESTIONS, CHANGED THEM ABOUT AND ASKED HIM THE SAME QUESTIONS AGAIN

9 - JACOB WALSH

AND HE WROTE THE ANSWERS, ABOUT THREE MONTHS AFTER QUESTIONING HIM THE FIRST TIME, AND HE ONLY VARIED IN ONE QUESTION; AND THAT WAS THE TIME JAKE SAID IT TOOK TO GO TO ADAMSVILLE AND BACK. THE FIRST TIME IT WAS THREE OR FOUR DAYS. THE LAST TIME, IT WAS FOUR DAYS. AT THAT TIME, IT WAS QUITE IMPORTANT, BUT AS THE STORY UNRAVELLED IT MADE NO DIFFERENCE.

RHINEY MADE SEVERAL TRIPS WITH THE PARTY AFTER THAT. SOMETIMES HERMAN WAS OFF ON HIS OWN HOOK, BUT THE OLD MAN PETRACH WAS ALWAYS ONE OF THE PARTY. SOMETIMES AFTER RHINEY HAD TOLD ME THE STORY, HELENA, THE OLD MAN, AND RHINEY CAME FROM OVER THE MOUNTAIN ON HORSES THAT WERE MISERABLY POOR AND ONE THEY WERE PACKING. THEY CAME TO THE RANCH ABOUT SUNDOWN AND ASKED IF THEY COULD GET SOME HAY AND GRAIN FOR THEIR HORSES AND SOMETHING TO EAT FOR THEMSELVES. THEY SAID THAT A DOLLAR AND SEVENTY-FIVE CENTS WAS ALL THE MONEY THEY HAD AND THAT THEY WOULD LIKE TO KEEP THAT SO AS TO BUY SOME GRUB AND HAY IN MESA, SOME THIRTY-FIVE MILES TOWARDS PHOENIX IN THE SALT RIVER VALLEY. I TOLD THEM THEY WOULD HAVE TO DO THEIR OWN COOKING, AS WE HAD HAD OUR SUPPER, AND THAT THE HORSE FEED WAS DOWN IN THE GRANERY, AND FOR THEM TO HELP THEMSELVES, WHICH THEY DID. THEY STAYED ALL NIGHT AND STRUCK OUT EARLY IN THE MORNING FOR MESA. THEY CAMPED THERE THAT NIGHT, AND DURING THE NIGHT THE HANDBAG THAT HELENA ALWAYS KEPT IN HER POSSESSION DISAPPEARED. WHO GOT IT, OR WHERE IT WENT, I DO NOT WISH TO SAY. THE HANDBAG CONTAINED JAKE'S PARCHEMENT MAP OF THE MINE, AND HOW TO GET TO IT, PERALTA'S NOTE TO THE TWO JAKES FOR SIXTY THOUSAND DOLLARS, BILL OF SALE TO THE TWO JAKES TO THE MINE AND JAKE WALSH'S NATURALIZATION CERTIFICATE. I SAW NO MORE OF THEM FOR A LONG TIME, BUT PETE SHOWED UP SHORTLY AFTER THAT, AND I HIRED HIM TO KEEP THE WATER HOLES CLEAN, ATTEND TO THE PUMPING, AND TO DO OTHER ODD JOBS, BUT NOT AS A COWHAND. I LET HIM HAVE LITTLE BILLY, A HORSE THAT WAS PERFECTLY SOUND, BUT NONE OF MY COWBOYS COULD MAKE HIM WORK. IF A COWBOY WAS RIDING BILLY AND TOOK AFTER A COW, IT WAS JUST SIMPLY IMPOSSIBLE TO MAKE HIM GET WITHIN ROPING DISTANCE OF THE ANIMAL YOU WERE AFTER. HE WAS THE GENTLEST HORSE I EVER SAW. YOU COULD CRAWL UNDER HIM

OR BETWEEN HIS HIND LEGS, OR LIE DOWN ON HIM WHEN HE WAS LYING DOWN, AND HE WOULD NEVER MOVE. NONE OF THE COWBOYS WOULD HAVE HIM IN THEIR MOUNT. PETE BEING SOME SIXTY YEARS OLD, I THOUGHT BILLY WOULD BE JUST THE HORSE FOR HIM TO RIDE BETWEEN WATER HOLES. PETE HAD BEEN WORKING FOR A WHILE AND I SUPPOSED HE AND BILLY WERE GETTING ALONG ALL O.K. AS I HAD SEEN PETE TAKE HIS SOAP BOX, GET UPON IT, FROM THERE PUT HIS FOOT IN THE STIRRUP AND GIVE A LUNGE AND LAND SOMEWHERE ON TOP OF BILLY, AND THEN SCRAMBLE INTO THE SADDLE, WHICH WAS ALL RIGHT WITH BILLY. I WAS WRITING IN THE HOUSE ONE MORNING, WHEN I HEARD A COMMOTION IN FRONT OF THE OPEN DOOR. I LOOKED UP AND THERE LAY PETE WITH THE BOX UPSET AND THE SADDLE ON TOP OF HIM; HIS NOSE WAS BEGINNING TO BLEED, AND BILLY WAS STANDING THERE WITH HIS HEAD TURNED LOOKING AT THE WRECK. PETE GOT UP AND THREW THE SADDLE TO ONE SIDE; AND OH, HOW MAD HE WAS! HE CAME TO ME AND SAID, "I QUIT". I SAID NOTHING, BUT HAD HARD WORK TRYING TO SUPPRESS MY LAUGHTER. "I WANT MY MONEY; I DID NOT HIRE OUT TO BREAK BRONCHOS, AND THAT IS ALL I HAVE DONE SINCE I HAVE BEEN HERE." I WAS SO FULL OF LAUGHTER THAT I STEPPED OUT OF THE DOOR TO WHERE BILLY WAS STANDING AND HE HAD NEVER MOVED. I LOOKED AT THE SADDLE AND PETE HAD FORGOTTEN TO CINCH HIM; HE DIDN'T EVEN PUT THE LATIGO THROUGH THE RING. SO I HAD TO TAKE BILLY AND GO DOWN THE CANYON TO THE WELL AND PUMP. THE FIRST LITTLE HILL I CAME TO ON THE TRAIL, BILLY POSITIVELY REFUSED TO GO DOWN, AND NO AMOUNT OF SPURRING WOULD FEASE HIM; HE WOULD KICK UP BEHIND, BUT COULDN'T BUCK A LITTLE BIT, SO I GOT OFF AND STARTED DOWN THE TRAIL TO A MESQUITE TREE, AND ALONG CAME BILLY FOLLOWING ME DOWN. I WALKED TO THE TREE AND GOT A GOOD CLUB AND BY THAT TIME WE HAD ARRIVED ON LEVEL GROUND. I GOT ON AND EVERYTHING WAS LOVELY, UNTIL WE CAME TO ANOTHER HILL THAT WE HAD TO GO UP. BILLY STOPPED AND HE WOULD NOT GO UP, POUND AS I WOULD; HE WOULD BUCK LIKE A COW. IF I HADN'T BEEN IN A HURRY, IT WOULD HAVE BEEN FUNNY. WHEN I GOT BACK TO THE HOUSE PETE WAS THERE AND I COAXED HIM TO STAY, THAT I WOULD GET HIM ANOTHER HORSE, A GENTLE ONE, BUT HE

11 - JACOB WALSH

REFUSED TO HAVE ANY OTHER AND THAT HE WOULD CONQUER HIM OR QUIT THE COW BUSINESS. SO PETE WORKED FOR ME ABOUT A YEAR, AND FINALLY GOT BILLY AS FAT AS A HOG; AND PETE HAD TO WALK BOTH DOWN AND UP HILL.

ONE EVENING WE WERE ALONE AND PETE WAS IN A PARTICULARLY GOOD HUMOR. HE ASKED ME IF I HAD EVER SEEN OLD JAKE'S NATURALIZATION PAPER, AND I TOLD HIM "NO". SO HE GOT UP, WENT TO HIS OLD TRUNK AND HANDED ME THE NATURALIZATION BERTIFICATE OF JACOB WALSH, A GERMAN, NATURALIZED IN NEW ORLEANS IN 1846.

QUITE SOME TIME LATER, HELENA MARRIED A MAN BY THE NAME OF SHAFFER, A RESIDENT OF SALT RIVER VALLEY, AND I DON'T THINK SHE MADE ANY MORE TRIPS TO LOOK FOR THE LOST DUTCHMAN. BUT PETE MADE FREQUENT VISITS TO THE SUPERSTITION MOUNTAINS UNTIL HE PASSED ON. I REMEMBER ONE BRIGHT MOONLIGHT NIGHT ABOUT ELEVEN O'CLOCK, I HEARD THE MOST UNUSUAL NOISE COMING DOWN THE TRAIL FROM OVER THE MOUNTAIN AND TOWARD THE HOUSE. I COULD NOT MAKE OUT WHAT WAS COMING, SO I PREPARED AS WARM A RECEPTION AS I COULD FOR WHATEVER IT WAS, AND WAITED. IT WOULD GO KLINTY BANG - KLINTY BANG, BANG, THEN ALL IN CHORUS, AND LOUDER AND CLOSER. WHEN THE HUMAN CALLIOPE ARRIVED, IT WAS OLD PETE, AND DRAPED ALL OVER HIM WERE FRYING PANS, COFFEE POT, TIN CUPS, PLATES, ETC. AND SOME BEDDING AND OTHER CAMP EQUIPMENT. THE EXPLANATION WAS THAT THE APACHE KID AND HIS SQUAW HAD BEEN SEEN BY PETE THAT DAY, AND THAT THEY HAD VISITED HIS CAMP, TAKEN WHAT LITTLE GRUB HE HAD AND THEN KICKED THINGS AROUND GENERALLY, AND PETE CONCLUDED TO IMMEDIATELY MOVE: BUT UNDER NO CONDITION COULD HE THINK OF LEAVING THAT FIVE DOLLARS WORTH OF CAMP EQUIPMENT TO THE TENDER MERCIES OF THE APACHE KID: HENCE THE EXTERNAL DECORATION OF OLD PETE. HERMAN, THE OLDER SON, HUNTED WITH HIS FATHER SOMETIMES: WHEN FUNDS RAN SHORT, HE WOULD GO WORK IN THE MINES AT GLOBE OR MIAMI, AND AFTER HIS FATHER PASSED ON, MOVED UP INTO THAT COUNTRY, MAKING A YEARLY TRIP FOR THE LOST DUTCHMAN. ALWAYS INDEPENDENT AND HONEST, I COULD DEPEND ON ANYTHING HE MIGHT TELL ME.

12 - JACOB WALSH

NOW, IN REGARD TO THE MONUMENTED TRAIL: I HAVE FOLLOWED IT OVER THE MOUNTAIN WITHOUT DIFFICULTY, AND THEN THE MONUMENTS COULD NOT BE FOUND FOR ABOUT TWO MILES, WHEN WE FOUND THEM ALONG A TRAIL WHERE CATTLE DID NOT TRAVEL, THEN OVER A DIVIDE INTO THE HORSE COUNTRY, WHERE THE MONUMENTS FAILED US AGAIN, BUT LED US TO A MUCH MORE SIGNIFICANT FEATURE OF THE STORY: THAT IS, THE CUT MESQUITE TIMBER. HELENA AND RHINEY SAID THAT JAKE TOLD THEM THAT THERE WERE TWO PITS AT THE MINE ABOUT 75 FEET DEEP AND A LIKE DISTANCE ACROSS THE TOP; THAT THEY WERE LINED WITH MESQUITE WOOD AND WOULD DROP ABOUT SIX FEET, THAN AN OFFSET OF ABOUT A FOOT, AND SO ON DOWN TO THE BOTTOM. UPON EVERY OFFSET, THERE WOULD BE A TOE LADDER, A STICK OF TIMBER WITH NOTCHES CUT IN IT THAT THE PEONS WOULD CLIMB BY PLACING THEIR TOES IN THE NOTCHES, TAKING HOLD OF THE TIMBER WITH THEIR HANDS AND FINALLY REACHING THE TOP WITH A RAWHIDE SACK ON THEIR BACKS AND A STRAP AROUND THEIR FOREHEADS, THE SACK FILLED WITH ORE. IT REALLY WAS ASTONISHING TO SEE HOW FAST MEN WHO WERE USED TO THAT KIND OF MINING COULD COME UP FROM THE BOTTOM WITH A LOAD OF ABOUT ONE HUNDRED POUNDS. NOW, THOSE PITS TOOK CONSIDERABLE MESQUITE, AND WE POSITIVELY HAVE FOUND WHERE THEY WERE CUT, AND THEY COULD NOT HAVE BEEN CUT FOR ANY OTHER PURPOSE THAT WE CAN CONCEIVE, AND THEY WERE ALL CUT WITH AN AX, A TOOL THAT APACHE INDIANS NEVER USED. THE CUT TIMBER MUST HAVE GONE DOWN, AS THE CUTTING IS ON TOP OF THE MOUNTAIN AND COVERS QUITE A LARGE AREA. WE HAVE BEEN WORKING ON THE CUT TIMBER FOR SEVERAL YEARS, AND WE HAVE FOUND WHERE THEY HAD QUITE A CAMP AT ONE TIME, JUST TO CUT AND PACK THE WOOD: THEY MUST HAVE LEFT IT IN A HURRY, AS WE FOUND A LOT OF SMALL HORSESHOES (NEW), SEVERAL KEGS, GONE TO STAVES AND HOOPES, AS THERE IS NO WATER THERE; AND OTHER EVIDENCES OF A CAMP. NOW, WE THINK WE KNOW VERY NEARLY WHERE THE WOOD LEFT THE MESA AND THAT WILL BE OUR NEXT ATTACK THIS WINTER. JAKE ALSO TOLD THEM THAT AT TIMES JUANQUIN MIRIETTA, THE BANDIT, WITH HIS GANG, WOULD JOIN THE PERALTA OUTFIT AND COME UP TO THE SUPERSTITION MOUNTAINS AND TO THE PERALTA CAMP AS A MUTUAL PROTECTION; AND THAT AT THE CAMP THEY HAD A MEXICAN RACE TRACK, AND THAT MURIETTA'S OUTFIT

13 - JACOB WALSH

WOULD RUN HORSES AGAINS PERALTA'S. HE SAID THEY WOULD SKIN THE PERALTAS OUT OF ALL THE MONEY THEY HAD, AND THAT MURIETTA HAD A SILVER MINE SOMEWHERE BEYOND AND THEY WOULD MOVE ON TO HIS MINE FOR A LOAD OF SILVER ORE. WE HAVE UNDOUBTEDLY FOUND THAT CAMP AND RACE TRACK.

I FORGOT TO STATE THAT UPON THE DEATH OF JAKE, THE GROUND AROUND HIS HOUSE WAS MOST THOROUGHLY DUG UP, BUT NOTHING FOUND.

THE FOREGOING IS PRACTICALLY THE STORY OF JACOB WALSH, WITH THE WHEAT SEPARATED FROM THE CHAFF AS NEAR AS IT IS HUMANLY POSSIBLE TO DO.

German translations of directions given to Reiney by Jacob as he lay ill in the storeroom.

The account of Jacob Waltz giving directions to his mine/cache to Julia Thomas and Rhinehardt Petrasch is true and verifiable. Gottfreid and Hermann Petrasch tried to make sense of the things Julia and Rhiney remembered Waltz saying. Either Waltz's remarks were too vague or Julia and Rhiney misinterpreted them, for whatever reason there just wasn't enough solid information to get them to the site.

Since Waltz spoke mostly German too Julia and Rhiney, most of the directions had to be interpreted from German to English. Hermann questioned both Julia and Rhinehardt about what they remembered Waltz saying. When both Julia and Rhiney agreed on what was exactly stated, Hermann would use that as a solid clue. When Julia and Rhiney differed on what Waltz said or meant, Hermann would consider those statements suspect. From everything that could be remembered Hermann compiled the directions he thought would lead them to Waltz's mine.

When Hermann realized the directions were too vague to get him to the mine he began to look for help. Shortly after the turn of the century Hermann approached Wm. Edwards and George Mathew Roberts (son of George R. Roberts). Edwards was a well known prospector and was rumored to be looking for mines in the "Superstitions". George M. Roberts had been a lifelong neighbor and friend to the Petrasch's. He was also known to frequent the Superstition, Pinal and Ancha ranges.

From this association Edwards and Roberts learned some of the things Julia and Rhinehart remembered Waltz saying.

<u>Julia and Rhinehart remembered Waltz saying :</u>

"Sie liegt uber de Salz fluss bergen".
It is in those Salt River Mountains.

"Ich kreuze der Grunen fluss unten am ein fort".
I cross the Green (Verde) River below the fort.

"Ich mache lager ubernacht in der Cottenwood wassenlauf".
I make an overnight camp in the (a) Cottonwood water course (wash).

"Ich kreuze der Salz fluss an der felsenseichter".
I cross the Salt River at the (a) rocky shallow.

"Ich folge das __?__ dem suden zu".
I follow it's __?__ to the south.

"Verlasse das wasser an der quelle".
Leave the water at the spring.

Hermann Petrasch (circa 1954)

Hermann Petrasch and a prospector friend beside Hermann's Superstition camp (circa 1979).

"Benutze der indianer weg der da folgt".
Follow the indian trail that leads (leaves) there.

"Drei kilometers ron der quelle wirst du mein spitzfelsen finden".
Three kilometers from the spring you find a (my) spire (pointed rock or uplifted rock).

"Dort oben wirst du sas Mexicaniscle lager finden".
Up there (or ,there above) you will find a (the) Mexican camp.

"Von meiner lagerstatte wen du im sattel sitzt kannst du in ferne im suden durch ein loch vom Berg sehen".
From my camp in it's (a or the) saddle you will see a peak in the distance to the south with a hole. (in it, or through it).

"Zu dem westen kannst du den sehen".
To the west you will see another peak.

"Meine miene ist nordlich aber ohne mich wirst du sie nicht finden".
My mine is to the north but you can not find (see) it without me.

"Die miene ist gut versteckt rom auge".
The mine is well hidden from sight.

"Ich versteckte die miene genau so wie die Mexicaner".
I hid the mine like the Mexicans. (had hidden the mine)

"Die Mexicaner lageren oben".
The Mexicans camp(ed) above (on high).

"Ich versteckte die werkzeuge ober den sattel".
I hid my tools (outfit) above the saddle.

"Schav fur die felsenspitze die Mexicaner benutzten die felsenspitze als ein anhalts punkt".
Look for the spire (s) the Mexicans used the spires as their markers.
Or; --- Look for the rock point(s) the Mexicans used the rock points as their stopping place (or point of reference).

"Ein hoher berg steht zu dem osten".
A high (or, higher) mountain is to the east.

"Ein tiefes tal lieft zu dem wester".
A low valley (or, a deep valley) lies to the west.

"Ich benutzte zwei von den Mexicaner lochern".
I used two of the Mexican's holes. * Does Waltz mean hole or cache ?

"Ich grub zwei eigene locher".
I dug two of my own. * Dug a mine, a hole or a cache ?

"Keine sind in sicht von einander und keine sind weit voneinander".
None are within sight of each other and none are far from the each other.
* Does "none" mean the holes Waltz dug ? How far does "far" mean ?

"Von der hohle kannst du sie alle sehen".
From the cave you can see all of them. * What cave ? Can see all the "holes" ?

"Komme riemals zuruck der selbe weg".
Never return (come back) by the route you arrived (or, the way you came).

"Die indianer haben mich einmal umstellt, Ich war gezwungen mein lager schnell zu verlassen".
The indians saw (found, trapped or surrounded) me there once, I had to leave my camp quickly.

West view of insane asylum where Reiney's father died and was buried (circa 1900).

From these few remembered directions Herman eventually ended up on Tortilla Mountain as did his father Gottfreid. Rhinehart ended up on Peters Mesa. With this same information Bark and Ely ended up on the North end of Bluff Spring Mountain and also across LaBarge on what some people call today (Black Mountain or Charley-Boy Mountain).

Wm. Edwards had already found a Mexican camp on "Peters Mesa" back in 1866 – 1867. The high ridge on Peter's Mesa is where both he and George M. Roberts ended up following these directions.

These directions, remembered by Rhinehart and Julia fit that ridge on Peters Mesa better than any other place in the mountains. Artifacts found over the years in that area fit with some of the things Waltz spoke of although no ownership could be positively established.*

The directions are to the best of Julia and Rhiney's memory. Interpretation is a big factor in the exchange of German to English.

Also things like Waltz saying a peak to the south with a hole in it can be seen. I know that on a clear day today the hole in Miners Needle can be seen to the south from atop Peters Mesa. However in Waltz's day I doubt if that hole could have been seen from there. I am not certain but I believe that hole was much smaller in Waltz's day and was enlarged later on. Also Waltz's eyesight was suspect, could he have seen a hole in a peak that far away? Possible explanations could be Waltz knew that peak had a hole and was not literally saying you could see the hole, only the peak with the hole in it. Also possibly Waltz had an old optical glass. Who knows ? There is so much open to interpretation.

My great aunt had occasion to talk with Rhinehart at the annual Pioneer Days picnic in 1933. Rhinehart told her the same story that Hermann had told her brother and Wm. Edwards almost 30 years before.

* Interesting note is that if Brownie's directions are not taken too literally they also lead one into that same area. Some people feel the Tortilla side is a stronger candidate because of some things Brownie said.

Special Report and Maps

Bob Corbin was the Maricopa County Attorney in the 1960s. His offices were in the courthouse in Phoenix, Arizona. During his tenure he became acquainted with Arnold Ortiz, a bailiff. Arnold was familiar with Bob's interest in lost treasure and told him the following story: "When my father was young, he hired an old Indian who came by our Chandler farm looking for work. The two became fast friends. Later, after the friendship cemented, he offered my father a letter and a map to a mine in the Superstitions. The Indian said years before he was hunting near Weaver's Needle when he came upon two white men. They started shooting at him. He fired back and one of them fell; the other ran away. After it was safe, the Indian approached cautiously to see if the man was still alive. He was dead. The Indian searched his pockets. The letter and the map were on the body."

For years following that incident, Arnold's father made trips into the Superstitions with some prospecting partners from Mesa. They looked for the mine but were unable to locate it. During one of those trips, the map and the letter disappeared. Mr. Ortiz never knew what happened to them.

In 1986, after Corbin became Attorney General, he called Arnold. Bob mentioned a map called the Ortiz map that had been given to him by Tom Kollenborn, his prospecting friend. He asked Arnold to come by his office and take a look at it. After arriving at the capitol complex and being handed the document, Arnold grinned and said, "Do you mind turning it upside down, Bob? I was a young boy when my father used to pore over that map, and I would stand with my nose table high on the other side. I memorized it upside down." After studying the document for a few moments, Arnold said there were some lines missing. He drew them lightly in pencil. He also explained that his father's map was drawn on a piece of leather and this map must have been copied from it.

Bob and Tom were unable to locate the mine from the map, even with the changes, because it had no explicit placement in the topography of the area.

In 1988 Corbin received a phone call from a woman in northern Arizona. She said she had heard he was interested in the Lost Dutchman Mine. She explained she had received docu-

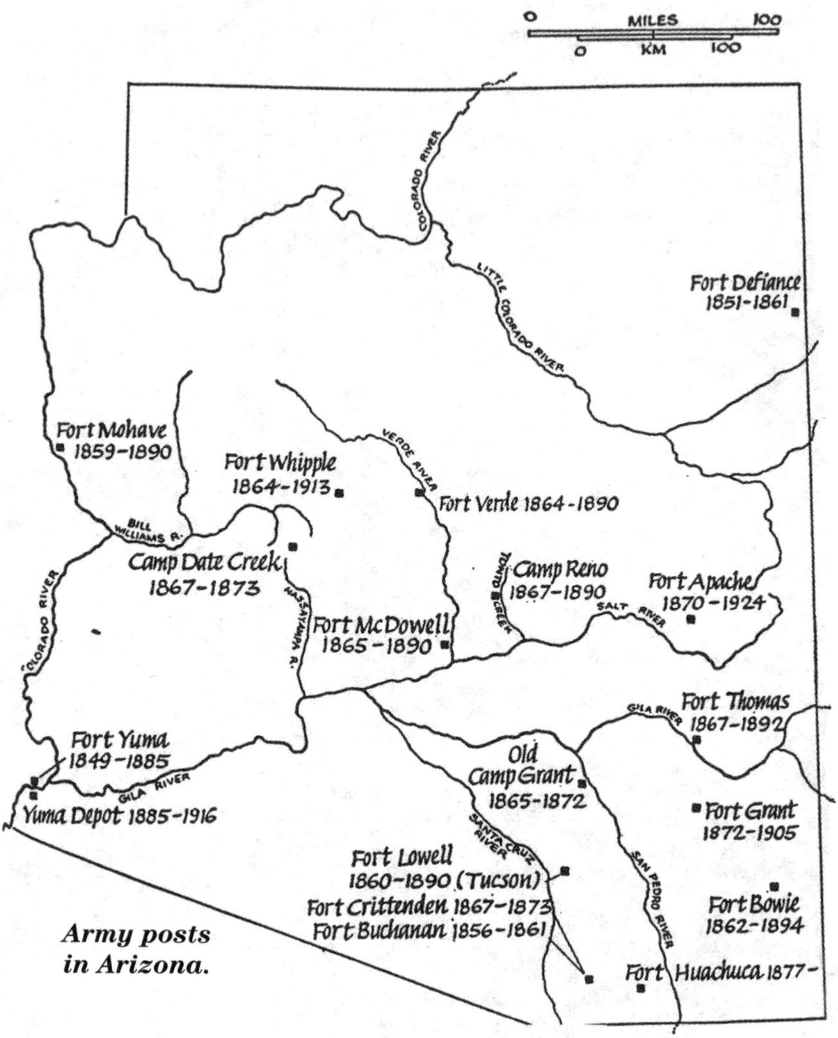

Army posts in Arizona.

ments that belonged to her husband's family in Mesa. Among the papers was a copy of a letter and a map pertaining to a lost mine. She wondered if he would like to have them. Of course, he did. This letter (shown on page 194), dated 1864, is addressed to Jacob Waltz and is signed by Don Peralta. The name is misspelled, but the information is so exact one cannot believe it is not Jacob from Württemberg. Even the casual observer can see it was written by an uncertain hand – almost a scribble – on very old, unusual paper. The information in the letter makes sense, and the man who wrote it was obviously having trouble with English.

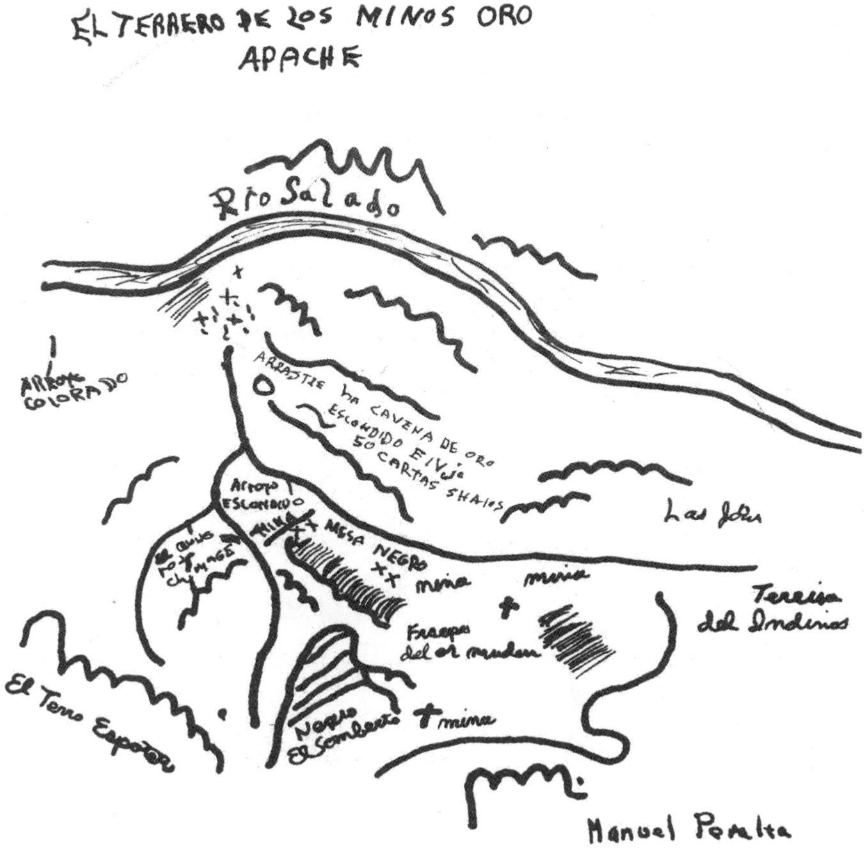

The Fish Treasure Map

Frank Fish was a successful contemporary treasure hunter. A number of years ago, while searching in Mexico, he met an American mining engineer who told him of a great gold mine in the Superstition Mountains of Arizona. Frank claimed he was eventually able to locate members of the Peralta family who gave him the map. They said the document came from Manuel Peralta. The full story of finding the map appears in a book,* Dead Men Do Tell Tales, *by Lake Erie Schafer. The original map is held by the Schafer family.

Bob believes the Germans may have headed into Mexico and, depending on their route, could have run into Peralta in the cantina in Arispa, saved his life and gone to the ranch where he encouraged them to join him. They went as soldiers to ride with the peons and later helped them dig up the ore. If that story is true, the letter may indeed be their rights to the mine that Peralta is supposed to have given them in return for their share of the ore.

* * *

People always ask Corbin what he would do if he found the mine. The reason for the question is all the U.S. Forest Service rules and regulations that govern the wilderness. The law essentially denies mining a new mine but not an existing mine.

Copy of the map sold by Richard (Dick) J. Holmes.

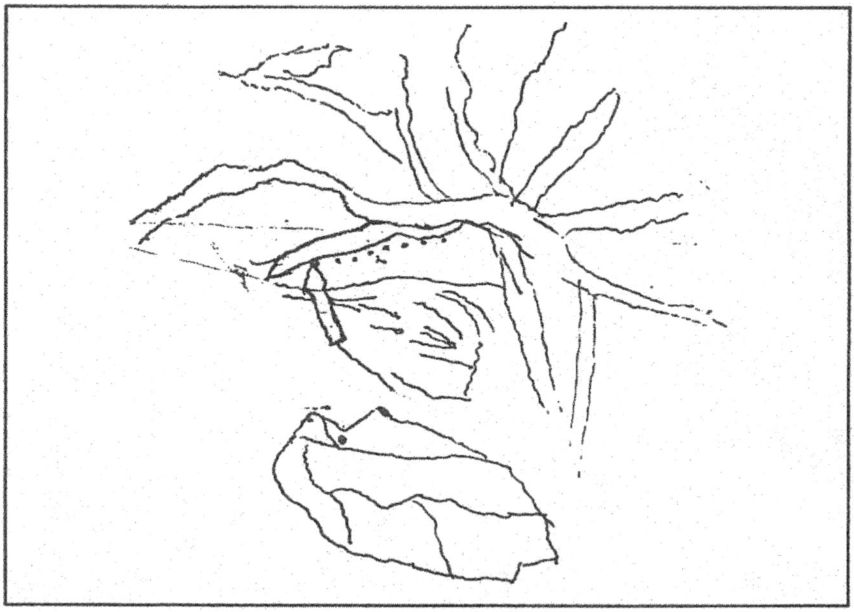

The Ortiz Map

THE ORTIZ LETTER

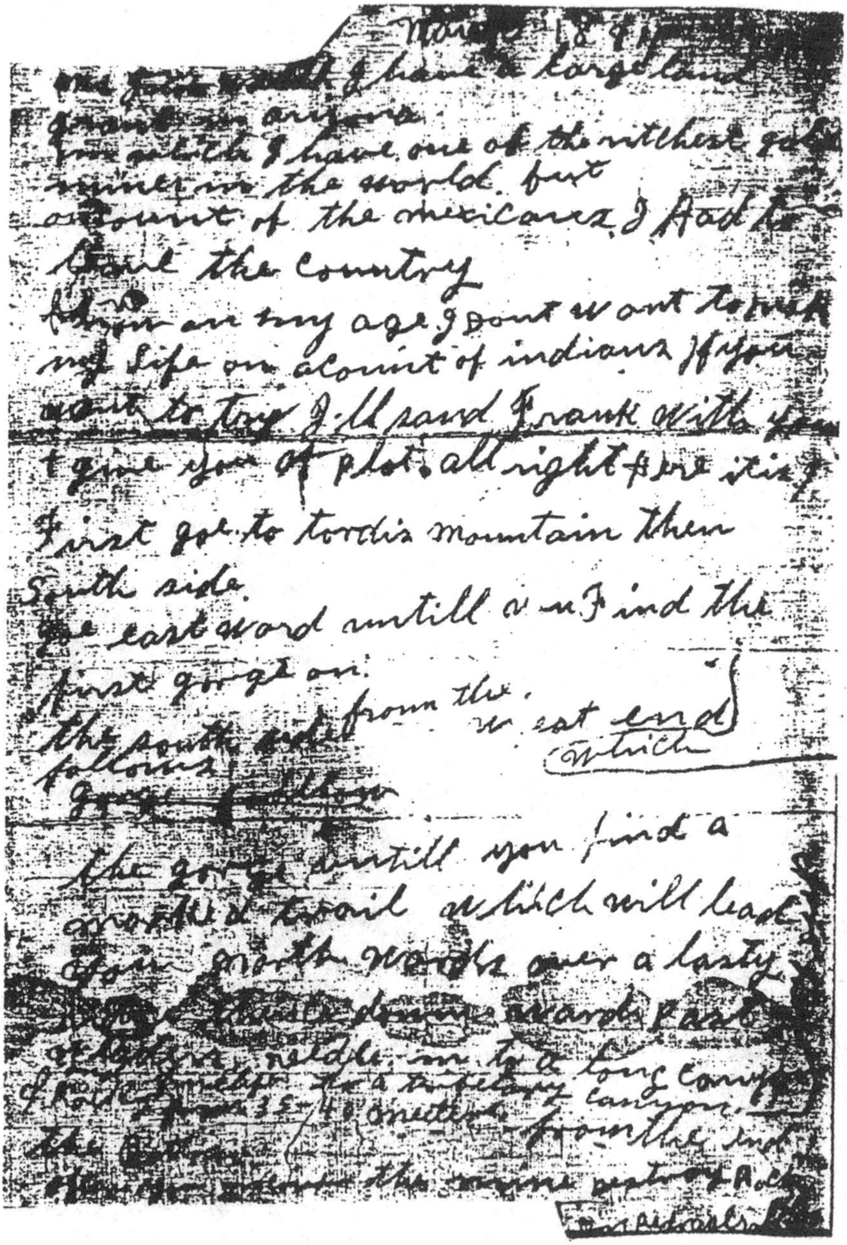

Nov [illegible] 18[illegible]

[illegible] I have a large land
grant in arizona
[illegible] I have one of the ritchest gold
mines in the world but
on acount of the mexicans I had to
leave the country
[illegible] at my age I dont want to risk
my life on acount of indians If you
want to try I'll send Frank with you
I give you a plot all right here it is
First go to tortilla mountain then
South side
go eastward untill you find the
first gorge on
the south side from the west end
follows [illegible] which
gorge [illegible]
the gorge untill you find a
worked trail which will lead
you northwards over a [illegible]
[illegible]
[illegible]
[illegible] canyon
[illegible] 35-40 meters from the [illegible]
[illegible] the mine [illegible] Rock
[illegible]

Letter given to Bob Corbin by Mesa woman.

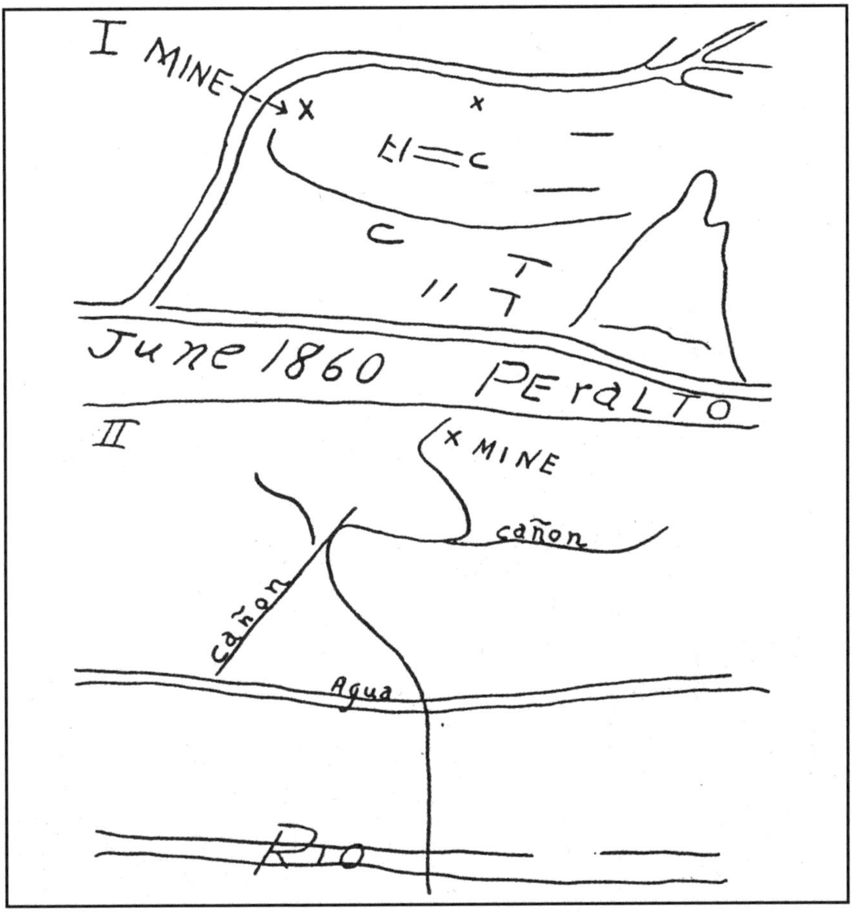

Peralta Locator Map

The Peralta locator map is one of the three maps that were received by Erwin Ruth from Senor Gonzales during his trip to Mexico. (It is claimed to be one of the maps to the Peraltas' lost gold mines.)

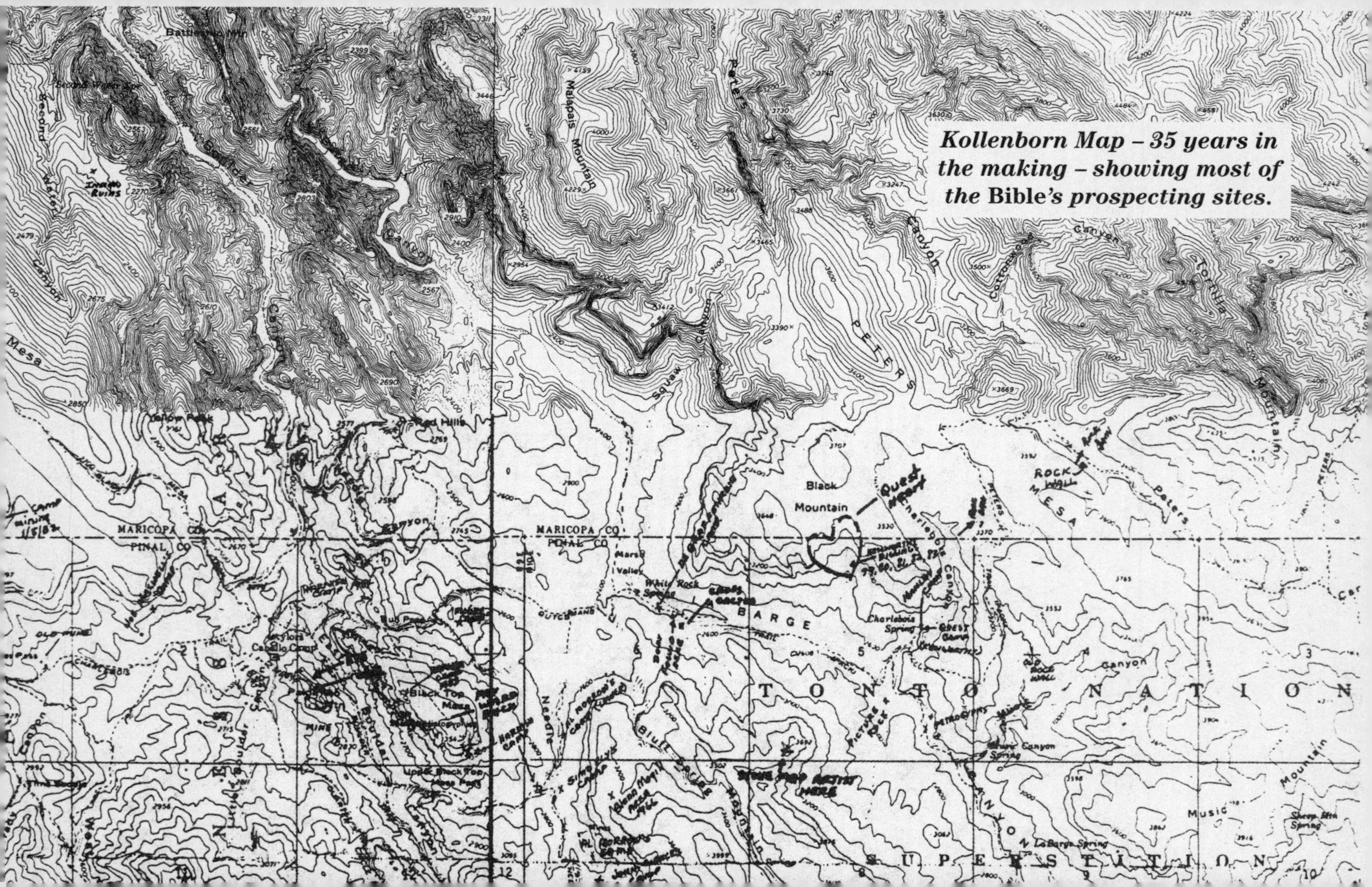

Kollenborn Map – 35 years in the making – showing most of the Bible's prospecting sites.

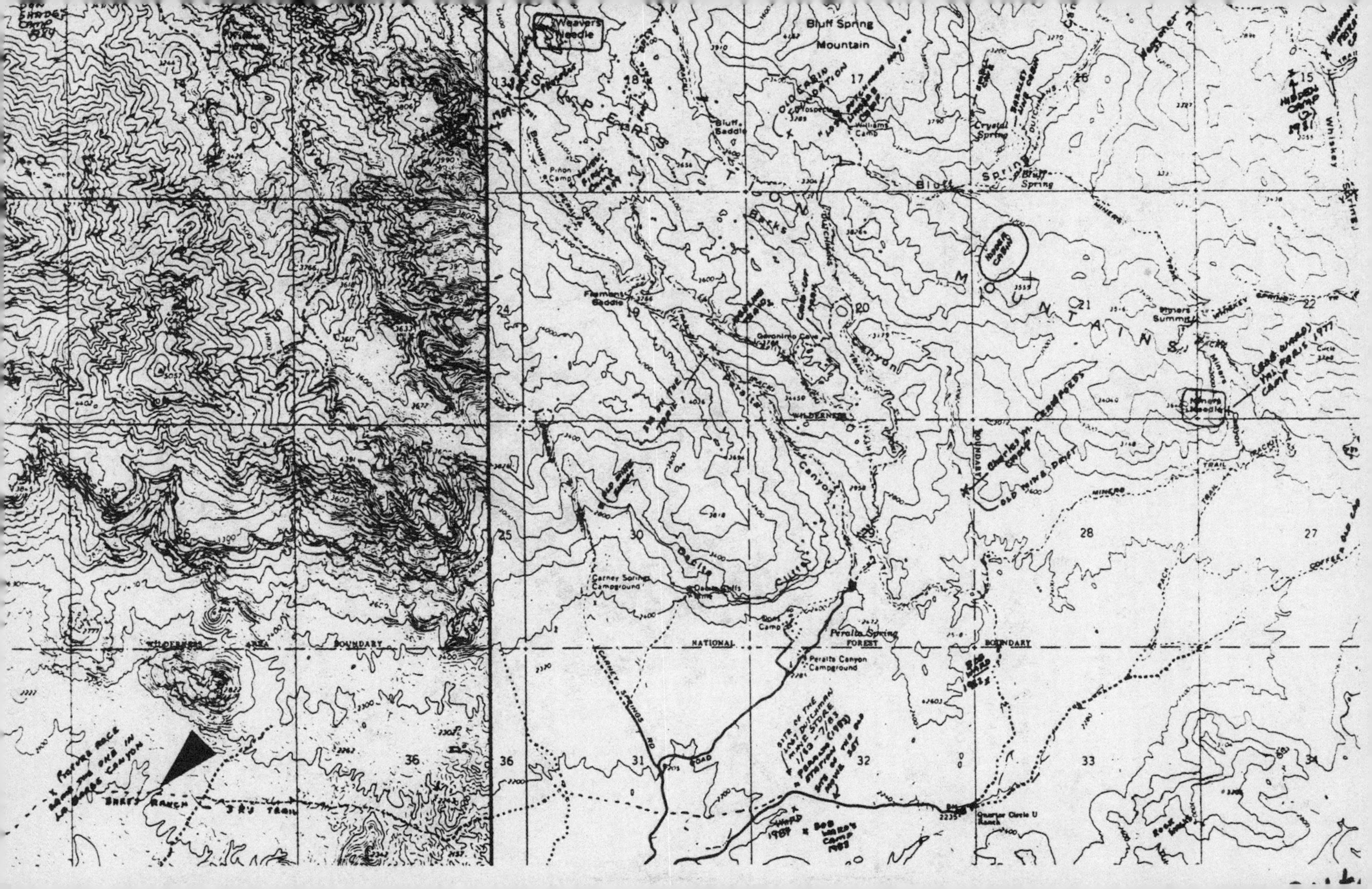
Weavers Needle
Bluff Spring Mountain
Bluff Saddle
Williams Camp
Crystal Spring
Bluff Spring
Pinon Camp
East Boulder
Fremont Saddle
Geronimo Cave
Miners Summit
Miners Needle
Peralta Canyon
Dutchmans Trail
WILDERNESS
Carney Springs Campground
Peralta Spring
Peralta Canyon Campground
Carney Springs Rd
NATIONAL FOREST BOUNDARY
WILDERNESS AREA BOUNDARY
Quarter Circle U Ranch
Barks Canyon
SUPERSTITION MOUNTAINS

Chapter 7

The Apaches and the Superstition Mountains

In the year 2001, it doesn't seem politically correct to repeat history as it occurred, but it is this author's opinion that no one should be allowed to do otherwise. Regardless of background, breed, ethnicity or tribe, the truth will out. The following explains who and what they were and what our invasion of their lands cost.

During our research we were privy to a manuscript that describes the Apache lore as it was during pioneer days. The information is especially significant because it alludes to the Superstition Mountains and the gods whose influence was so pervasive in their way of life. It also gives some explanation for the brutality that followed.

The Anglos, Blacks and Mexicans who came West had an expression that explained succinctly their feelings regarding the savages: "The only good Indian is a dead Indian" – harsh words indeed. When one reads the notes of the early pioneers there is complete redemption of that statement, but in all fairness to the red men, one needs to hear both sides of the story. Therefore, in an attempt to explain to the uninitiated what really transpired in the early West, we will print verbatim the Roberts documents on the Apache warriors. Later we will include excerpts from other sources regarding retaliation that was done.

As the reader absorbs the information, he needs to remember the wagons that came bearing families, prospectors, farmers, cattlemen, foreigners and the dregs of humanity sounded the death knell to the life of the native population. Their hunting grounds were denuded, their land usurped, their people captured and killed and finally surrender and relocation. As in all wars, there is the ultimate loss.

The situation in the Superstition Mountains became so intolerable by the mid-1800s that several leaders banded together and raised a following of some 300 warriors. These warriors were a mixed group of Chihenne, Bedonkohe and Yavapai Apaches. Three chiefs led these groups, who acted in union to defeat a common enemy. Many battles ensued, and the final end came when by sheer numbers the Apaches forced the invaders to make a run for it. While several smaller groups harassed and herded the Mexicans, a larger group set a trap and lay in wait for retreating miners. This final battle occurred on the northwest slope of today's Superstition Mountains. There, amid a group of boulders on a hillside, the last of the Mexicans were caught in the Apaches' trap and were forced to make a final stand. They were slaughtered to a man according to Apache accounts and their bodies left to the animals. This particular story as given by Indians relates to the famous massacre that is constantly questioned by researchers, miners and Dutchman hunters but is obviously true and is hereby verified

Apaches living today at San Carlos Reservation in Arizona, the Mescalero Reservation in New Mexico and at Lawton, Oklahoma, are descendants of Apaches who took part in the battles with the Mexicans in the mid-1800s. Some of these present-day Apaches have spiritual ties to the Superstition Mountains as a result of those battles with the Mexicans.

One family in particular are descendants of a Bedonkohe Apache who was known as Ghost Horse, who was born around 1840 near the headwaters of the Gila River, about where the town of Clifton, Arizona, is today. Ghost Horse was a fierce warrior who fought to the end with Geronimo and surrendered with him in September 1886. Ghost Horse and Geronimo were brothers but only in the Apache sense. (The tribe has no word for grandfather, grandmother or cousin. Grandfathers and great-grandfathers and so on were referred to as "father." It was confusing to the white eyes but quite logical to the Indians. Apaches considered first cousins as brothers.) So fierce and warlike was Ghost Horse that even Geronimo feared him. After their final surrender, Geronimo's group was sent to Florida, Alabama and finally to Fort Sill in Oklahoma. All the warriors were allowed to live under house arrest except Ghost Horse. He never bowed to a white man and remained imprisoned for the entire captivity, then died in his cell after going into a trance-like coma following days of chanting.

Ghost Horse had strong spirit ties to the Superstition Mountains. These ties involved the burial sites of his ancestors and his involvement with the return to Yusen of gold and silver taken in raids.

In Eve Ball's biography of James Kawaykla (grandson of Cochise), she gives an account whereby a Bedonkohe chief (Nana) tells young James of a time when he (Nana) and Ghost Horse went into a cave in the mountains where they had stacked a large amount of gold and silver taken in raids. Nana would not tell even Cochise's grandson the cave's location.

As the Apache Geronimo grew old and sick, he attempted to buy his freedom with the location of a cave laden with gold and silver hidden deep in the Arizona mountains. Geronimo wanted to die and be buried in the sacred mountains in the spiritual ritual of his Apache beliefs. To die and be buried in a white man's cemetery would have been a weakening of his spirit body, and his next life would have doomed him to untold hardships. His request was denied; he took the treasure's location with him to the grave.

Shortly before Ghost Horse's death, he was brought up on charges of attempting to kill a guard. At his trial, Lt. Britton Davis chastised him for not accepting his captivity and adapting to the white man's ways as other warriors had done. George Wratten acted as interpreter for Ghost Horse, whose answer became part of the official tribal record:

> You tell me I should teach my people how to live? By that you mean that I should teach them the ways of the white people. It is the duty of one who has been robbed of all that was his to accept the thief as his schoolmas-

> ter and spiritual guide. . . . Could you expect one who has been humiliated and broken to set up the author of this degradation as his ideal and pattern? . . . For gold your people destroyed the forests, desecrated our mountains, dried up our watering places, laid waste to the grass and brought death to all life. For gold you would rob, enslave and kill every race that is not of your blood. . . . By this same gold that the Apache people have been destroyed shall the Apache be revenged, for by that gold shall the destroyers themselves in their turn be destroyed. . . . The spirit of *na-deh-ja* cannot be killed by your evil power. Ta-neah will destroy all who enter the *kunh-ga-nah-hay*. When you kill me, my spirit will join Ta-neah there. There is nothing left for me to do but to die. I Ghost Horse have spoken.

Those were his last words. Refusing to eat or drink, he chanted constantly for three days – and then died. Toclanay, an Apache interpreter, stated Ghost Horse was in better health when he died than any Apache being held in captivity.

The meaning of the words at his trial were lost on the white man and his court of laws, but to the Apaches his words were a prophecy with a deeply solemn omen. The warrior chief said, "The spirit of *na-deh-ja* cannot be killed by your evil power . . ." He was telling his captors and all white men who ever desecrated the Apaches' sacred ground that even though they held his mortal body, they could not imprison the spirit body. When he said, ". . . Ta-neah will destroy all who enter the *kunh-ga-nah-hay*," he was referring to Ta-neah (an Apache *gan* – ghost spirit) called to guard the sacred places (*kunh-ga-nah-hay*) in the Superstition Mountains. Ghost Horse believed he would die, after which his spirit would join Ta-neah (his ancestor spirit) and stay in the Superstition sacred places to protect the ritual site.

Ghost Horse was believed by his fellow Apaches to have been a reincarnated shaman of war who could predict future events and summon the ghost spirit Ta-neah who guarded the spiritual centers. Toclanay believed Ghost Horse had summoned Ta-neah to come in four days and take his spirit to the mountains where he would gain

Apache War Chief Geronimo (on right) and Nachez at the San Carlos Reservation in Arizona (circa 1884).
Photo courtesy of Arizona State University.

A group of Apaches, under sentence of death for murder (circa 1880s Arizona).

revenge over the white men for their desecration and dishonor.

The direct descendants of this Apache warrior live today at Mescalero, New Mexico, and on the San Carlos Reservation in Arizona. Members of this family frequent the Superstition Mountains. They are looking after the sacred sites and performing rituals that are required by the age old *na-deh-ja*. No other Apache family of today follows the rituals of the old religion of Yusen as closely as this family.

There are two infamous incidents necessary for the reader to know regarding the other side of the brutality in the pioneer West.

Some of the Indian chiefs were wisely exhibiting to their people a careful understanding of the facts as they existed. One such chief was the Apache Eskiminzin. The warrior, realizing the futility of fighting against the growing odds, made peace with his enemies to preserve his people. They were offered a camp at a rancheria near Tucson while awaiting a visit from First Lieut. Royal Emerson Whitman. Upon his arrival, the two men settled into a comfortable discussion with coffee at the chief's wickiup.

Eskiminzin expounded on the perceived safety of the settlement after a compliment from the soldier regarding the efficient layout of the compound. Whitman asked the chief to provide some warriors to help round up the hostile Apaches still marauding in the mountains. They talked of peace. Each man was appreciative of the efforts of the other. Whitman commented on how few men were in camp. The chief explained the braves were out on hunting parties because the army rations that were donated were not sufficient to sustain his tribe. It was spring in the year 1871.

Unbeknown to the lieutenant, a group of Mexicans, Papago Indians and a few whites had been meeting in nearby Tucson planning an attack on the rancheria with the intention of murdering all who were there. In fact, they had already begun their march. The underlying reason for the attack was greed. A group of men, known as the Tucson Ring, controlled selling of supplies in that area. They had good reason to keep the war going. The army was a lucrative purchaser; therefore, peace was not an option. And, it was not difficult to convince the settlers the Indians were a threat and needed to be destroyed. Every family had an Indian story. There had been too much bloodshed, and fear was a constant companion.

Much later, Whitman made his way back to Camp Grant along Aravaipa Creek; then spent the evening in his quarters, having dinner and writing to his wife and sons in Maine. Afterward he poured a whiskey, lit a cigar and went out into the brisk wind soughing in the canyon to admire the brilliantly lit night sky. The soldier's comely face evidenced his satisfaction of the matter at hand. He was about to show everyone that the Indian situation could be accomplished with diplomacy.

Under that same dusky sky, the forces of evil were creeping toward the rancheria, awaiting the dawn. There were 92 Papago Indians, 42 Mexicans and 6 Americans in the party.

The expedition took the Cebedilla Pass route. It was infrequently used, and the heavy growth on either side of the Rillito afforded protection from prying eyes on the west side of the Catalinas. The San Pedro was on the east. Near dawn they made their way into a grassy valley and made camp near a stream about 16 miles below the army camp and what they believed was the location of the Apache rancheria. They had chosen early morning for their attack.

A scout returned to say they had misjudged the distance; their quarry was camped much farther away than they had estimated. It took many more hours to approach the village, and night was again upon them. The unsuspecting Apaches slept peacefully. The Indians, in their wisdom, had moved their village farther upstream to be certain the "waters always flowed."

In the biting, predawn chill, the interlopers crept into the cactus forest just beyond the camp. Two men went forward to kill the two sentries sitting by a fire. One, an old woman, was dealing cards to pass the time. The other was a young brave who was enjoying the card game. They were both sleepy. Within seconds their throats were cut, and they were hidden in the cactus without a sound. The leaders instructed part of the party to go up onto the bluff overlooking the camp and the river and to kill all who tried to run away. Having planned carefully, they carried war clubs and knives for a silent attack. Once the melee was underway, they would use their guns.

After the signal, the carnage began. An old chief, who slept fitfully, was the first to be alerted. His head came out of the wickiup and was bashed in. In the next tepee, a young girl lay sleeping. A Papago grabbed her, tore off her clothes and raped her. Two other men entered and ravaged the trembling girl, after which they crushed her skull, slashed her breasts with their knives and later riddled the body with bullets. Silently the enemy moved into each wickiup. War clubs crushed the skulls of sleeping squaws, and razor-sharp knives slashed the throats of sleeping children.

Dogs barked; they were beaten to death. Aravaipas and Pinals awakened into the living hell; they died immediately.

A frenzied mother attempted to shield her baby only to watch in horror as it was murdered and one of its legs

Geronimo (fourth from left) and Chiricahua Apaches waiting to be shipped to a Florida reservation.

hacked off. An old woman, pierced by an arrow, stumbled helplessly out of her wickiup before crumbling into a deathly heap. A nearby Papago clubbed her until her brains spilled onto the dirt. Some of the tribe tried to escape toward the bluff; they were cut down by the steady reverberation of gunfire.

Eskiminzin came awake early on. He forced his family out, shouting instructions, before grabbing the youngest child and running toward the mountains. In the turmoil he lost the others. He decided to go back would be risking the child's life. Those behind him were brutalized. He lost his entire family except for the daughter he saved.

Bodies were stripped, beaten and mutilated. Within a half hour the deed was done. The terrorists set about burning all the wickiups and firing bullets into any body they came across. There were some small children who had not been killed and were wandering aimlessly, their

Al-Che-Say – Chief of the White Mountain Apache Indians holding a Winchester Model 1886 (circa 1886) .

terrified cries searing the dawn. They were rounded up, forced up onto the saddles, then abducted. They were later given to the Papagos as slaves.

When Lieutenant Whitman received the news, he took a squad of men to the site with wagons filled with medical supplies and shovels for burials. Appalled by the sight he witnessed, he dispatched the soldiers to start a burial detail. There would be no need for medical supplies. He believed that act would show his sincerity to the chief. Sixty-eight bodies were buried and many others were missing, he reported to the command. Eskiminzin was in deep grief but agreed with Whitman that the army was not responsible. He knew there were many who did not want peace. He even agreed to try once more and began to rebuild the rancheria. During the rebuilding, a troop of cavalry riding through the valley came upon the Indians and were surprised. A battle ensued. Fortunately, no one was killed, but the incident forever damaged the relationship between the white eyes and the Indians. Eskiminzin went on the warpath, and there was untold bloodshed for the next decade.

The other outstanding event occurred in New Mexico near the Santa Rita copper mine. This massacre involved the Mescalero Apaches under similar circumstances. The Mexicans had put a bounty on Indian scalps, and a group of white men decided it might be a lucrative project to act upon. A party was planned by the people of the nearby town, and the Indians were invited. There was to be food, drink, games, toys for the children, music and camaraderie. The braves, who did not trust the white eyes, went out hunting. Only women, children and elders arrived for the fiesta. Shortly, the party commenced and half-way into it the scalp hunters swooped into the town square and slaughtered the Indians to a person.

One of the braves was Mangas Coloradas, who joined forces with Cochise after the slaughter of his people. He was a fierce adversary and was feared by his own as well as the enemy. There was one factor not apparent to the people of the town: All supplies had to be brought in by wagon and mules. From that day on, they were left to starve; Mangas Coloradas attacked and stole every wagon coming into the area. Eventually it became a ghost town.

Open warfare with the Chiricahua Apaches began in February 1861 after the army rode into Cochise's stronghold at Apache Pass. Cochise was accused of kidnapping a young boy from a ranch near Sonoita Creek. The chief denied the accusation. The soldiers tried to capture him, but he slit the sides of the wickiup and escaped with hostages, whose return the army demanded. The soldiers took a bunch of the braves as hostages and sent word to Cochise they would be killed. They were hanged, and among the captives were some of the chief's relatives. The Indian had told the truth about the abduction of the boy, but the fat was in the fire. He murdered his captives, mutilated their bodies and left them where the army would find them.

From then on angry bands of savages under the leadership of Chief Mangas Coloradas and the great Chief Cochise combined into a formidable force that terrorized southern Arizona for the next 10 years.

* * *

The Apache of the early 1800s were a loose knit organization of four separate groups. The white man gave the Apache and their groups and subtribes many different and varied names, but the Apache always referred to themselves as the *Inday* or *Shis-Inday*, which translated means men of the forest. To the Apache their four groups were known as the Chihenne, the Bedonkohe, the Chokonen and the Nedni. The word Chiricahua is a word used to describe the Chokonen group of Apache. Cochise was known as a Chiricahua chief, but in reality he was a Chokonen Apache. Geronimo and his band were Bedonkohe Apache.

Yavapai Apache were not true Apache; they were a combination of Navajo, Apache and mostly Yumen Indian. The true Apache looked upon the Yavapais as being *tagoon* or fools. They referred to the Yavapai as "the Biendene," which literally translated means brainless people. The true Apache and the Yavapai were enemies by their nature and culture but banded together at times to defeat a common enemy. To understand the Apache and Yavapai and how they related to the area known as the Superstition Mountains, one must forget everything the white man has misconstrued about them over the past 150 years.

The true Apache did not use the Superstitions as their home. Those mountains were in the far northwest boundary of their land; the Bedonkohe and Chihenne groups had the closest ties to the Superstitions. The true Apache used the mountains as a stronghold and for religious and spirit rituals. The Superstitions were a burial place for important members of the Apache. The Yavapai occasionally used the Superstition Mountains as their home but did not have the religious and spirit ties to the mountains as did the true Apache. The Four Peaks area held the Yavapai's sacred sites.

The true Apache of old practiced the religion of Yusen, their creator. (It's also their word for nature and the realm of the heavens that included the spirit world.) The part of the religion that dealt with the spirit world was known as the *na-deh-ja*. It is a deeply serious and sacred part of the Apache belief and existence. Its significance cannot be underestimated! It is this part of the old religious ways that affected the area of the Superstition Mountains the greatest.

To understand the Apache spirit connection one must first understand how the Apache lived and what they believed about life, death and the hereafter.

The white man believes that when you die your spirit goes to heaven to dwell there forever with God, and your human remains are buried in a "sacred" cemetery. The Apache of old believed that life and death were like a wheel, always turning. When an Apache died, his spirit went to dwell in the spirit world, a place that represented the universe, the stars, the clouds, the rain, a tree, a bird such as a hawk or a raven or even the wind. The spirit world to the Apache encompassed all that existed and everything that ever was. The Apache believed they would return to the living again as an Apache. This reincarnation would occur over and over again like a wheel always turning.

While alive the Apache practiced what was known as the "high laws" and the "low laws." The high laws were dominated by strength, courage, cunning, resourcefulness, treachery, ability to endure hardship and pain and the ability to persuade others to follow your bidding. The low laws were represented by laziness, cowardice, ignorance and generally being inept. To the Apache the strict adherence to the high laws would strengthen your spirit body so when you died you would be a little stronger in the next life – and stronger in the next and in the next and so on until finally you became so strong in your spirit that you did not have to return to the living and all its hardship and pain. Your spirit would dwell with Yusen in the spirit world forever. Conversely, an Apache who exhibited the low laws would have his spirit weakened with each successive life and be doomed to return to the living over and over again each time to face an even more miserable and painful existence.

The death of an Apache and the manner of his passing was of great importance to him. To die bravely in battle was the highest achievement and strengthened his spirit more than any other deed. This is why the Apache would endure any hardship without complaint, why they would not cry out in pain when captured and tortured. To endure torture and death at the hands of your enemy was also a great spirit achievement. This is why the Apache tortured and mutilated their captives so horribly. To force an enemy to cry out in pain and beg for mercy weakened that enemy's spirit, and when death finally came, his spirit would come back in the next life much weakened and be less of a threat to the Apache. This is what they believed and was the basis for many of their actions and deeds.

Someone once asked, "Why would the Apache choose *that* certain place for a sacred site? It makes no sense to me." The answer to why it might seem to make no sense is because in fact the sacred site is not chosen by the Apache at all. Sacred sites, also know as *chi-hah-agan* (places of power), are the will of Yusen and are chosen for the Apache by Yusen and the spirits of the Apache dead. This is a difficult thing to understand and should be explained:

When an Apache dies, he will one day be returned to the living, but that is not to say it will happen in a set amount of time. It could be in a day or a week or a year or a century. Only Yusen can choose the time of their return. After death the Apache spirit dwells near the place of his death or by the will of Yusen at sacred places of religious or ceremonial significance. Yusen can give the spirit any form, and through a complex ritual the living Apache can communicate with the spirit of the dead. This communication is the essence of the *na-deh-ja* religion and is the most significant and potentially deadly of all Apache beliefs.

When an Apache dies bravely, especially in battle with his enemy, the site of his death is a sacred place of power. The site of his chosen burial also becomes sacred and a place of power. Apache burial sites are chosen when a medicine man (shaman) or a close family member receives a sign or "vision" from Yusen directing him to the sacred site. Sometimes these visions take many days to occur; while the shaman awaits the vision he deprives himself of food, water and comfort. At times Yusen will send the shaman on a journey to unknown destination, the shaman wandering for days and many miles until he reaches the place of power and the vision is revealed to him. This is how the Superstition Mountains became a sacred stronghold and place of power to the Apache. Long before the days of Geronimo, the Superstitions were already sacred, the Apache Mah-ko, who was the grandfather of Geronimo, spoke of the power of those mountains, and their sacred origin predated even him. Over the years the Superstitions, and in particular some special areas, have been considered sacred and places of ceremonial power to the Apache.

The living Apache of today believe they can communicate with their spirit ancestors at these sacred places of power using a complex ritual. These spirit sites are not just bare rock and earth. Sacred items of power are found at these places; sometimes the places are the burial sites of the Apache. It is of *vital* importance to the Apache that these sites are not disturbed or altered, or the power of the spirit place will be destroyed and with it the living Apaches' connection with their ancestors. This point cannot be stressed enough! Once a site is lost, their ancestor's spirit and their own spirit is lost along with it.

Keeping in mind everything that has been said, it is now possible to put things into a historical perspective and explain how and why these things affected the Superstition Mountains and the white man's presence in that area.

In the late 1700s and very early 1800s, Spanish explorers pushed into the area of the Gila and Salt Rivers in central Arizona. They were looking for silver, gold and mercury (cinnabar). It should be noted that historically the period of 1780 – 1816 was the time of greatest peace between the Apache and the Anglos in northern New Spain (Mexico) and the Arizona Territory. The Spaniards discovered rich deposits of all three ores from present-day Nogales to Flagstaff. Mines were established and a half-friendly trade existed between the Anglos and the Apache. The areas directly west and north of the Superstition Mountains were mined for silver and gold, and the interior of the mountains produced mercury in the form of cinnabar. The drift tunnel in Pistol Canyon was a rich deposit of cinnabar along with realgar, orpiment, some enargite gold and late stage development of native silver.

One must consider this time frame: the early 1800s. Back then the Apache sacred sites were openly marked with their ceremonial markers in place and in full view. And why shouldn't they have been? The area then was completely remote; the Apache never had to worry about desecration and plunder of their sites because all other Indian cultures held each others' religious sites in superstitious awe. They went out of their way to avoid these sites fearing their power and the spirits that guarded them. When the first Spaniards arrived, they were drawn immediately to the sites by the elaborate ceremonial markers. Artifacts and items of power placed at these sites were stolen by the Spanish and the sites desecrated.

This desecration broke the spirit connection for the Apache, a tragedy of the highest order as far as they were concerned. Nothing the Spanish could have done would have been more damaging to the Apache. To the Apache the situation was intolerable, and their custom required revenge for these acts. It was this situation and not the mining that began the conflicts between the Spaniards and the Apache in the area of the Superstition Mountains. Mining was merely the thing that brought the Spanish into their sacred area. Had the Spanish left the sacred spirit sites alone, there may never have been a problem.

From these earliest desecrations, the Apache learned to hide their sacred sites from the Anglos. No longer did they display in full view the ritual poles and displays that marked the spirit places. But the damage had already been done; by the desecration the war had effectively begun.

It must be understood that the events in the Superstition Mountains concerning the Anglos and the Apache were not one-dimensional problems centered around sacred sites. It was a complex set of issues that also included trade, use of the land, treatment of the Apache people and gold itself. The Apache believe that gold is the possession of Yusen and belongs to him alone. The Apache did not use gold even in their jewelry. Copper was of highest value among the metals of the Apache. The Apache did not at first mind the Spanish mining the gold for they believed Yusen would destroy the Spanish for them in revenge for stealing "his gold." Gold was avoided by the Apache, and it was of no value whatsoever to them. Later when the Apache raided the Anglos and took gold and silver in the raids, they followed a practice of taking the metal to a remote cave or crevice and placing it there for Yusen. A prayer was offered to Yusen and a spirit ritual was held to place a spirit at the site to guard the metal for Yusen.

Later on, as more "metal" was taken in raids, the same caves would often be used to place the gold in safe keeping for Yusen. A ritual would be strictly followed for the entering of these spirit sites as even the Apache were forbidden from entering the sites once the spirit had been called. The Apache even today stay a safe distance from the sites when they visit them. A shaman would say a ritual prayer to Yusen and, placing a blue stone to the right of the entrance and a white stone on the left, would be allowed to enter the site and place new objects (gold or religious items) at the spirit site. Over the years with the many successful raids, the Apache acquired much gold and silver from the Anglos. The Superstition Mountains, a stronghold and deeply religious area to the Apache, became a major place to "return the gold to Yusen."

In the early 1800s there was no word in the Apache language for gold. All metal was called *pesh*. It was not until the Anglos came that the Apache learned the differences of the metals and their value to the Anglos. The Apache then adopted words for each of the metals; the word *pesh* remained to denote metal, but a second word was added to differentiate the metals. Gold became *pesh kitzo*, and silver became *pesh lickoye*, literally meaning yellow metal and white metal. An <u>extremely</u> important point must be made here. The Apache had no differentiation <u>in their language or in their thinking</u> to tell a mine from a cave filled with gold. Both were the <u>same</u> to them!

It is at this point, and understanding the aforementioned knowledge, that the significance of the Apache traditions and beliefs begin to blend and make sense of the stories and legends that have arisen concerning the Superstition Mountains over the past 150 years. One story in particular is the account by an elderly Apache who told a white man, ". . . a spirit guards the mine there and it only sleeps four hours a day." This story while not <u>precisely</u> what the Apache ritual involves is an <u>extremely close</u> accounting. First, the Apache by stating the word "mine" may well have been describing a spirit cave, and in reality that is more probably the case. Second, the number of hours the spirit sleeps is significant as the number four is the sacred number of the Apache. Ritual and ceremony are accomplished strictly by the use of the number four – four items, four days, four moons and so on. The Apache day was measured from the point when the evening star (Venus) was at its highest arc in the night sky, their equivalent to our midnight. There were no "hours" to the Apache; they did not know the concept of minutes and hours. The Apache day was divided into <u>four</u> parts, which would have equated to six white-man hours. When an Apache says, ". . .sleeps four hours," he means sleeps one-fourth of a day. <u>Everything</u> to the Apache, even parts of the day, were measured using the number four.

To an Apache the white man's word "mine" meant literally a place where gold is found. A gold deposit yet to be excavated was as much a "mine" to an Apache as was a tunnel and a shaft. This critical point cannot be stressed too emphatically!

The areas of today's Tortilla Mountain and Peter's Mesa were areas where the Apache and the Spanish (and later the Mexicans) clashed. The battles to drive out the Spanish went on for decades and at times were not much more than harassment and raiding. Neither side held the clear upper hand, and the balance of power shifted back and forth. The Apache are not a united, democratic society led by a single leader and fighting for a singular cause. While some groups of Apache raided and harassed the invaders, others traded with them. Each Apache is an independent entity; they give their allegiance to whomever they choose and fight and die only by their own will. No chief or leader can order another Apache to follow him. A chief's power was in his ability to persuade his followers to do his bidding or fight and die with him.

The past 114 years have changed the ways of the Apaches' religious beliefs. Over the years fewer and fewer Apache have followed the old ways, and as they became more and more modernized, they became as the white man: lazy in their day-to-day life and no longer were the ways of the high laws practiced. As the Apache became "Americanized," they cast aside their old beliefs and adopted the white man's religions. Today, less than one-half of one percent of all the true Apache living at Mescalero, San Carlos and Lawton, Oklahoma, practice the old religion of the high laws and the *na-deh-ja*.

The Apache never were a people with a shared history or knowledge of their past. They were composed of many loose-knit groups whose history was known to its members but never shared with outside groups. Some groups shared pieces of the same history, but much of this history was lost after the final surrender and imprisonment of 1886. Not until 1914 was an attempt made by certain Apache families to put down some of this history and traditional ways. The families and groups that suffered the most during the final days of freedom were the ones who lost most of their past.

The families who had the strongest ties to the land and the spirits were the ones who retained the most knowledge and history. Apache such as James Kawaykla, Jason Betzinez, Asa Dakluge, Edgar Perry, Eva Geronimo and Margaret Kleghorn literally saved the Apache past from extinction. If it were not for those people and others like them, an entire culture and its history would have been lost forever to the Apache of today. Even with their best efforts, much of the old history and the old ways slipped away into the fog of the ages.

In the Superstition Mountains between the years of 1920 and 1965, at least a dozen or more bodies were found of prospectors and mine hunters who had been killed and beheaded by unknown persons, and for unknown reasons.

In the latter days of the Apache wars, the U.S. Army paid a bounty on the lives of certain Apache leaders. To prove one had dispatched the right Apache, the U.S. Army required his head be brought in for identification. Army records show at least 100 heads were taken during this period. Some were of the desired Apache, but most were "mistaken identity." It is without question, this author can state that mysterious beheadings in the Superstition Mountains were the work of Apache revenge for the practice of the U.S. Army.

It is an age old Apache tradition to seek revenge on the race of people who committed an injustice or an atrocity on an Apache. It is the duty of the family of the injured Apache to right the wrong by revenge.

The closer to the source of the injustice the revenge can be wreaked, the better; but if none of the principal offenders can be sought, any member of that race is eligible for retribution. And traditionally that retribution is swift and brutal; often many lives were required to atone for the life of one Apache. This author can further attest that some of those who met that similar fate in those mountains have yet to be discovered. To say that this old and violent practice no longer continues, or could ever happen again in those mountains, would be a foolish assumption to make.

The Apache who frequent the Superstition Mountains today know well the locations of the sacred ritual and sites of power. They keep a close eye on these places, and they watch all that goes on in these areas. As recently as this past decade, the Apache have lost some of their sacred sites to thieves and vandals, both in the Superstitions and on their own reservation at San Carlos. These desecrations are intolerable to the Apache who still follow the religion of Yusen. To them the destruction of their spirit sites today requires revenge as surely as it did in the days of the early Mexicans. As some of these sites have been exposed to contact with outsiders (cattlemen, mine hunters and hikers), the Apache have been forced to evoke an age old spirit ritual to move some of these sites to places of greater safety and hiding.

With more and more outsiders coming into the formerly remote reaches of the mountains the possibility that more of these sites will be discovered is high. This fact is a great cause of distress to the Apache, and by no means are they standing by in idle frustration. The families who have the greatest stake in those sites have formed a plan to guard those sacred places, with their lives if necessary. There is a great danger in those mountains today to those who would destroy a site or even happen upon one accidentally. To be in these areas, even while minding one's own business with no ill intentions, brings one into suspicion. One can rest assured that to be in these areas for even a short time brings you directly under the watchful eye of these Apache.

The legends and stories told throughout the years by white man and Indian alike are based in truth and fact, a truth and fact remembered and held sacred to this day by only a very few.

Chapter 8

The Peraltas – Real or Imagined?

Throughout history there have always been rumors and myths concerning stories revolving around treasure. Many claim in every myth there is a kernel of truth enlarged upon until one wonders if there is any fact left. The matter of the Peralta family is just such a story and has been rumored for over a century.

Jacob Waltz is supposed to have told Julia (Helena) Thomas and Reiney Petrasch, at Christmas dinner outside his adobe, the story of saving Miguel Peralta from a dishonest dealer during a gambling episode in Mexico. The result of this chance meeting was a deal involving a rich gold mine on a land grant given to Miguel's father by the King of Spain. Apparently, Jacob and his partner Jacob Weiss agreed to the deal. After making one trip into the Superstitions in the Arizona Territory with Miguel and 18 peons, they turned over their share of the gold ($60,000 worth) to the gambler in exchange for the rights to the mine. (The story is hearsay from Julia to Sims Ely and is in his book, but many people believe it is untrue.)

Jacob Weiss, as he lay dying, is supposed to have told Doctor Walker that the parchment he had was a map to a really rich mine in the mountains to the north. (That would be the Superstitions.) He also said he and Waltz had been soldiers, and a Mexican named Peralta wanted some trained men to accompany them to the mine where his father had been killed by Apaches. (This story is hearsay from Tom Weedin to Sims Ely. It also appears in Ely's book. Jim Bark's notes confirm part of the story too.)

Because Sims said the Germans had been in the Civil War, which ended in 1864, everyone assumed that was correct. Of course, we now know the two men had been conscripted to the Prussian army at age 20 for three years and that they immigrated to the United States in 1839 when they were 28 and 29 years of age. So when the two Jacobs were quoted as having said they were soldiers, this was exactly correct.

It is important for the reader to understand the history of the Mexicans who came north from Sonora during the Civil War. The military had been called away from the territory to be used in the war. This gave Mexican miners the chance to extract precious minerals from mines previously owned by relatives having land grants from the King of Spain. They had been mining the area since the 1700s. The Arizona Territory was a wilderness occupied mainly by Apaches. Of course, much of the territory had been part of Mexico until the Hidalgo Treaty following the Mexican War of 1846. Then in 1852 the Gadsden Purchase set the present border between Arizona and Mexico.

The Mexicans rarely lived in the territory. Most of them lived in Ures, Fronteras and Hermosillo in the state of Sonora. However, they were used to running cattle, farming the fertile washes and river beds and mining in the territory.

The Peraltas were literally run out of the country following the war with Mexico. Their mines and ranches were stolen from them with no compensation ever given. Santa Rita Del Cobre mines in Silver City, New Mexico, rich in silver and copper, were owned by the Peralta family of Janos. The Peralta family of Sonora was said to have rich mines to the north, which were carefully concealed. Small mines were often covered over in an effort to preserve what minerals might be needed some time in the future. This was common practice in the years after 1848.

Mexicans would drift with a vein of ore deep inside the earth. The opening to the mine would be so small as to allow one man entry. Inside would be just enough room to work. The waste rock, which we would call the dump, was often carried off for long distances and hidden in a fault seam. After they were finished working, the two-foot opening would be walled up and plants or brush put in front of it. One of these lost drift mines is purported to be the Lost Dutchman. (Anglos rarely drifted. They learned from Welsh and Cornish miners in Europe, who dug tunnels and shafts, did timbering and often carved out caves in rocky terrain. When one came upon their mines, they were wide open and easily found.)

Mathew Roberts at a walled up drift mine.

Now, who was Miguel Peralta? Born in Ures, Sonora, in 1838, this young Mexican became a prospector and miner. In 1864 he and his father owned and worked a mine in the Bradshaw Mountains of the Arizona Territory. (Legal papers to prove this statement are included.) This bit of information begs the question: If they had a rich mine in the Superstitions, why weren't they mining it? Perhaps they had several land grants. It was common knowledge there was a lot of gold to be found in the Bradshaws, the vast range behind Prescott. And, perhaps, the Indians weren't as fierce as they were in the Superstitions. That same year Miguel and his father sold their land grant to Dr. George Willing for $1,000. It is possible they went back to their ranch in Mexico, planned to reopen the mine in the Superstitions whenever they needed gold and actually made their initial trip. The father was killed. Miguel, for whatever reason, was skittish about returning to the range for gold, but he had heavy gambling debts. That was when he accidentally met the two Germans.

Miguel Peralta went on to become a respected pioneer. In 1870 he ran a mercantile store in Wickenburg. Eventually he built and ran the Phoenix post office along with a mercantile store located on the same block on which George Roberts ran a saloon and drugstore. (The two men had been friends since 1863.)

The Peraltas were known to be honest, hard-working people. In 1880 Joseph Goldwater, the uncle of Senator Barry Goldwater, convinced Miguel to sell his business and work with him at his store in Fairbank, Cochise County. (The Goldman brothers, who ultimately bought the gold that was under Waltz's bed when he died, purchased the Phoenix mercantile store from Peralta.)

If Miguel Peralta did indeed give the mine to the two Jacobs to repay his gambling debt, then perhaps he waited until Waltz was dead and tried to relocate the mine or was looking for the

Mining Claim and land transfers of the Magdalena group which includes the "Valencia" mine. Yavapai County Arizona Territory 1864 - 1869

Grantor		Grantee	Date	Bk/Pg	Item
Ignacio A. Garcia	to	Jas. McMullen	19 Jan 1865	1/153	Claim#2 Valencia
Francisco Ochoa	to	Bernarbe Robles	17 Mar 1865	1/168	300' Valencia
Ramon Rodriguez	to	Bernarbe Robles	17 Mar 1865	1/170	100' Valencia
Florentine Otero	to	Chas. Johnson	1 May 1865	1/170	300' Profidencia
Pablo Peralta	to	Wm. Flower	22 May 1865	2/125	1000' Valencia
Adolph Arvizu	to	Wm. Flower	22 May 1865	2/126	1600' Valencia
Ambrocio Quintero	to	Wm. Flower	22 May 1865	2/126	1600' Valencia
Bernardo Salcido	to	B.H. Wunderlich	29 May 1865	1/367	100' Valencia
Ramon Rodriguez	to	Joseph Allen	30 May 1866	3/213	100' Valencia
Antonio M Peralta	to	Geo. Willing	20 Oct 1864	1/74	Land

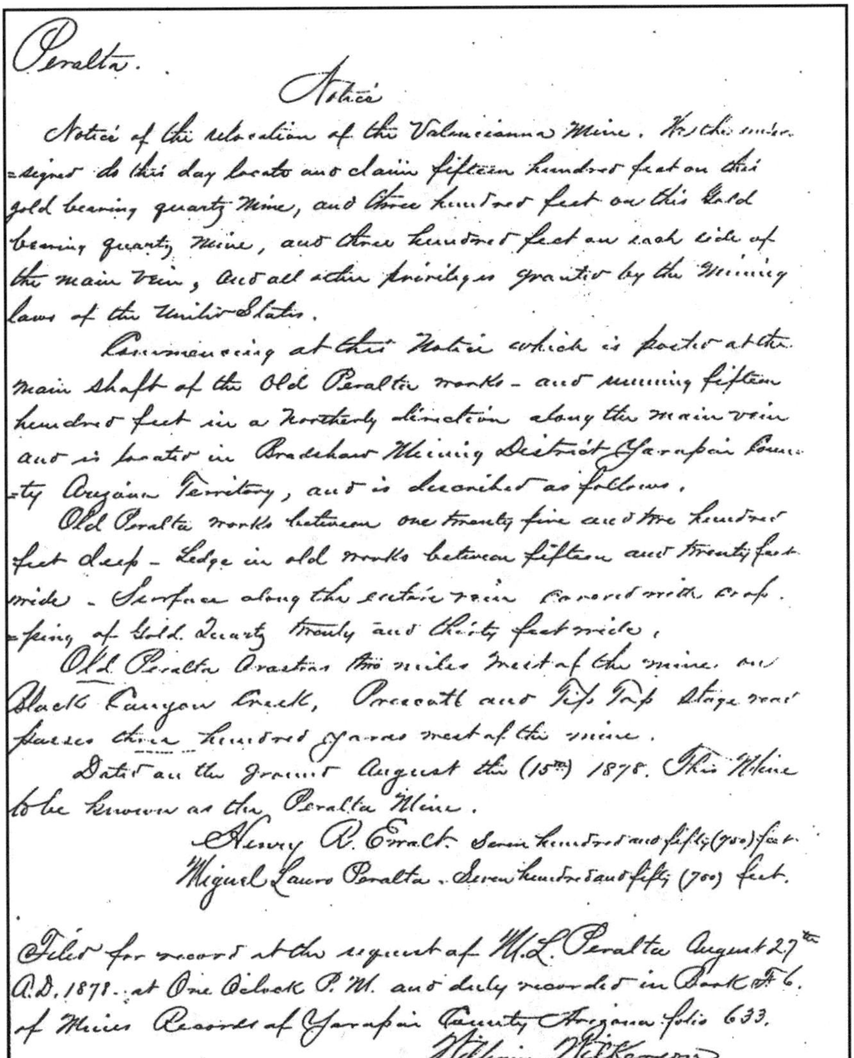

Peralta.

Notice

Notice of the relocation of the Valenciana Mine. We the under-signed do this day locate and claim fifteen hundred feet on this gold bearing quartz Mine, and three hundred feet on this Gold bearing quartz mine, and three hundred feet on each side of the main vein, and all other privileges granted by the Mining laws of the United States.

Commencing at this Notice which is posted at the main shaft of the Old Peralta works - and running fifteen hundred feet in a Northerly direction along the main vein and is located in Bradshaw Mining District Yavapai County Arizona Territory, and is described as follows.

Old Peralta works between one twenty five and two hundred feet deep - Ledge in old works between fifteen and twenty feet wide - Surface along the entire vein covered with cropping of Gold Quartz twenty and thirty feet wide.

Old Peralta Arastras two miles west of the mine on Black Canyon Creek, Prescott and Tip Top Stage road passes three hundred yards west of the mine.

Dated on the ground August the (15th) 1878. This Mine to be known as the Peralta Mine.

Henry R. Ewalt. Seven hundred and fifty (750) feet.
Miguel Lauro Peralta.. Seven hundred and fifty (750) feet.

Filed for record at the request of M.L. Peralta August 27th A.D. 1878.. at One O'clock P.M. and duly recorded in Book F 6. of Mines Records of Yavapai County Arizona folio 633.

William Wilkerson
County Recorder
by W.C. Foster
Deputy

caches. If Miguel were as honest as his reputation indicated, he would not have trespassed onto the Waltz mine but would have been as good as his word. Since his father had been killed by the Apaches, it is not hard to accept that he might have been afraid to take his sons into the Superstitions.

TOPO! map printed on 10/26/00 from "Phoenix.tpo" and "Untitled.tpg"

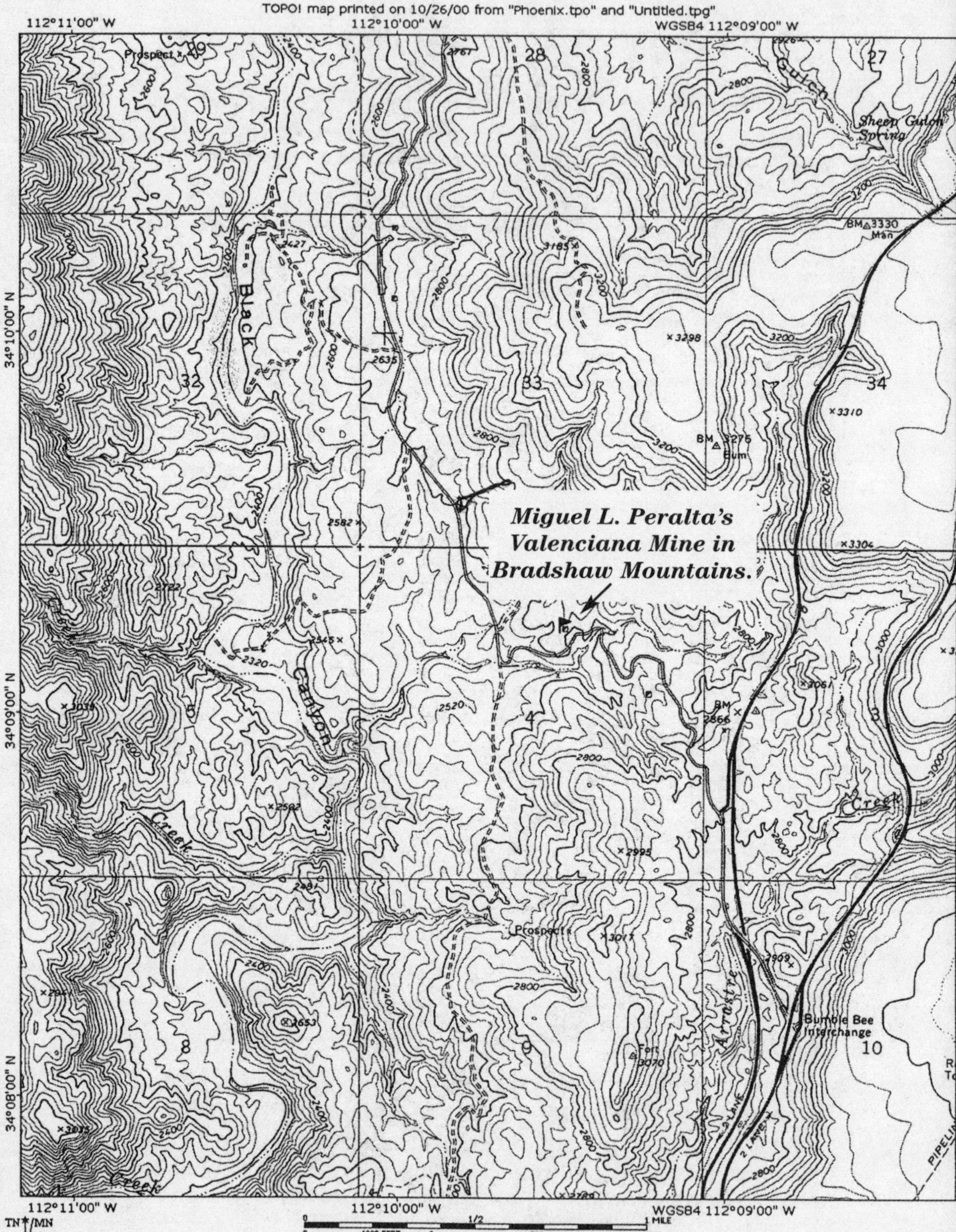

Printed from TOPO! ©1999 Wildflower Productions (www.topo.com)

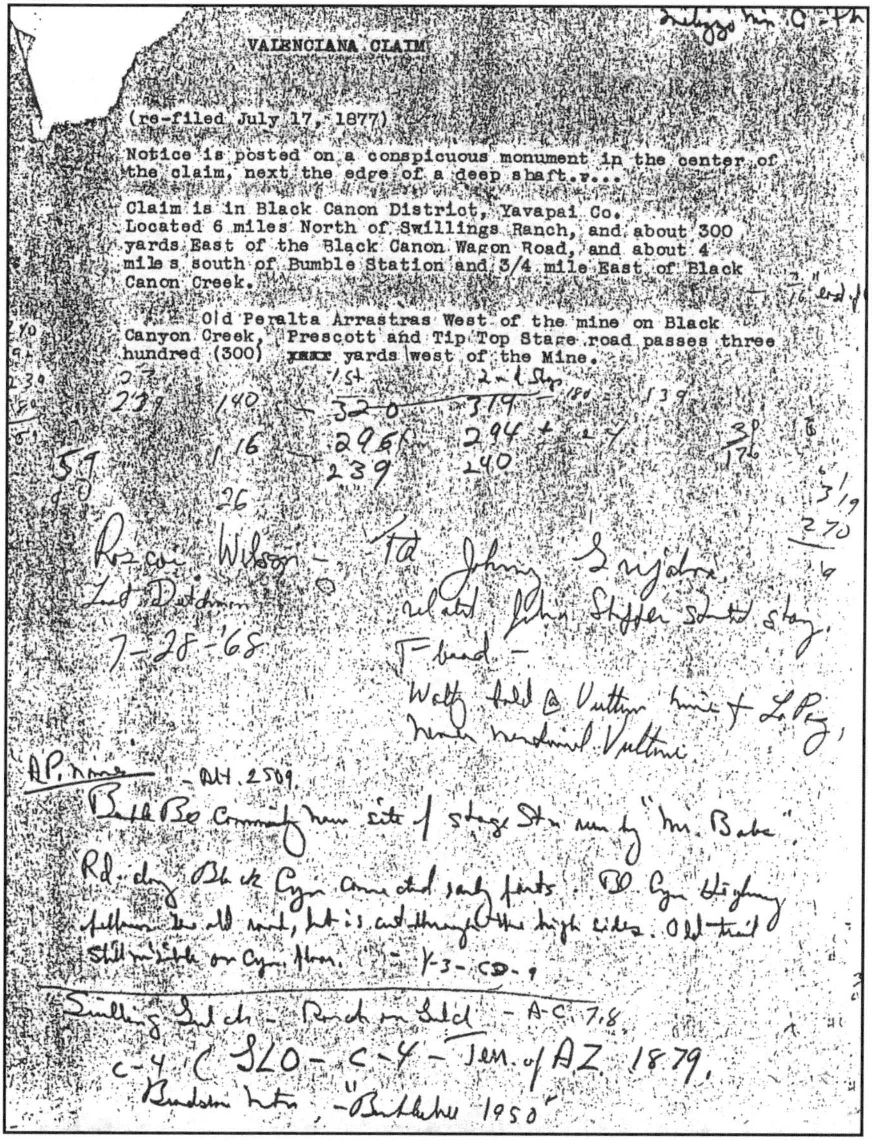

VALENCIANA CLAIM

(re-filed July 17, 1877)

Notice is posted on a conspicuous monument in the center of the claim, next the edge of a deep shaft.....

Claim is in Black Canon District, Yavapai Co.
Located 6 miles North of Swillings Ranch, and about 300 yards East of the Black Canon Wagon Road, and about 4 miles south of Bumble Station and 3/4 mile East of Black Canon Creek.

Old Peralta Arrastras West of the mine on Black Canyon Creek, Prescott and Tip Top Stage road passes three hundred (300) ~~yaxx~~ yards west of the Mine.

Peralta's Valenciana Claim (circa 1877).

In 1895 Miguel moved to Nogales. He and his sons had prospected the mountains and canyons along the Salt River and the Sierra Madre Mountains of Mexico without success. In later years, his son Raymondo said of his father, "He was searching for the gold left behind in those mountains along the Salt River." On November 6, 1897, Miguel Peralta committed suicide while sitting at his desk in Nogales, Arizona Territory. He had a pistol in one hand and a Mexican dollar in the other. He left a note that read: "I had the money; I have lost it: good bye."

The Valenciana Mine, Arizona.

We have had an opportunity of examining a report made by Henry A. Bigelow, of Prescott, Arizona, to the President of the California and Arizona Mining and Real Estate Agency, on the Valenciana mine, in Black Canyon district, Yavapai county, Arizona. We make such extracts from the report as may be of interest. The mine is about 40 miles southeast of Prescott and 60 miles north of Phœnix, near the stage road between the two towns. The vein is in the great gold-bearing slate belt which traverses the country east of Prescott and the Sierra Prieta and Bradshaw ranges of mountains, in which are many clusters of gold veins known as the Black Hill mines, the Bigbug mines, the Black Canyon (in which the Valenciana is the most conspicuous) and the New River and Cave Creek mines.

The Black Canyon mines were first discovered in 1864 by Mexicans, who were working placers in Turkey creek, on which Black Canyon is situated, and they at once built arastras and worked the mine for two years, during which time they extracted, according to Peralta's books, over $35,000 in gold.

Three times the Apaches stole from the Peralta party all their horses and mules and harassed the miners so much that they abandoned the mine, which then laid unworked till 1870, when a party of Americans worked it for several months, taking out about $11,000 by arastras until the silver excitement in the Bradshaw mountains lured them away. In the fall of 1877, J. R. Deroche sorted over the ore left on dump by previous workers as worthless and worked 10 tons which paid $46 per ton. During the present year M. L. Peralta & Co. worked three tons from the mine which paid in the Phœnix mill a trifle over $100 per ton—selected ore from the bottom of the old workings.

The ledge runs nearly north and south and crops out boldly above the surface for several hundred feet, where the principal work is done, showing at intervals for 3,000 feet. The main shaft is worked down in the old Mexican style by benches to about 90 or 100 feet, and is not timbered, not being in a safe condition for work. The main opening is made where the vein is 20 feet thick and the quartz has been taken out from wall to wall and 30 feet along the vein. South of the shaft is a good location for running a tunnel in on the vein which would underrun the old workings and admit of stoping out all the ore to the surface. The quantity of ore in sight is enormous. Second-class ore on dump is estimated at 250 tons, which will pay from $12 to $15 per ton. The quartz in chimney in shaft for about 200 feet in length on surface will mill $20. Walls of vein are soft slate. Vein runs down nearly perpendicular and shows every indication of permanency. Quartz is of a grayish white in which gold shows plainly; through richest portion are copper stains in which not so much gold is visible. A good site for a mill can be had for the choosing on Turkey creek anywhere from a mile or two from the mine, from which to the creek the hauling would be down hill.

At the mine there is no timber. All along the creek, however, is a good growth of cottonwood, willow and ash suitable for firewood. A heavy growth of pine timber crosses the Bradshaw mountains six miles west of the mine. Lumber and sawed timbers can be bought at mills in Prescott for $30 per thousand.

The Nigger, Brown, Gillespie, Swilling, Mason and other gold mines, which have been worked by arastras in years past, are near the Valenciana and could furnish ore to any mill erected on Turkey creek that would work custom ore. Mr. Bigelow, in conclusion, states that all of the facilities of access and working considered, he regards the Valenciana mine as one of the most desirable gold mines yet discovered in Yavapai county.

Article regarding the Valenciana Mine in the **Mining and Scientific Press** ***(circa 1879).***

IN NOGALES P5 col 2

The Oasis

Saturday, Nov. 6, 1897.

A Suicide.

Having some business yesterday morning at the room of Mr. Miguel Peralta, who had been the bookkeeper at his store, Mr. A. F. Paredes went to the room of the gentleman, on International street, and was shocked and horrified to find him stretched in death upon the floor; a bullet wound in his head and a pistol in his hand with a discharged cartridge told that the fatal wound had been self inflicted. The authorities were notified, and Justice Duffy impanelled a coroner's jury who took the testimony of several who knew of some facts bearing upon the case. Mr. Thomas Harrison testified that about three o'clock in the morning he took the deceased (who had been drinking and gambling) to his room. Mr. E. J. Pellegrin, in the adjoining room, heard the shot, but thought it a piece of furniture falling. Mr. S. D. Piper, who lives in the vicinity also heard the shot. The two witnesses agreed that the time was between three and four o'clock. Mr. Paredes introduced a note from the deceased which was found on the table with a Mexican dollar resting upon it which read: "I had the money; I have lost it: good bye."

Deceased was an old and well known citizen of Pima county, having resided many years in Tucson and was also at one time in business at Crittenden. He was a native of Ures, Sonora, but was raised and educated in California.

Newspaper article reporting on the suicide of Miguel Peralta (circa 1897).

Chapter 9

Jim Bark's Recollections of Those Who Searched

Preface from Jim Bark's Notes

I wish to say that the following statements are in no sense fiction, they are a collection of the stories relating to what is probably the richest deposit of gold ore ever worked on by man. I have not included many stories which have come to me from different individuals, who would not or could not offer substantiating proof.

The reader's first thought will be, "If Jim Bark knows so much about where the mine is located, why doesn't he find it himself?" I wish to say that the vicinity in which I now think the mine must be, I have not nearly exhausted; that I have hunted for the mine on and off for nearly thirty years, and I am more certain than ever that the Superstition Mountains of Arizona conceal one of the great gold mines of the world. For many years I have dreamed of coming on the hidden opening of the mine. All the accounts agree that it is so cunningly concealed that one could walk within a few yards, or even feet, and miss it.

It is possible, of course, that someone may come upon the Dutchman by accident, but I should like to think the fortunate wanderer who discovers the lost treasure will do so because of the material I am giving him in the following stories. They each contribute their part to the evidence that the mine exists, in a definite locality in the Superstition Mountains.

Old prospectors who sit on park benches in all of our Western Towns and are filled with hope, exaggeration, specimens, and nicotine, must step aside and let the younger

James E. Bark (1858-1938). Photo courtesy of the Arizona Historical Foundation.

Sims Ely at the Relief Gold Mine (circa 1908).

ARIZONA DAILY HERALD

OCTOBER 5. 1899.

P. 1 Col. 4

JIM BARK'S FIND

Old Olla Found on Superstition Filled With Aztec Arrowheads

Jim Bark has one of the finest cattle ranges in Arizona, situated over in the Superstition mountains. While spinnitticating over the mountains for several years past he has several times been rewarded by finding some excellent mineral prospects that will some day astonish the mineral world.

Jim has never been much of a hunter after ancient knowledge, believing in the old maxim that "sufficient for the day is the evil thereof," but he keeps his eyes and ears open for anything new or strange in the line of values.

Recently there was a Boston expert prospecting up about Jim's place and he was principally after evidences of ancient tribes. He regaled the boys in camp with lectures about beads, ornaments of all kinds, arrow-heads and war implements of different dead and gone tribes, and particularly about Aztecs.

Some time after the Bostonian had gone away a party of surveyors from the geological bureau chained from the highest peak of the Superstitions and they put up for a visit at the always open door of Jim Bark. Speaking of their experience on the mountain they displayed quite a numbe of arrow heads of moss-agate which they had picked up on the sides of the old peak.

After the geological survey party left Jim Bark went up that peak. It was a hard climb but he reached the top and prospected.

Yesterday, W. A. Kimball of Mesa was in Phoenix and told the story of Jim Bark's find. It seems that Kimball and some friends had learned about the arrow-heads from the geological crew, and they at once started out prepared to find the place and make a thorough search for what they felt sure would be a large quantity of these reminders of the ancient inhabitants of the mountains of superstitions.

They found the place and picked up 106 arrow-heads. Then they descended and called at Jim Bark's. Whether they had agreed to keep silent as to their find or not Mr. Kimball did not say, but he did say that he and his friends were completely astonished when Jim Bark brought out a sack and rolled out onto the bread table 887 of as perfect moss-agate Aztec arrow-heads as were ever seen in the world They handled them and counted them, and counted them and talked about them until bed time became a matter of news of the day before, and they came away impressed with the thought that Jim Bark did not need a telegraph wire nor a long distance telephone in order to keep up either as to current events or ancient values.

All of Jim Bark's find were in an old olla hidden between a crevice in the peak and it required acrobatic skill to get to it. The olla came to pieces as soon as Jim touched it, but there were arrow-heads by the hundreds and some beautiful turquoise strings of beads.

These arrow-heads are valued at $2.50 apiece in Boston and Mr. Kimball was offered $3 for one of his yesterday.

The* Arizona Daily Herald *reports on Jim Bark (circa 1899).

ARIZONA DAILY GAZETTE OCT 14, 1893 P. 5 col 2

valley.

Jim Bark came in last night from his cattle ranch, where, he says, grass is good and cattle fat. Mr. Bark has confined mining, or rather prospecting, with his many other avocations, and says he has really struck the trail of some very good quartz.

The* Arizona Daily Gazette *reports on Jim Bark (circa 1893).

generation hunt for the Lost Dutchman – chew their own tobacco, tell their own lies and buy or steal their own specimens. Hunting for the Lost Dutchman is not for old men.

Someone someday will fit the parts together more successfully than we have done. Good luck to him!

* * *

The earthquake in the late 1880s was indeed a strong one. According to Bertie Roberts, it shook the buildings in downtown Phoenix and broke some windows but mostly did only minor damage around town. A hay barn at Smith's Station (between Phoenix and Fort McDowell) collapsed, so the quake had to be fairly substantial. Fort McDowell received major damage to all its buildings. The roads to Superior, Florence and Globe were all blocked by falling rocks and took days to clear.

A lot of rock fell in Pinal County and in the Superstition Mountains. It may have covered camps and mines dug into the sides of some canyons. Since the area was so sparsely populated back then, the damage was considered light, but early ranchers knew that enormous damage had been done to the environment. Three springs on the west bank of La Barge, where it makes the bend at Marsh Valley, stopped flowing after this earthquake. Only White Rock Springs continued to flow, and according to William Edwards, it had only a fraction of its normal output. Edwards noted several other springs and seeps stopped flowing, and within a few years the trees and vegetation around those spots were gone.

DR. THORN

IN MY INTERVIEW SOME YEARS AGO WITH JOHN MONTGOMERY, BOB GROOME, JOE FUGIT, AND JOHN HEWITT, ALL OF WHOM HAD BEEN WITH THORN ON ONE OR MORE OF HIS EXPEDITIONS, HUNTING FOR WHAT IS KNOWN AS THE THORN MINE, AND WHICH I FEEL CERTAIN, IS ALSO CALLED THE LOST DUTCHMAN, THE ABOVE MENTIONED FOUR MEN PRACTICALLY AGREED UPON THE FOLLOWING FACTS TOLD THEM BY DR. THORN:

THAT THORN, WHO WAS AN ARMY SURGEON, AND THREE OTHER OFFICERS (NAMES FORGOTTEN) WERE ORDERED FROM SANTA FE, NEW MEXICO, TO REPORT AT YUMA, ARIZONA. THIS WAS ABOUT 1866 OR LATE IN 1865. THEY WERE FURNISHED AN ESCORT OF FOURTEEN CAVALRY, AND AN EXTRA MAN ON THE SEAT WITH THE DRIVER, WHO DROVE FOUR MULES. SOMEWHERE EAST OF MARICOPA WELLS THEY WERE ATTACKED BY THE APACHE INDIANS. ALL WERE KILLED EXCEPT TWO OF THE OFFICERS, THORN, AND ONE OTHER OFFICER IN THE STAGE. DR. THORN HAD HIS CASE OF INSTRUMENTS IN THE STAGE, AND WHEN THE APACHE CHIEF LOOKED IN AND SAW THEM, HE MADE IT KNOWN BY SIGNS THAT HE WANTED TO KNOW WHO THE INSTRUMENTS BELONGED TO, AND DR. THORN CLAIMED THEM. THE OLD CHIEF SIGNED FOR HIM TO GET THEM AND FOLLOW HIM, WHICH HE DID. THE CHIEF LED THORN TO A MESQUITE TREE, UNDER WHICH SAT AN APACHE WARRIOR WITH ONE ARM BADLY SHATTERED AT THE ELBOW. THE CHIEF MOTIONED THORN TO FIX IT. THORN AMPUTATED THE ARM, AND THEY ALL STARTED NORTH ON HORSEBACK. HIS FELLOW OFFICER WAS WITH THEM UNTIL AFTERNOON, THEN HE DISAPPEARED AND WAS SEEN NO MORE. THERE WAS A PARTY OF ABOUT FORTY WARRIORS; THEY TRAVELLED NORTHERLY UNTIL THEY CAME TO THE JUNCTION OF THE SALT AND VERDE RIVERS, AND THERE ON THE NORTHEAST CORNER OF THE JUNCTION WAS THE APACHE CAMP. THORN WAS GIVEN IN CHARGE OF A YOUNG APACHE WARRIOR, AND THE WARRIOR WAS GIVEN TO UNDERSTAND THAT HE WOULD BE RESPONSIBLE FOR THORN'S PRESENCE AND SAFETY. AS THORN HAD BEEN SUCCESSFULLY TREATING THE INDIANS FOR A SCALP DISEASE AND OTHER AILMENTS, HE WAS GIVEN FULL LIBERTY IN THE CAMP, BUT TO GO ANYWHERE OUTSIDE THE CAMP, HE MUST GET HIS KEEPER TO GO WITH HIM, BUT IF HE REFUSED, THEN THORN COULD NOT GO. THORN SAID HE NOTICED ON

2 - DR. THORN

SEVERAL OCCASIONS THAT A SMALL PARTY OF INDIANS WOULD BRING IN QUITE A QUANTITY OF GOLD, AND EVERY TIME THEY DID, SOON THEREAFTER, THERE WOULD BE TWO WHITE MEN WITH BURROS CAMPED UP ON A COTTONWOOD CANYON TOWARD FOUR PEAKS, SEVERAL MILES EAST OF THE INDIAN CAMP, AND THE INDIANS WOULD TAKE THEIR GOLD AND TRADE IT TO TWO WHITE MEN FOR BULLETS, CAPS, AND POWDER, JUST SO MUCH OF EACH, FOR A LIKE WEIGHT OF GOLD, AS THE WHITE MEN HAD ONLY A SET OF BALANCES. HOW THE INDIANS KNEW THAT THE WHITE MEN WERE COMING TO TRADE, HE NEVER KNEW, BUT THE INDIANS ALWAYS KNEW. THORN AND THE INDIANS WERE GETTING QUITE FRIENDLY, AS HE HAD ALREADY HANDLED AN OUTBREAK OF MEASLES; ALSO, AN EPIDEMIC OF SORE EYES, AND MANY INDIVIDUAL TROUBLES. HE PRESUMED TO ASK ONE OF THE CHIEFS WHERE THEY GOT THE GOLD. HE REPLIED BY POINTING UP SALT RIVER - "WAY UP THERE". HE ASKED THEM TO TAKE HIM WITH THEM THE NEXT TIME THEY WENT FOR GOLD, AND THEY FINALLY PROMISED HIM THEY WOULD, BUT THAT HE WOULD HAVE TO BE BLINDFOLDED (AS THAT WAS THE ONLY WAY THEY COULD OVERCOME THE BAD MEDICINE). SO ONE DAY IN AUGUST, 1865, OR 66, THEY TOLD HIM THEY WERE GOING AFTER SOME GOLD AND HE COULD GO ALONG. THAT EVENING, AFTER SUNDOWN, HE WAS TOLD TO GET ON HIS HORSE; THEY BLINDFOLDED HIM, AND EVIDENTLY CIRCLED THE CAMP SEVERAL TIMES, AS HE COULD SEE THE REFLECTION OF THE CAMP-FIRE IN HIS BLINDFOLD. THEY RODE INTO OR ACROSS THE WATER ELEVEN TIMES DURING THE NIGHT, AND WHICH AFTERWARD PROVED TO BE SALT RIVER, AND ARRIVED AT THE PLACE WHERE THEY GOT OFF THEIR HORSES ABOUT EIGHT O'CLOCK IN THE MORNING; THEY DID NOT TAKE THE BLINDFOLD OFF UNTIL THEY HAD WALKED QUITE A WAY, AND WHEN THEY REMOVED IT THE SUN WAS SHINING VERY BRIGHT. "I WAS STANDING ON THE EDGE OF A DEEP AND VERY STEEP CANYON, HIGH UP. TO MY LEFT WAS A SPUR, AND ON THE SPUR WAS A LIGHT-COLORED RUIN OF AN OLD STONE CORRAL--NOT AN INDIAN RUIN, BUT VERY OLD, AND IT LOOKED AS THOUGH IT WAS AZTEC. WE WENT CAREFULLY DOWN THE TRAIL ON THE SIDE OF THE CANYON, AND WHEN WE GOT DOWN ONTO A LITTLE FLAT THERE WAS AN INCLINE PIT; QUITE A LOT OF WORK HAD BEEN DONE, AND GOLD COULD BE SEEN LYING AROUND IN THE QUARTZ THAT HAD BEEN BROUGHT UP OUT OF THE PIT AND EVIDENTLY DISCARDED AS NOT BEING RICH ENOUGH. THERE WAS NO SCRAMBLING FOR IT; IN FACT, I WAS THE ONLY

3 - DR. THORN

ONE WHO SEEMED AT ALL EXCITED. I ASKED THE CHIEF IF I COULD TAKE SOME, AND HE SAID, "TAKE ALL YOU WANT, AS YOU WILL NEVER COME HERE AGAIN". I DID NOT KNOW JUST HOW TO TAKE HIS STATEMENT, BUT NEVERTHELESS, I GATHERED SOME OF THE BEST SPECIMENS, AND GOT SOME OF THE INDIANS TO CARRY THEM FOR ME, AS THEY DID NOT OBJECT. I BELIEVE I COULD HAVE LOADED A SIX MULE TEAM RIGHT FROM THAT DUMP. IN MY JUDGMENT, IT WAS ABOUT NINE O'CLOCK IN THE MORNING; AS I LOOKED OFF TOWARD THE SOUTH, AND ABOUT FIVE MILES AWAY (JUST A GUESS AS TO THE TIME AND DISTANCE) WAS A SHARP PEAK THAT LOOKED LIKE A MEXICAN SOMBRERO; WE STARTED BACK, BUT THEY DID NOT BLINDFOLD ME UNTIL WE GOT TO THE HORSES. IN THE MEANTIME, WE HAD WALKED THROUGH A HOLE OR CAVE, AND IT HAD A LOT OF OLD MEXICAN MINING TOOLS, SUCH AS SCREENS, ETC. HORSES COULD NOT GO THROUGH; THERE THEY AGAIN BLINDFOLDED ME, AND WE RODE UNTIL ABOUT TWO O'CLOCK WHEN I HEARD THE HORSES TO INTO WATER. I ASKED THE CHIEF TO STOP AND LET ME GET A DRINK, AND HE SAID THEY WOULD CROSS OVER AND ALL STOP AND GET A DRINK. THEY TOOK THE BLINDFOLD OFF, AND I GOT DOWN AND TOOK A DRINK. AS I LOOKED UP, I COULD SEE THE SAME SOMBRERO PEAK THAT I SAW FROM THE MINE, GUESSING IT TO BE ABOUT FIFTEEN MILES TO THE SOUTH. WE RESUMED OUR JOURNEY, ARRIVING AT CAMP AT ABOUT THE SAME TIME WE LEFT IT THE NIGHT BEFORE. I PROCEEDED TO COLLECT MY GOLD, AND GUESSED I HAD ABOUT FIFTEEN HUNDRED DOLLARS. I BURIED IT UNDER A MESQUITE TREE RIGHT IN CAMP, AND NO ONE BOTHERED IT. SOMETIME AFTER, AN INDIAN CAME GALLOPING INTO CAMP, HIS HORSE ALL WHITE WITH SWEAT, AND SAID THE SOLDIERS WERE COMING. THERE WAS SHOUTING, DOGS BARKING AND CRYING CHILDREN. WICKUPS WERE TORN DOWN AND HITCHED TO PONIES, AND HERE CAME THE COMPANY OF SOLDIERS -- NOT AFTER THE INDIANS, BUT TO SELECT A LOCATION FOR A POST TO BE NAMED FORT MCDOWELL, IN THE EXCITEMENT, THE INDIANS SEEMED TO HAVE FORGOTTEN ALL ABOUT ME, SO I WENT AND DUG UP MY GOLD AND JOINED THE SOLDIERS. I DID NOT HAVE MUCH OF A JOIN, AS THEY SELECTED A SPOT JUST ADJOINING WHERE THE INDIANS HAD CAMPED, AND CALLED IT CAMP MCDOWELL. THE SOLDIERS CAMPED THERE ABOUT A YEAR, THEN MOVED UP THE VERDE RIVER ABOUT EIGHT MILES, AND ACTUALLY ESTABLISHED FORT MCDOWELL, OCCUPIED FOR MANY YEARS BY THE

4 - DR. THORN

GOVERNMENT TO HOUSE THEIR SOLDIERS WHILE DRIVING THE INDIANS ON TO THEIR RESERVATIONS. FORT MCDOWELL A FEW YEARS AGO WAS ABANDONED, AND THEN MADE INTO AN INDIAN RESERVATION, AS THERE WAS ALREADY A DITCH TAKEN OUT OF THE VERDE RIVER. THE GOVERNMENT MOVED QUITE A NUMBER OF THE OLDEST APACHES FROM WHITE MOUNTAIN AND SAN CARLOS, AND GAVE THEM EACH A FEW ACRES TO CULTIVATE, AS THE CLIMATE IS MUCH MILDER AT MCDOWELL THAN UP ON THE OTHER RESERVATIONS, AND THE INDIANS APPEAR QUITE CONTENTED.

(ON WITH THE STORY): I SPOKE TO THE COMMANDING OFFICER, AND TOLD HIM IF HE WOULD GIVE ME A SMALL DETAIL OF SOLDIERS, THAT I WOULD BE GONE BUT A SHORT TIME AND WOULD RETURN JUST AS SOON AS I FINISHED LOCATING. HE TOLD ME THAT HE COULD NOT DO THAT, BUT HINTED THAT HE WOULD AFFORD ME ANY PROTECTION THAT WAS WITHIN HIS POWER, INDIRECTLY.

I THEN WENT TO PRESCOTT, AND THERE, AFTER TELLING MY STORY AND SHOWING MY SPECIMENS, I ORGANIZED A COMPANY OF PROSPECTORS AND MINERS, SOLD MY GOLD AND BLEW IT ON MY ORGANIZATION. WE WENT WELL ARMED AND FULLY EQUIPPED TO TAKE POSSESSION OF THE MINE. AT THAT TIME, I HAD NO DOUBT WHATEVER BUT WHAT I COULD GO RIGHT TO IT. MY GREATEST WORRY WAS HOW TO RESTRAIN MY FOLLOWERS FROM PLACING TOO MUCH GOLD UPON THE MARKET AT ONE TIME. WELL, WE STARTED AND WENT TO SOMBRERO BUTTE AND LOOKED AROUND THERE, AND THERE WAS CONSIDERABLE MUTTERING AMONG MY FOLLOWERS AS THE FORMATION ALL AROUND (AS FAR AS ONE COULD SEE) WAS TUFA, AND NON-MINERAL BEARING. THE SOLDIERS WERE GIVING US ALL THE PROTECTION, INDIRECTLY, POSSIBLE. IF IT HAD NOT BEEN FOR THE GOLD I HAD SHOWN IN PRESCOTT, I BELIEVE I WOULD HAVE BEEN IN A HAZARDOUS POSITION. GRUB WAS RUNNING LOW, HORSES WERE TIRED, MEN WERE DISGUSTED, AND SO WE VOTED TO RETURN TO PRESCOTT.

I IMMEDIATELY BEGAN TO ORGANIZE ANOTHER EXPIDITION; A GREAT MANY OF MY FORMER FOLLOWERS WERE TELLING ALL KINDS OF STORIES ABOUT ME, ABOUT WHAT A BUNCO ARTIST I WAS; THAT I JUST WANTED TO GO PROSPECTING IN THE APACHE COUNTRY, AND WAS AFRAID TO GO ALONE (INTIMATING THAT THEY WERE NOT), AND AFTER A GOOD WHOLESOME

5 - DR THORN

DRUNK, A GREAT MANY OF THEM JOINED MY SECOND EXPIDITION. WE STARTED SOUTH, AND DID NOT TAKE IN THE OLD TUFA COUNTRY, BUT HUNTED FOR ANOTHER SOMBRERO BUTTE. WE FOUND ONE THAT WOULD DO, BY A SLIGHT STRETCH OF THE IMAGINATION, UP ON CHERRY CREEK, BUT COULD FIND NO MINE, NOR A COUNTRY THAT LOOKED LIKE WHERE THE MINE WAS LOCATED, SO WE GAVE UP AGAIN AND RETURNED TO PRESCOTT. I HAD WRITTEN MY BROTHER, WHO WAS APHYSICIAN IN ILLINOIS, SHORTLY AFTER MY FIRST ARRIVAL IN PRESCOTT, MENTIONING MY POSSESSION OF THE GOLD AND HOW I CAME BY IT. I WROTE HIM AGAIN AT THE END OF MY SECOND TRIP AND TOLD HIM THAT I HAD ABOUT EXHAUSTED MY FUNDS; THAT I HAD BEEN OUT TWICE, GIVEN A THOROUGH SEARCH EACH TIME, AND THAT I HAD CONCLUDED TO GIVE IT UP. HE WROTE ME TO WAIT AROUND PRESCOTT, UNTIL I COULD HEAR FROM HIM AGAIN (WHICH WOULD BE VERY SHORTLY). HE SOON AFTER SENT ME A LETTER CONTAINING FIVE THOUSAND DOLLARS TO BE EXPENDED IN THE FURTHER SEARCH FOR THE MINE; THAT HE HAD RAISED THE MONEY AMONG HIS DOCTOR FRIENDS, AND FOR ME TO GO TO IT. I IMMEDIATELY WROTE HIM THAT I WOULD HOLD THE MONEY INTACT; THAT I WAS NOT AT ALL CERTAIN THAT I COULD FIND IT(IN FACT, DID NOT THINK I COULD); THAT IT TOOK QUITE A STRONG AND WELL-EQUIPPED PARTY, ON ACCOUNT OF INDIANS, AND, IN GENERAL, I WAS DISCOURAGED, AND THAT I WOULD BE GLAD TO RETURN THE MONEY. HE WOULD NOT HEAR OF IT, AND INSISTED THAT I GO AHEAD, WHICH I DID, AND MADE TRIP AFTER TRIP, UNTIL FINALLY I HAD ONLY ELEVEN FOLLOWERS WITH ME. GLOBE HAD JUST BEEN DISCOVERED, AND WE STARTED FROM THERE, TOWARD SALT RIVER. IT HAD BEEN RAINING AND THE GROUND WAS VERY SOFT, AND WHEN WE REACHED SALT RIVER, IT WAS OUT OF ITS BANKS AND COULD NOT BE CROSSED, SO WE DISBANDED, THREW UP OUR HANDS, AS IT WERE, AND QUIT; ALL I CAN SAY IS, THAT IT WAS THERE AND I KNOW IT STILL IS.

JOHN HEWITT, A NEIGHBOR CATTLEMAN, TOLD ME THAT HE HAD BEEN ONE OF DR. THORN'S EXPEDITION MEMBERS, AND GOT QUITE WELL ACQUAINTED WITH HIM; THAT IN HIS OPINION, THERE WAS NO QUESTION ABOUT THE TRUTH OF THE STORY, AND THAT HE, HEWITT, RECEIVED A LETTER FROM THORN SEVERAL YEARS BEFORE (THAT WAS AT LEAST TWENTY YEARS AGO). I ASKED HEWITT IF HE WOULD WRITE THORN AND REQUEST HIM TO

6 - DR. THORN

COME OUT TO HIS PLACE FOR A VISIT, AS HE, HEWITT, HAD A FRIEND, A NEIGHBORING CATTLEMAN, WHO WAS LOOKING FOR WHAT HE THOUGHT WAS THE SAME MINE, FROM AN ENTIRELY DIFFERENT SOURCE OF INFORMATION, AND HE WAS HAVING NO BETTER SUCCESS THAN THORN HAD; THAT IF HE WOULD COME, ALL HIS EXPENSES WOULD GLADLY BE PAID. THAT HIS FRIEND WAS VERY ANXIOUS TO HAVE HIM IDENTIFY SOME OF THE LAND MARKS THAT HE (THE CATTLEMAN) HAD GAINED FROM THE OTHER SOURCES. HE REPLIED TO HEWITT THAT HE WAS OVER EIGHTY YEARS OF AGE, ALMOST BLIND, HAD A LITTLE DRUG STORE AND POST OFFICE IN A LITTLE TOWN ON THE RIO GRANDE RIVER IN NEW MEXICO; THAT HE WAS JUST MAKING A LIVING, BUT WAS TOO FEEBLE TO DO ANY MORE TRAVELING; THAT HE HAD GIVEN THE BEST YEARS OF HIS LIFE HUNTING FOR THE MINE, HAD RUN HIS BROTHER AND HIS FRIENDS IN DEBT IN SEEKING THE MINE; THAT HE HAD ONLY A SHORT TIME TO LIVE AND WISHED TO LIVE THAT LITTLE TIME IN PEACE; AND THAT IF HEWITT AND HIS FRIENDS HAD INFORMATION FROM ANOWHER SOURCE, HE THOUGHT THAT BETWEEN THE TWO, THEY SHOULD FIND IT, AND THAT HE CERTAINLY HOPED THEY WOULD.

TWO SOLDIERS

SOME TIME IN THE EARLY DAYS OF SILVER KING, THERE WERE TWO SOLDIERS DISCHARGED AT MCDOWELL. THEY REFUSED TRANSPORTATION BACK EAST TO THEIR PLACE OF ENLISTMENT AS THEY WANTED TO STAY OUT WEST AND GO TO WORK IN THE MINES. THE WAGES WERE MUCH HIGHER THAN THEY COULD GET IN THE EAST FOR ANY WORK, SO THEY STRUCK OUT FOR THE KING AFOOT. WHEN THEY GOT TO PINAL THEY WENT INTO THE SUPERINTENDENT'S OFFICE AND APPLIED FOR WORK. ARON MASON WAS SUPERINTENDENT. HE ASKED IF THEY WERE MINERS. THEY TOLD HIM THEY WERE NOT. "WELL," HE SAID, "I SUPPOSE YOU CAN MUCK?" AND THEY WANTED TO KNOW WHAT THAT WAS. HE TOLD THEM IT WAS SHOVELING ORE, FILLING BUCKETS, AND ANY KIND OF WORK AROUND A MINE EXCEPT ACTIVE MINING. THEY SAID THAT WAS WHAT THEY EXPECTED TO DO, AT LEAST FOR A WHILE. MASON TOLD THEM TO GO UP TO THE KING AND REPORT TO THE FOREMAN, BOB BOWEN, AND HE WOULD PUT THEM TO WORK. ONE OF THEM PUT HIS HAND IN HIS COAT POCKET, PULLED OUT A HANDFUL OF NUGGETS AND ASKED MASON WHAT THAT WAS. HE TOLD THEM IT WAS GOLD, AND THEY SAID, "OH, YOU ARE FOOLING". MASON TOLD THEM THAT THEY COULD TRADE IT IN AT THE COMPANY STORE UP AT THE KING FOR SIXTY DOLLARS AN OUNCE FOR ANYTHING IN THE STORE. SO THEY WENT UP, AND SURE ENOUGH, THEY COULD CASH IT IN FOR ANYTHING THE COMPANY HAD. THEY BOUGHT TWO BURROS AND PACK SADDLES, BEDDING, GRUB, ETC. TO THE TUNE OF EIGHTY-FOUR DOLLARS; PACKED UP AND STARTED BACK FOR THEIR MINE. IN PASSING THROUGH PINAL, THEY STOPPED AT THE SUPERINTENDENT'S OFFICE AND ASKED MASON IF HE WOULD KEEP THIS GOLD THAT THEY HAD WRAPPED IN A RAG FOR THEM UNTIL THEIR RETURN. MASON WEIGHED IT, PLACED IT IN A FRUIT JAR, PUT A SLIP OF PAPER WITH THEIR TWO NAMES AND THE AMOUNT, EIGHT HUNDRED AND SOME DOLLARS, AND PUT IT IN THE SAFE. (MY AUTHORITY FOR THIS WAS WILEY HOLMAN, WHO SOMETIMES LIVED AT PINAL, FOLLOWING MINING, AND THEN HE WOULD CHANGE AND GO TO COW PUNCHING. HE WOULD CONSIDER AS TRUTHFUL AS A COMBINED MINER AND COW-PUNCHER COULD BE). WILEY TOLD ME THAT HE HAD SEEN THE GOLD AND HAD DISCUSSED IT WITH MASON ON DIFFERENT OCCASIONS. YEARS AFTER, HE HAD WRITTEN MRS. MASON IN NEW YORK IN REGARD TO THE GOLD, AND SHE REPLIED

2 - TWO SOLDIERS

THAT SHE HAD SUCH A JAR WITH THE TWO NAMES IN IT; THAT A MINING FRIEND HAD TOLD HER THERE WAS ABOUT THREE HUNDRED DOLLARS WORTH OF GOLD IN THE JAR, AND THAT IT WAS SUBJECT TO THE OWNERS CLAIM. MASON ASKED THE TWO SOLDIERS IF THEY DIDN'T WANT A PARTNER, AND THEY THEN TOLD HIM THE FOLLOWING: THAT THEY WERE DISCHARGED SOLDIERS FROM MCDOWELL AND HAD DECIDED TO COME OVER TO THE KING TO SEE IF THEY COULD GET WORK IN THE MINE. SO THEY STRUCK OUT TOWARD THE KING, CROSSED SALT RIVER, STRUCK A TRAIL WHICH THEY HAD BEEN TOLD WAS THE PROPER ONE TO TAKE, AND WHICH IS NOW CALLED THE APACHE TRAIL, BUT IT IS NOW AN AUTOMOBILE ROAD. THEY FOLLOWED IT FOR SEVERAL MILES TO A CREEK CROSSING, WHERE THERE WAS WATER. THE TRAIL AFTER THAT APPEARED TO RUN NEARLY NORTH, AND THE KING WAS NEARLY SOUTH, SO WHILE THEY FELT CERTAIN THAT THE TRAIL WOULD EVENTUALLY LAND THEM AT THE KING, IT MUST BE A LONG WAY AROUND AND THEY WERE TIRED. THEY DECIDED TO MAKE A SHORT CUT, WENT UP THIS CREEK FOR SOME DISTANCE, CAME TO A WATERFALL AND COULD GO NO FURTHER. THEY CAME BACK DOWN THE CREEK, AND FINALLY GOT OUT ON THE SIDE OF THE CREEK TOWARD THE KING, AND UP ON A VERY ROUGH AND HIGH MOUNTAIN. THERE WAS NO TRAIL. THEY STRUCK OUT, ALWAYS TRYING TO WORK TOWARD THEIR DESTINATION, BUT MAKING VERY SLOW PROGRESS. THEY RAN ON TO A TRAIL, AND SUCH A QUEER PLACE FOR A TRAIL; THEY CONCLUDED TO FOLLOW IT AND SEE IF IT WOULDN'T LEAD THEM OUT OF THAT GOD-FORSAKEN COUNTRY. THEY FOLLOWED IT BUT A SHORT DISTANCE AND WERE IN HIGH HOPES WHEN THE TRAIL LED THEM THROUGH A CAVE BETWEEN THE PEAKS. IT MUST HAVE BEEN A FOOT TRAIL, AS ANIMALS COULD NOT GO THROUGH THE CAVE. THEY WENT ON A LITTLE FURTHER, AND CAME TO A TUNNEL THAT HAD BEEN WALLED UP, WITH WORKINGS ABOVE AND OVER. THEY SAID THAT THEY DID NOT BELIEVE WHAT THEY SAW WAS GOLD, AS THERE WAS SO MUCH OF IT, THEY SAID THEY CERTAINLY COULD LOAD THEIR BURROS DOWN TO THE WATER LINE. THEY TOLD MASON THAT THEY HAD NEVER TAKEN OUT THEIR NATURALIZATION PAPERS AND IF THEY WOULD GO OVER TO TUCSON WITH THEM AND HELP THEM GET OUT THEIR PAPERS, UPON THEIR RETURN, WHICH WOULD NOT BE MORE THAN TEN DAYS, HE COULD GO WITH THEM TO THE MINES: THAT THEY WOULD TAKE JUST ONE CLAIM AND HE COULD LOCATE ALL THE BALANCE. MASON SAID ALL RIGHT, THAT HE WOULD LOOK

3 - TWO SOLDIERS

FOR THEIR RETURN. UP AT THE KING WAS A FELLOW WHO HAD A CROOKED FOOT AND WAS KNOWN AS "CLUB FOOT", AND EVERYONE SPOKE OF HIM AS CLUB-FOOT. HE WAS A CASE KEEPER, HANGER-ON AND GENERAL ROUSTABOUT FOR THE GAMBLING GAMES. WHENEVER MEAL TIME CAME, THE DEALER AT WHATEVER GAME CLUB-FOOT HAPPENED TO BE WORKING WOULD REACH IN THE DRAWER AND TOSS HIM TWO OR FOUR BITS, ACCORDING TO THE FEELINGS OF THE DEALER, AND TELL HIM TO GO GET SOMETHING TO EAT. IN EACH SALOON IN THE MINING CAMPS WAS A LUNCH COUNTER, OWNED AND PRESIDED OVER BY A CHINAMAN, AND THE FOOD SERVED WAS VERY MUCH SUPERIOR TO THAT SERVED AT THE LUNCH COUNTERS IN THE MINING TOWNS OF TODAY. IT WAS CONSIDERED AN AWFUL THING FOR ANY OF THE EMPLOYEES OF ONE SALOON TO EAT EVEN ONCE AT THE LUNCH COUNTER OF ANOTHER. THERE WAS VERY SELDOM A REASON FOR SO DOING AND IF IT WAS REPEATED, THAT CHINAMAN SOON FOUND A BUYER AND THERE WAS A CHANGE, OR ELSE THE EMPLOYEE FOUND A JOB SOMEWHERE ELSE. THE CHINAMAN WAS THE SOURCE WHERE THE EMPLOYEES OF THE HOUSE WOULD GO WHEN THEY WERE IN NEED OF A FEW DOLLARS, BUT HE WOULD NOT LOAN A MAN THE SECOND TIME IF HE BORROWED THE MONEY TO GAMBLE. HE ALWAYS FIGURED THAT HE WAS CAPABLE OF GAMBLING HIS OWN MONEY, AND IN REALITY HE WAS. IF HE DID NOT APPROVE OF A MAN'S CONDUCT, HE WOULD BE BRUTALLY FRANK IN HIS METHOD OF COLLECTION. HE WOULD WATCH HIS OPPORTUNITY AND GET HIS DEBTOR ALONE, AND SAY "WHEN YOU PAY?" "I AM GOING TO WORK AT SUCH AND SUCH A GAME MONDAY JIM (ALL CHINAMEN WERE NAMED CHARLEY OR JIM) THEN I PAY YOU". "YOU ALL TIME SAY THAT, NO GOOD." THAT GAMBLER KNEW HE HAD BETTER BUCK UP AND CUT OUT HIS DRINKING AS HE PROBABLY OWED QUITE A GRUB BILL, AS WELL AS CASH, BUT ALL GAMBLERS WOULD RATHER BORROW OF CHARLEY OR JIM, AS THEY DID NOT CONSIDER THEY WERE UNDER ANY OBLIGATIONS TO THE CHINAMAN EXCEPT TO PAY IT BACK, AND THEY DID NOT HAVE TO HUMILIATE THEMSELVES BY ASKING A FELLOW GAMBLER FOR A LOAN, THEREBY DISCLOSING THEIR FINANCIAL DISTRESS. A CHINAMAN HELD TRANSACTIONS OF THE KIND AS SACRED. THE LINES OF SOCIETY WERE DRAWN JUST AS TIGHTLY AMONGST THE GAMBLERS AND THEIR FAMILIES AS THEY EVER WERE OR CAN BE IN OUR TOWNS OF TODAY.

Muckers with dynamite caps. Muckers gathered the loosened rock after the miners had picked and broken it.

200 feet underground – mining gold.

4 - TWO SOLDIERS

TO CONTINUE: THE TWO EX-SOLDIERS THEN STRUCK OUT FOR THE MINE, AND AGAIN TOLD MASON THAT THEY WOULD NOT BE GONE OVER TEN DAYS. SOON THEREAFTER CLUB-FOOT DISAPPEARED FROM THE KING, AND NO ONE KNEW WHERE HE HAD GONE OR WHY. TEN DAYS WENT BY, AND NO SOLDIERS; FIFTEEN AND NO SOLDIERS; THEN CLUB-FOOT SHOWED UP. EVIDENTLY HE HAD PLENTY OF MONEY, AS HE WAS BETTING TWENTY-DOLLAR PIECES AT THE DIFFERENT GAMES. THE MINERS AND GAMBLERS BECAME VERY SUSPICIOUS, AND THEY GOT TOGETHER, HELD A SECRET MEETING AND AGREED THAT AT THE NEXT SHIFT CHANGE THEY WOULD HANG CLUB-FOOT AND MAKE HIM TELL WHERE HE GOT HIS GOLD. SOMEHOW, IT GOT TO CLUB-FOOT'S EARS AND HE DISAPPEARED. JUST THEN IT REACHED THE KING THAT THE BODY OF ONE OF THE SOLDIERS HAD BEEN FOUND OVER IN THE MOUNTAINS, LYING ALONG SIDE OF THE TRAIL, BY SOME CATTLEMEN. THEIR MANNER OF IDENTIFICATION WAS HIS HAT, WHICH THEY SENT TO THE SUTTLER AT MCDOWELL; HE SAID IT HAD BEEN PURCHASED BY ONE OF THE DISCHARGED SOLDIERS WHO STARTED AFOOT FOR THE SILVER KING MINE ON A CERTAIN DATE, GIVING HIS NAME, WHICH WAS CORRECT. HE HAD A BULLET HOLE IN HIS BACK; HAD BEEN DEAD LONG ENOUGH SO THAT THE BODY WAS PRETTY WELL DECOMPOSED. OLD MAN WHITLOW, WHO WAS ONE OF THE MEN WHO FOUND THE BODY, SAID THAT THERE WAS A SHOD HORSE TRACK ON THE TRAIL LEADING OVER THE MOUNTAIN INTO THE SUPERSTITION. HE SAID THEY JUST DUG A HOLE, ROLLED THE BODY INTO IT, COVERED HIM WITH WHAT LITTLE DIRT THEY HAD, PILED ROCKS ON THE GRAVE AND LET IT GO AT THAT. THIS GRAVE WAS SHOWN ME BY ONE OF THE WHITLOW BOYS. IT WAS ABOUT A QUARTER OF A MILE ABOVE MY RANCH HOUSE. HUSE WARD AND I DUG HIM UP, AND FOUND NOTHING BUT BONES AND BUTTONS, SO WE RE-BURIED HIM, AND COVERED THE GRAVE WITH ROCK. THE GRAVE IS STILL THERE AND UNDISTURBED. BILL KIMBALL OF MESA, SAID THAT AT ABOUT THAT TIME HE SAW THE BODY OF A MAN (NAKED) LYING NEAR BLUFF SPRING, SO HE PULLED OUT FOR MESA IMMEDIATELY. WHETHER THIS WAS THE OTHER SOLDIER IS ONLY CONJECTURE, BUT I AM INCLINED TO THINK THAT IT WAS, AS THERE WERE VERY FEW WHITE MEN TRAVELING IN THE SUPERSTITION MOUNTAINS AT THAT TIME. ONE THING CERTAIN, NEITHER OF THE SOLDIERS WAS EVER SEEN AFTERWARD.

BLANK

NOT MANY YEARS AGO, I ONE DAY MET A MAN IN PHOENIX, BY THE NAME OF BLANK, WHO REMARKED TO ME: "DO YOU KNOW, MR. BARK, THAT THERE OUGHT TO BE A RICH GOLD MINE UP NEAR YOUR COW RANCH". OF COURSE, I ASKED HIM WHAT MADE HIM THINK SO. HE THEN PROCEEDED TO TELL ME THE FOLLOWING STORY: THAT IN 1905 AND 1906 HE WAS FUEL AGENT FOR THE N.C. CO., OPERATING A LINE OF STEAMBOATS ON THE TANANA RIVER IN ALASKA. ONE DAY, A MAN CAME TO HIM SAYING THAT HE WAS AN OLD ARIZONA MAN, WAS BROKE AND WANTED TO GO UP TO FAIRBANKS. BLANK TOOK HIM TO THE MATE AND ASKED HIM TO PUT HIM ON AS A DECK HAND, WHICH HE DID. BLANK HELPED HIM SEVERAL TIMES TO GO UP THE RIVER. (ONE COULD MAKE A RAFT AND COME DOWN THE RIVER ALL RIGHT, BUT GOING UP WAS SOMETHING ELSE). IN SEPTEMBER 1906, HE WANTED TO GO UP AGAIN AND WAS TOLD THAT HE HAD BETTER MAKE HIMSELF A GOOD DECK HAND, AS BLANK WAS GOING TO THE STATES AND WAS NOT COMING BACK TO ALASKA. HE WAS A VERY GOOD WORKER, THOUGH LAME IN ONE FOOT. HE APPEARED TO BE VERY GRATEFUL FOR WHAT BLANK HAD DONE FOR HIM, AND SAID, "IF I TOLD YOU WHERE THERE WAS A RICH GOLD MINE IN ARIZONA, WOULD YOU GIVE ME PART OF IT?" TOLD HIM "YES, BUT WHY ARE YOU UP HERE HUNTING A MINE IF YOU KNOW OF ONE IN ARIZONA"? HE REPLIED THAT HE WAS NEVER GOING BACK TO ARIZONA. THAT HE WAS GOING TO STAY IN ALASKA UNTIL HE DIED. HE DREW A MAP FROM MEMORY IN A BOOK BELONGING TO BLANK AND TOLD HIM THAT WHEN THEY WERE NEAR MCKINLEY CROSSING THAT HE WOULD SHOW HIM A MOUNTAIN THAT LOOKED LIKE THE ONE IN ARIZONA THAT THE MINE IS ON. THE MOUNTAIN HE POINTED OUT RESEMBLED AN OXBOW IN SHAPE WITH VERY PERCIPITOUS CLIFFS, EACH CLIFF SITTING BACK OF THE ONE IT WAS RESTING ON, LIKE A GIANT STAIRWAY. IT COULD NOT BE CLIMBED FROM THE NORTH, EAST OR SOUTH SIDES BUT WAS QUITE ACCESSIBLE FROM THE WEST SIDE. HE ALSO TOLD HIM THAT WHEN THE STEAMER GOT TO FAIRBANKS HE WOULD GO TO HIS CABIN AND GET A MAP SHOWING THE TRAIL FROM THE FOOT OF THE MOUNTAIN TO THE MINE. WHEN HE GOT TO FAIRBANKS, HE BROUGHT ME A MAP, OLD AND NEARLY WORN OUT. BLANK COPIED IT, AND IT WAS THEN TORN UP. THE MAP HE DREW IN BLANK'S BOOK HAD ITS STARTING POINT FROM THE SILVER KING MINE, THEN IN A WESTERNLY DIRECTION (PASSING CERTAIN LAND MARKS) FIFTEEN MILES TO THE MINE: HE SAID IT CERTAINLY WAS VERY DIFFICULT TO FIND.

2 * BLANK

BLANK REMARKED, "IT PROBABLY HAS BEEN FOUND AND WORKED OUT BY THIS TIME". HE SAID, "DON'T FOOL YOURSELF. I AM CERTAIN THAT IT HAS NOT BEEN FOUND, AS I SURELY WOULD HAVE HEARD OF IT AND SO WOULD YOU." NOW, AT THAT TIME, BLANK HAD NOT HEARD OF ANY TWO SOLDIERS OR ANY CLUBFOOT,--IN FACT, HE NEVER GAVE IT A THOUGHT UNTIL HE HEARD THAT I HAD BEEN LOOKING FOR A LOST MINE UP IN THAT COUNTRY. AS HE TOLD ME THE FOREGOING STORY, I IMMEDIATELY CAME TO THE CONCLUSION THAT HIS FRIEND IN ALASKA WAS CLUB-FOOT, AND THEN I TOLD BLANK THE TWO SOLDIER STORY. BLANK LET OUT ANOTHER LINK BY SAYING, "HE GAVE ME THE ADDRESS OF HIS BROTHER IN THE STATE OF WASHINGTON, WHO WAS POSTMASTER IN A SMALL TOWN", AND ASKED ME TO SEND HIS SHARE TO THAT ADDRESS. I WAS CHANGING MY PLACE OF BUSINESS FROM PHOENIX TO MAYER IN YAVAPAI COUNTY, AND REQUESTED MR. SIMS ELY, SR., TO WRITE TO THE ADDRESS IN WASHINGTON, WHICH HE AND BLANK DID, AND THEY RECEIVED A SHORT AND RATHER CURT REPLY, SAYING THAT HIS BROTHER HAD BEEN DEAD FOR FIVE YEARS (WHICH BLANK KNEW WAS NOT TRUE), AND WROTE AGAIN TO WASHINGTON TO THAT EFFECT. THEY THEN RECEIVED A MORE COURTEOUS REPLY, SAYING THAT HE HAD BEEN MISINFORMED AND WISHED TO APOLOGIZE; THAT HE HAD WHAT HE SUPPOSED WAS RELIABLE INFORMATION, THAT HIS BROTHER HAD DIED ABOUT A YEAR AGO. BLANK THEN BECAME A CONFIRMED SEEKER OF THE LOST DUTCHMAN MINE, BUT ONLY FROM THE INFORMATION HE OBTAINED IN ALASKA. HE HAD MADE SEVERAL TRIPS UP THERE, AND STILL CONTENDS THAT WHERE I AM LOOKING IN THE MAIN RANGE OF THE SUPERSTITION MOUNTAINS IT IS AT LEAST TWENTY MILES FROM THE SILVER KING MINE; AND HIS ALASKA FRIEND SAID THAT IT WAS ABOUT FIFTEEN MILES IN A WESTERNLY DIRECTION.(SEE MAP). I FIRMLY BELIEVE THAT I HAVE LOCATED THE MOUNTAIN; ALSO HIS ZIGZAG TRAIL. I HAVE FOLLOWED IT TO THE TOP OF THE MOUNTAIN, AND THERE THE MOUNTAIN IS FLAT AND SHOWS NO TRAIL. IF ONE COULD FOLLOW THAT TRAIL, I BELIEVE IT WOULD LEAD TO EITHER THE PERALTA CAMP OR THE MINE, POSSIBLY BOTH. BLANK HAS NEVER TOLD ME HOW TO FIND THE MINE AFTER REACHING THE TOP OF THE MOUNTAIN, OR HAVE I SEEN HIS MAP. WHETHER IT WOULD BE ANY MORE EXPLICIT AS TO DIRECTIONS THAN THE MAPS I HAVE, I DO NOT KNOW, BUT HE WILL NOT LOOK BEYOND THE FIFTEEN-MILE LIMIT.

JOE DEERING

ONE DAY IN THE LATE 80'S THERE CAME TO THE RANCH A MAN, AND CAMPED DOWN AT THE WELL. HE HAD A BURRO AND A PUP. THAT EVENING, HE CAME UP TO THE HOUSE, AND WE WERE TALKING ON GENERAL SUBJECTS, WHEN HE SAID THAT THE KING HAD SHUT DOWN AND THAT HE DID NOT WANT TO GO TO TOWN. HE WAS SPEAKING OF THE SILVER KING MINE WHICH WAS ABOUT FIFTEEN MILES SOUTHEASTERLY FROM US. IT HAD PRODUCED BETWEEN FOURTEEN AND FIFTEEN MILLION DOLLARS IN SILVER, BUT NOW HAD SHUT DOWN FOR AN INDEFINITE TIME AND THAT HE THOUGHT HE WOULD REST UP FOR A WHILE. I ASKED HIM TO STOP WITH ME AT THE RANCH AND HIS GRUB WOULD COST HIM NOTHING. I WAS BUSY CLEANING WATER HOLES. THE NEXT MORNING I WAS PUMPING WATER AND LETTING IT RUN DOWN A PIPE ABOUT FIVE HUNDRED FEET TO A STRING OF TROUGHS FOR CATTLE (WE HAD BEEN WITHOUT RAIN FOR QUITE A WHILE AND WATER WAS GETTING LOW IN THE WELL AND ALL THE WATER HOLES). THIS MAN WHOSE NAME WAS JOHN CHUNING, SUGGESTED THAT WE SINK OR DRIFT FROM THE BOTTOM OF THE WELL AND UNDER THE CREEK, AFTER PUMPING IT DRY AND MAKING AN EXAMINATION, AND THAT HE WOULD DO THE WORK GLADLY AND NOT CHARGE ANYTHING. SO HE PUMPED AND KEPT THE WATER DOWN AND I MADE AN EXAMINATION, AND WE CONCLUDED TO DRIFT. WHAT TIME I COULD SPARE AWAY FROM THE OTHER HALF DOZEN WATER HOLES ON VARIOUS PARTS OF THE RANGE, I PUT IN HELPING JOHN. I THINK HE WAS ONE OF THE MOST EFFICIENT MEN I EVER SAW AND ONE OF THE HARDEST WORKERS, AS WELL AS ONE OF THE MOST TRUTHFUL. WELL, HE DRIFTED IN THE WELL AND ABOUT DOUBLED THE FLOW OF WATER. ONE EVENING AFTER SUPPER, I SAID, "JOHN, WHAT WILL YOU WORK FOR PER DAY UNTIL YOU GET ALL THE WATER HOLES FIXED UP IN GOOD SHAPE; THOSE THAT HAVE TO BE KEPT CLEAN FROM THIS RANCH?" HE THOUGHT A MOMENT, AND SAID, "TWO DOLLARS A DAY AND BOARD." I HIRED HIM AND TOLD HIM THAT HE WOULD FREQUENTLY HAVE TO DO HIS OWN COOKING, WHICH HE SAID WAS ALL RIGHT. HE FIXED ALL THE WATER HOLES AND THEN BUILT A ROCK GRANARY, THE WALLS OF WHICH ARE IN PERFECT SHAPE TODAY. ONE EVENING, JOHN SAID, "WELL, MR. BARK, THE GRANARY IS ABOUT FINISHED AND I DON'T SUPPOSE YOU HAVE ANY MORE WORK FOR ME." I SETTLED UP WITH JOHN AND GAVE HIM AN ORDER ON GOLDMAN'S STORE, AT PHOENIX, AS AT THAT TIME THERE WERE NO BANKERS. JOHN HANDED THE ORDER BACK TO ME, AND AS I WAS GOING TO PHOENIX FOR SUPPLIES THE NEXT DAY, JOHN SAID HE WOULD MAKE OUT A LIST OF GRUB, AND FOR ME TO HAVE IT SENT UP WITH MY SUPPLIES, AND TO LEAVE THE BALANCE

2 - JOE DEERING

OF THE ORDER TO HIS CREDIT IN GOLDMAN'S. THAT EVENING, WE WERE SITTING OUTSIDE THE HOUSE WHEN HE SAID, "MR BARK, DO YOU KNOW THAT THERE IS A RICH GOLD MINE OVER THE MOUNTAIN?" AND THEN HE LAUNCHED FORTH AS FOLLOWS: THAT WHILE HE WAS WORKING AT THE SILVER KING, A FELLOW BY THE NAME OF JOE DEERING ASKED FOR WORK. THE FOREMAN, WHO WAS BOB BOWEN, ASKED HIM IF HE WAS A MINER, AND HE SAID THAT HE WAS; SO HE WAS TOLD THAT HE COULD GO ON AT THE NEXT SHIFT, AND TO REPORT TO JOHN CHUNING, WHO WAS A SHIFT BOSS IN THE SILVER KING. THIS DEERING HAD ONE POOR RATTY BURRO AND A VERY SMALL AND DISREPUTABLE CAMPING OUTFIT. HE HAD BEEN WORKING BUT A SHORT TIME WHEN HE SPOKE TO JOHN AND SAID, "I AM ONLY GOING TO WORK HERE ABOUT THIRTY DAYS". JOHN DID NOT REPLY, AS THAT WAS A COMMON OCCURRENCE, AND IN A LITTLE WHILE, HE SAID, "I GOT A RICH GOLD MINE OVER IN THE SUPERSTITION MOUNTAINS AND IT IS SOME MINE". JOHN SAID, "IF YOU GOT A RICH MINE OVER THERE, YOU DON'T NEED TO WORK HERE AT ALL, AS YOU CAN GET A PARTNER AND HE WILL PUT UP THE MONEY AND WORK IT". DEERING REPLIED THAT HE DID NOT WANT A PARTNER, AS HE HAD HAD ONE AND THAT WAS THE CAUSE OF HIS WORKING NOW. HE SAID THAT A MERCHANT UP IN COLORADO HAD STAKED HIM TO PROSPECT ABOUT ELEVEN MONTHS AGO, AND THE LAW OF COLORADO SAYS THAT ANYTHING FOUND BY A MAN WITHIN ONE YEAR AFTER ACCEPTING A PROSPECTING STAKE, ONE-HALF OF SUCH FIND SHALL BELONG TO THE MAN WHO FURNISHED THE STAKE, AND AS THE YEAR HAD ABOUT THIRTY DAYS TO RUN YET, I WILL WORK HERE UNTIL THE TIME EXPIRES AND THEN LOCATE THE MINE. JOHN SAID, "SUPPOSING SOME ONE SHOULD LOCATE THE MINE WHILE YOU ARE WORKING HERE?" "OH, THERE IS NO DANGER OF ANYONE'S FINDING IT". JOHN SAID, "YOU FOUND IT, WHY COULDN'T SOME ONE ELSE DO THE SAME?" "THERE IS NO DANGER, AS THE MINE IS IN THE MOST UNGODLY PLACE YOU EVER SAW OR HEARD OF." DURING ALL OF THIS CONVERSATION, THEY WERE WORKING AWAY, SIDE BY SIDE; JOHN SAID THAT DEERING KEPT REFERRING TO HIS FIND, AND SAID THAT IF ANYONE SHOULD FIND THE MINE, THAT HE HAD PUT UP A LITTLE MONUMENT WITHIN A HUNDRED FEET OF IT, BUT THAT ONE COULD NOT SEE THE MINE FROM THE MONUMENT AND HE DID NOT PUT ANY NOTICE ON IT. JOHN ASKED WHY. AND HE REPLIED THAT IF ANYONE SHOULD FIND IT HE COULD SWEAR THAT HIS MONUMENT WAS THERE AHEAD OF HIM, AND THAT HE HAD DESTROYED THE LOCATION NOTICE. JOHN SAID, "HOW DID YOU COME TO FIND IT?" "WELL, I WAS CAMPED

3 - JOE DEERING

AT A SPRING IN A BIG CANYON, AND WAS HEADED FOR THE KING. I HAD MY BREAKFAST, TOOK MY CANTEEN FULL OF WATER AND WENT TO LOOK FOR MY BURRO. I SAW HIM ABOUT HALF WAY UP THE SIDE OF THE MOUNTAIN AND ABOUT A MILE AND A HALF UP THE CANYON AND ABOVE THE CAMP. AS I STARTED UP TOWARD HIM, I SAW A DEEP WORN TRAIL AND IT WAS SO MUCH LARGER AND WORN SO MUCH DEEPER IN THE ROCKS THAN ANY TRAIL I HAD SEEN ON THOSE MOUNTAINS THAT IT EXCITED MY CURIOSITY. I LEFT THE BURRO AND STARTED TO FOLLOW IT, AND BELIEVE I FOLLOWED IT SIX OR SEVEN MILES AND CAME TO THE WORST PLACE I EVER SAW. THERE WAS A TUNNEL AND IT HAD BEEN WALLED UP." JOHN SAID IT PROBABLY HAD CAVED IN. DEERING SAID, "NO, I AM ENOUGH OF A MINER TO TELL WHEN A TUNNEL HAS BEEN WALLED UP. THE WALL HAD SETTLED ABOUT EIGHT INCHES, AND I DON'T KNOW HOW DEEP THE TUNNEL WAS. ABOVE THE TUNNEL, AND FURTHER OVER, IT LOOKED AS THOUGH THERE HAD BEEN TWO BIG SHAFTS, BUT THEY WERE PRETTY WELL FILLED". HE RETURNED TO HIS CAMP, OVER THE SAME TRAIL WHICH HE TOOK GOING IN, AND ON HIS WAY BACK, THERE WAS A WILLOW TREE GROWING JUST AT THE LOWER EDGE OF THE TRAIL. HE RESTED UNDER IT AND TOOK THE LITTLE HAND AX HE WAS CARRYING AND CUT A CROSS IN THE TREE (I FOUND THE WILLOW TREE AND SURE ENOUGH, THE CROSS WAS THERE). THERE WAS AN OLD HATCHET PICKED UP ON THE SAME TRAIL BY A MAN BY THE NAME OF WRIGHT. THE HEAD OF THE HATCHET WAS ALL BATTERED OUT OF SHAPE FROM POUNDING ROCK; AND THE TRAIL WAS MONUMENTED WITH TWO LITTLE ROCKS, UNTIL IT LEFT HAVALINA CANYON AND DROPPED OVER INTO THE HORSE COUNTRY. DEERING SAID THAT HE WOULD NOT HAVE STAYED ALL NIGHT ALONE AT THE MINE FOR ALL THE MONEY IN THE WORLD. JOHN ASKED HIM WHY NOT. HE SAID IT WAS SO GHASTLY AND CREEPY. I CANNOT UNDERSTAND HOW A MAN LIKE DEERING, WHO HAD COME ALL THE WAY THROUGH THE MOUNTAINS FROM COLORADO, ALONE EVERY NIGHT, WITH NOT EVEN A DOG, COULD GATHER SUCH FEAR FROM THIS PARTICULAR SPOT. THE NEXT MORNING, AFTER THINKING IT OVER, HE CONCLUDED THAT HE WOULD VISIT THE MINE ONCE MORE, SO HE WENT OVER THE SAME TRAIL, BUT IN GOING BACK TO HIS CAMP FROM THE MINE HE SAID THAT HE CAME DOWN OFF THE BIG MOUNTAIN INTO A CANYON AND THERE HE BUILT FOUR MONUMENTS. HE SAID THEY WERE NOT MUCH MONUMENTS -- JUST A LONG SLIM STONE STOOD ON END WITH FOUR OR FIVE SMALL STONES LAID AROUND ITS BASE, WITH NO PARTICULAR ATTENTION PAID TO DISTANCES OR DIRECTIONS. "WHAT IN THE WORLD WAS YOUR

4 - JOE DEERING

IDEA OF DOING THAT?" JOHN ASKED HIM, AND HE REPLIED THAT "I THOUGHT IF ANYONE SAW MY MONUMENT AT THE MINE AND THEN SAW THESE THEY WOULD THINK THAT CRAZY PROSPECTOR HAD BEEN HERE ALSO, OR IF THEY SAW THESE FIRST, AND THEN THE ONE I BUILT AT THE MINE, THEY WOULD SAY THE SAME THING AND PAY NO ATTENTION TO EITHER. (NOW I HAVE FOUND THE MONUMENTS, WITHOUT A DOUBT, THAT DEERING BUILT IN THE CANYON). DEERING SAID THEN THAT HE WENT UP THE CANYON TO HIS CAMP, AND THE NEXT DAY CAME OVER TO THE KING. JOHN ASKED HIM HOW FAR IT WAS TO SALT RIVER. THE REPLY WAS, "OH, A MILE--MILE AND A HALF--TWO MILES". JOHN SAID THAT DEERING COULD NOT HELP TALKING ABOUT IT, AND WOULD KEEP BREAKING FORTH, SAYING, "JOHN, THERE IS A TRICK ON THAT TRAIL; OH, IT IS NO TRICK EITHER, BUT YOU HAVE TO GO THROUGH A CAVE OR HOLE, AND SAY, JOHN, IT IS HIGH UP AND YET YOU GOT TO GO DOWN TO IT. SAY, JOHN, WHEN I WORK THAT MINE, I AM GOING OUT THE OTHER WAY, TOWARD THE DESERT. OH, YOU DON'T THINK I HAVE A MINE--YOU THINK I AM JUST STRINGING YOU, JUST TO BE TALKING. YOU MEET ME AT JESSE BROWN'S SALOON TONIGHT AND I WILL SHOW YOU SOME OF THE ORE THAT I PICKED UP IN FRONT OF THE TUNNEL AND EVIDENTLY WAS THROWN AWAY AS WASTE; AND IF IT WAS, YOU CAN JUDGE WHAT THE ORE THEY PACKED OUT MUST HAVE BEEN". JESSE BROWN KEPT A SALOON IN PINAL, WHERE THE MILL FOR THE SILVER KING WAS LOCATED, AND THE ONLY TIME JESSE COULD GET AWAY WAS WHEN CHUNING WOULD COME DOWN AND TEND BAR FOR HIM, WHICH HE FREQUENTLY DID WHEN OFF SHIFT. SO DEERING SHOWED UP THAT EVENING AND HE SHOWED JOHN AND JESSE BROWN ABOUT FIVE POUNDS OF GOLD ORE, AND JOHN SAID IT WAS QUITE RICH. THE GOLD IN THE DIFFERENT PIECES WAS FROM SPLIT PEAS TO COFFEE KERNELS IN SIZE, (SO WHILE I DID NOT IN THE LEAST DOUBT JOHN'S STORY, I WENT ON A HUNT FOR JESSE BROWN AND FINALLY LOCATED HIM IN NOGALES, ARIZONA. UPON INTERVIEWING HIM HE VERIFIED WHAT JOHN HAD TOLD ME). A FEW DAYS AFTER THAT, DEERING WAS WORKING IN A STOPE IN THE KING WHEN A CAVE TOOK PLACE AND HE WAS QUITE BADLY HURT. A FELLOW WORKMAN, ED JONES OF MESA, PUT A BOX ON THE ELEVATOR, SAT UPON IT AND HELD DEERING IN HIS LAP TO THE TOP. HE WAS PUT UPON THE OPERATING TABLE, EXAMINED AND IT WAS FOUND THAT HE HAD BOTH LEGS BROKEN, BESIDES OTHER INJURIES. THE LEGS WERE SO BADLY INJURED THAT THE COMPANY DOCTOR DECIDED THAT AMPUTATION WAS NECESSARY. DEERING NEVER RECOVERED CONSCIOUSNESS.

5 - JOE DEERING

I went to the county seat and found the record of where they had held an inquest ON ONE JOE DEERING; THE VERDICT WAS ACCIDENTAL DEATH, SO THAT PROVED THAT PART OF THE STORY. JOHN SAID, "I WILL GO OVER AND MAKE A HUNT FOR THE MINE, FOR A WHILE AT LEAST. I TOLD JOHN THAT I WOULD BE PLEASED IF HE WOULD MAKE THE RANCH HIS HEADQUARTERS, AND ANY TIME HE NEEDED ANY GRUB TO COME TO THE RANCH AND GET IT, OR ANY TIME THAT HE WANTED ANYTHING FROM TOWN, I WOULD SEE THAT IT WOULD BE SENT OUT TO THE RANCH. JOHN THANKED ME, AND SAID UNDER THOSE CONDITIONS HE OUGHT TO FIND IT. HE CAME OVER TO THE RANCH ABOUT ONCE A MONTH WITH HIS BURROS AND HIS LITTLE CUR DOG. I REMEMBER ONE AUGUST NIGHT, WE WERE SITTING OUT IN FRONT OF THE HOUSE. THE MOON WAS SHINING FULL AND IT WAS HOT. JOHN WAS SITTING ON OUR RODEO TABLE WHICH WAS LEANING AGAINST THE FRONT OF THE HOUSE. JOHN HAD HIS SHOES AND STOCKINGS OFF, AS HIS FEET WERE SWOLLEN FROM CLIMBING, AND I WAS SITTING ON A CHAIR FURTHER OUT FROM THE HOUSE TO GET MORE BREEZE. JOHN STARTED TO GET OFF THE TABLE WHEN HIS LITTLE DOG RUSHED FROM UNDER THE TABLE AND TOOK HOLD OF ONE OF JOHN'S BARE FEET. HE SAID, "WHAT AILS THAT DOG?" AND SETTLED IN HIS SEAT ON THE TABLE AGAIN. THE DOG QUIETED DOWN, AND IN A FEW MINUTES JOHN MADE A MORE DETERMINED EFFORT TO GET DOWN. THE DOG WENT THROUGH THE SAME PERFORMANCE OF TAKING HOLD OF JOHN'S FOOT, AND WHEN HE SAW THAT JOHN MEANT TO STEP DOWN, HE LET GO OF HIS FOOT AND RUSHED AT A RATTLER COILED RIGHT UNDER JOHN'S FOOT. THE DOG KILLED THE RATTLER, BUT WAS BITTEN ON THE NOSE AND THE SIDE OF THE HEAD. THERE WAS SOME WATER DOWN IN THE CREEK AND THE DOG WENT DOWN, BURIED HIMSELF ALL BUT THE TIP OF HIS NOSE IN THE MUD AND WATER, AND IN THE MORNING WAS STILL THERE. HIS HEAD WAS THE SIZE OF THREE HEADS; HE STAYED THERE FOR THREE DAYS AND CAME UP TO THE HOUSE AS SORRY A LOOKING DOG AS ONE EVER SAW. HE NEVER AMOUNTED TO MUCH AFTER THAT; ALL HE WANTED TO DO WAS TO LIE AROUND IN THE SHADE. UNLESS JOHN KEPT CALLING HIM, HE WOULD NOT GO OVER THE MOUNTAINS WITH HIM. WELL, JOHN HUNTED FOR ABOUT A YEAR, AND WENT BROKE AND WANTED TO QUIT. BUT I WOULD NOT HEAR OF IT AND TOLD HIM THAT HE HAD HUNTED SO LONG, THAT THIS WAS NO TIME TO QUIT AS HE MUST HAVE COVERED A WHOLE LOT OF COUNTRY, AND THAT WE WOULD STAY WITH IT UNTIL WE RAN DOWN. JOHN LOADED UP AND WENT BACK OVER THE MOUNTAINS. THE DOG FOLLOWED A MEXICAN FREIGHTER, WHO HAD JUST delivered a load at the ranch and we saw him no more. John continued to hunt, and AT TIMES HE WOULD HAVE TO TAKE A LAYOFF AT THE COW RANCH AND KEEP OFF HIS FEET AS MUCH AS POSSIBLE AS BOTH HIS KNEE JOINTS WOULD SWELL TO DOUBLE THEIR NORMAL SIZE. JOHN HAD TOLD ME THAT HE HAD WORKED FOR JOHNNY AYRES, WHO HAD A HORSE RANCH UP NEAR THE GRAND CANYON. HE HAD BEEN DEVELOPING WATER FOR AYRES. ONE TIME WHEN JOHN WAS AT THE RANCH, I RETURNED FROM PHOENIX AND THE FIRST THING JOHN WOULD GRAB WOULD BE THE ARIZONA REPUBLICAN. IN IT HE READ OF JOHNNY AYRES COMMITTING SUICIDE. "WELL", JOHN REMARKED, "I MIGHT JUST AS WELL TEAR THIS NOTE UP. POOR JOHNNY". JOHN HAD GONE INTO THE HOUSE WHERE HE KEPT HIS GO-AWAY BAG AND DUG A PAPER FROM WITHIN. I ASKED HIM WHAT IT WAS THAT HE WAS GOING TO TEAR UP. HE REPLIED THAT WHEN HE GOT THROUGH WORKING FOR JOHNNY HE TOOK A NOTE FROM HIS PAY, AS JOHNNY HAD NO MONEY. THE NOTE WAS FOR FOUR HUNDRED AND SOME DOLLARS. I REMARKED, "DON'T TEAR IT UP, JOHN, LET ME HAVE IT AND I WILL SEE WHAT I CAN DO WITH IT THE FIRST TIME I GO TO PHOENIX, I MADE A FEW INQUIRIES IN PHOENIX, AND SOME ONE TOLD ME THAT E.J.BENNETT WAS A SILENT PARTNER OF JOHNNY. SO I TOOK A CHANCE, WENT TO E.J.BENNETT'S OFFICE, TOLD HIM THAT CHUNING WAS HUNTING FOR A LOST MINE UP IN THE SUPERSTITION MOUNTAINS, AND WAS WORSE THAN BROKE; THAT HE WAS GETTING ALONG IN YEARS, AS HE WAS THEN OVER SIXTY; THAT HE HAD A NOTE OF JOHNNY AYERS FOR WORK HE HAD DONE ON THE HORSE RANCH, ETC., E.J. SAID, "HAVE YOU THE NOTE WITH YOU?" AND I PRODUCED IT. HE LOOKED IT OVER, TOOK A PENCIL, FIGURED THE INTEREST, REACHED FOR HIS CHECK BOOK AND ASKED ME WHO HE SHOULD MAKE THE CHECK TO. I SAID, "TO ME". HE MADE IT TO ME, AND IT CALLED FOR SIX HUNDRED AND SOME DOLLARS. WHEN I HANDED JOHN THE CERTIFICATE OF DEPOSIT, HE THANKED ME AND FROM THAT DAY ON HE WOULD NOT TAKE A DOLLAR FROM ME. HE CONTINUED TO HUNT FOR THE MINE, AND STILL MADE THE RANCH HIS HEADQUARTERS. HE BUILT NUMEROUS ROPE LADDERS TO LET HIMSELF DOWN OVER CLIFFS IN WHICH COULD BE SEEN CAVES, AND WHICH WERE ABSOLUTELY UNAPPROACHABLE IN ANY OTHER WAY. ROPE LADDERS THAT I WOULD NOT TRUST MYSELF AT ALL UPON HE WOULD GO DOWN TWENTY-FIVE AND THIRTY FEET, AND IF ANYTHING GAVE WAY IT WAS "KATY BAR THE DOOR FOR JOHN". HE ALSO RAN A TUNNEL ONE HUNDRED AND FIFTY FEET, ALONG A CRACK IN A HILL NEAR THE PAINT MINE, NORTH OF SOMBRERO BUTTE. THE CRACK WAS ABOUT EIGHTEEN INCHES WIDE AND FILLED

Typical prospectors packing supplies.

7 - JOE DEERING

WITH BOULDERS. HE RAN THE ENTIRE DISTANCE WITHOUT ANY TIMBERING, AND ALL ALONE. EVERY FOOT OF IT HAD TO BE BLASTED, WITH NO POSSIBILITY OF THEIR BEING A MINE OR ANY ORE. I THINK IT WAS THE MOST DANGEROUS PIECE OF WORK I EVER SAW ACCOMPLISHED BY MAN. WHENEVER JOHN RAN OUT OF MONEY, HE WOULD QUIT HUNTING AND GO TO WORK GENERALLY FOR BILL KIMBALL, WHO KEPT A HOTEL AND LIVERY STABLE IN MESA. JOHN WOULD TAKE CARE OF THE LIBERY PART OF IT, AND HE WAS A NUMBER ONE WORKER. KIMBALL TOLD ME THAT HE WAS ALWAYS GLAD TO GET HIM.

ONE DAY COMING FROM TOWN, I FOUND JOHN RESTING AT THE RANCH; HE SHOWED ME A PIECE OF ORE AND I COULD SEE A LITTLE FREE GOLD IN IT. I ASKED HIM WHERE HE GOT IT AND HE SAID OVER IN THE BOX CANYON OF SALT RIVER, ABOVE WHERE FISH CREEK COMES IN. HE ASKED ME IF I THOUGHT IT WAS VERY GOOD, AND I TOLD HIM THE ORE WAS GOOD ENOUGH; IT DEPENDED UPON THE QUANTITY. I ASKED HIM IF HE FOUND IT IN PLACE, AND HE SAID HE DID. I ALSO ASKED HIM HOW WIDE AND LONG IT WAS, AND HE SAID HE DID NOT KNOW, BUT THAT HE DID NOT THINK MUCH OF IT. I ASKED HIM IF HE COULD SHOW IT TO ME, AND HE SAID THAT HE COULD AND WOULD GO ANY TIME I WAS READY. SO THE NEXT DAY WE TOOK A PACK HORSE AND TWO SADDLE HORSES AND RODE TO WITHIN ABOUT TWO MILES OF THE VEIN, BUT THAT TWO MILES WAS A HOLY TERROR TO GET OVER, EVEN ON FOOT, BUT WE GOT THERE WITH MOST OF OUR CLOTHING TORN OFF AND MANY SKINNED PLACES. AFTER EXAMINATION OF THE VEIN, I TOLD JOHN THAT WE MIGHT RUN INTO SOMETHING PRETTY GOOD BY TUNNELING AND I THOUGHT THAT WE HAD BETTER SPEND A LITTLE MONEY ON IT. HE SAID, "ALL RIGHT, THAT THE MONEY SPENT ON IT WOULD NOT HAVE MUCH OF A WE TO IT, AS HE WAS IN HIS USUAL CONDITION FINANCIALLY, BROKE, BUT THAT HE WAS WILLING TO WORK ON IT. SO I LOCATED IT IN HIS AND MY NAME AND THEN TOLD HIM THAT THE FIRST THING WAS A TRAIL. WE WENT BACK TO THE COW RANCH, PACKED JOHN'S THREE BURROS WITH GRUB AND TOOLS, AND ESTABLISHED CAMP ON THE RIVER ABOUT TWO MILES FROM THE PROSPECT. WE HAD POWDER, PICKS AND SHOVELS AND FINALLY FINISHED OUR TRAIL. SOME OF IT, TO PUT IT MILDLY, WAS VERY DANGEROUS, AS IT WAS BUILT ALONG A CLIFF ABOUT FIVE HUNDRED FEET ABOVE THE RIVER, AND FIFTEEN HUNDRED FEET ABOVE US TO THE TOP OF THE CLIFF. THE FIFTEEN HUNDRED FEET ABOVE US DID NOT CONCERN US MUCH AS THERE WAS NO DANGER IN THAT

8 - JOE DEERING

DIRECTION UNLESS THE POWDER HAPPENED TO GO OFF. WE THEN WENT BACK TO THE COW RANCH AND LOADED THE THREE BURROS WITH MINING SUPPLIES AND GRUB WHICH I HAD SENT UP, AND STRUCK OUT FOR OUR FIRST TRIP OVER THE TRAIL WITH THE ANIMALS. WE GOT ALONG ALL RIGHT UNTIL THE TRAIL GAVE AWAY WITH JENNY, THE OLD BURRO, AND JOHN'S SPECIAL FAVORITE. WHERE IT GAVE WAY, IT WAS AT LEAST FIVE HUNDRED FEET STRAIGHT DOWN AND JUST BELOW JENNY. THERE WAS A MOUNTAIN MAHOGANY TREE GROWING OUT OF A CRACK IN THE CLIFF, AND JENNY LAID THERE ON HER SIDE, LOADED WITH A BLACKSMITH BELLOWS, FIFTY POUNDS OF POWDER AND SIX BOXES OF DYNAMITE CAPS AND FUSES. FIRST OF ALL, WE DROVE THE OTHER TWO BURROS TO A PLACE OF SAFETY, UNPACKED THEM, TOOK THEIR LASH AND SWING ROPES, AND WENT BACK AND TRIED TO FASTEN JENNY TO ANOTHER TREE THAT WAS GROWING ABOVE THE TRAIL. WE RAN A LINE FROM THAT TREE TO A POINT OF ROCK ABOVE. WE MADE HER FAST TO THIS ANCHORAGE, BOTH BOW AND STERN. EVERY TIME ONE OF US WOULD STEP DOWN ON JENNY, THE TREE WOULD GIVE SOMETHING AWFUL, BUT JENNY, (BLESS HER HEART) WOULD WATCH US AND NEVER MOVE, AND WE WORKED FAST. JOHN SAID, "JENNY, IF YOU DON'T LIE STILL, IT WILL BE MOLLY, TOLL THE BELL". AND IT APPEARED AS THOUGH THAT BURRO UNDERSTOOD EVERYTHING JOHN SAID TO HER. WE STARTED TO GET THE PACK ON TO THE TRAIL. JOHN WAS CERTAINLY A GENIUS. IF WE LOOSED THE LASH ROPES, EVERYTHING WOULD BE OVER, AND PERHAPS US WITH IT. OH, FOR AN EXTRA SUPPLY OF ROPE. BUT FIRST, WE MANAGED TO GET THAT CIGAR BOX OF DYNAMITE CAPS, AND WHEN WE GOT THAT UP AND PLACED IN SAFETY THE SUN SHONE BRIGHTER. PIECE BY PIECE WE FINALLY GOT THE PACK UP AND IN SAFETY. AS SOON AS WE GOT THE POWDER UP, I WANTED TO THROW THE BALANCE OF THE PACK OVER, BUT JOHN WOUDD NOT HEAR OF IT. THE NEXT PROBLEM WAS HOW TO GET JENNY BACK UP ON THE TRAIL. IT REALLY LOOKED IMPOSSIBLE, AS SHE WAS IN THE TREE AND ABOUT SIX FEET BELOW THE TRAIL. I CAN'T DESCRIBE HOW WE DID IT, BUT I WILL SAY THAT JENNY WAS THE BEST MAN IN THE PARTY. WHEN WE WOULD PLACE ANY OF HER FEET WHERE WE WANTED THEM, SHE WOULD NOT PUT ANY WEIGHT ON THEM UNTIL WE GOT HOLD OF THE ROPES AND GAVE THE SIGNAL. WE GOT DOWN TO THE PROSPECT AND MADE CAMP AND HEAVED A SIGH OF RELIEF.

9 - JOE DEERING

AFTER JOHN AND I HAD BEEN WORKING SEVERAL MONTHS RUNNING A BUNNEL AND SORTING THE ORE WE CAME TO THE CONCLUSION THAT WE WOULD BUILD AN ERASTRA, AND TRY AND GET SOME OF THE GOLD OUT OF THE ROCK. JOHN SAID HE COULD BUILD IT. I KNEW I COULD NOT AS I HAD HAD NO EXPERIENCE, BUT I COULD DO SOME OF THE HEAVY WORK. AN ERASTA IS A CIRCLE OF STONE ABOUT TEN FEET IN DIAMETER, TWO FEET HIGH, WITH THE BOTTOM LAID IN STONE, AND CLAY, OF GOOD QUALITY, TAMPED IN THE CRACKS. TO PREVENT THE GOLD FROM GOING DOWN BETWEEN, AN UPRIGHT POST IN THE CENTER WITH A CROSS BAR RUNNING HORIZONTALLY FROM THE POST, AND ABOUT THREE HEAVY ROCKS AROUND ON THE BOTTOM ROCKS ARE SO ARRANGED ON THE CROSS BAR THAT NO ONE ROCK FOLLOWED IN THE TRACK OF ANOTHER, THE DRAG ROCKS TO COVER THE ENTIRE SPACE IN THE BOTTOM AS THEY WENT AROUND AND AROUND AND AROUND. THEN A LONG SWEEP WAS FASTENED TO THIS CENTER POST, AND A BURRO WAS HITCHED TO IT. HOW FOXY A BURRO WOULD GET, -- THE MOMENT YOU WERE OUT OF SIGHT PROCEEDINGS AND THE BURRO STOPPED. IN A LITTLE WHILE HE WOULD STOP IF YOU WERE NOT YELLING AT HIM, THEN HE WOULD STOP UNLESS YOU WERE CUSSING HIM AND THROWING ROCKS AT HIM. ONE PRONOUNCED TRAIT OF A BURRO HAS, IF YOU THROW ANYTHING AT HIM WHEN HE IS PACKED, OR HITCHED TO ANYTHING, HE WILL STAND PERFECTLY STILL AND CLINCH HIS TAIL TIGHTLY BETWEEN HIS LEGS. JUST WHY, I DO NOT KNOW, AND FINALLY, TO KEEP THE BURRO MOVING YOU WOULD HAVE TO RIDE THE SWEEP AND THREATEN HIM WITH SOME KIND OF A WHIP. I HAVE DIGRESSED IN MY DESCRIPTION OF THE ERASTRA PROCESS OF EXTRACTING GOLD. YOU BROKE THE ORE YOU WERE GOING TO ERASTRA AS FINE AS POSSIBLE (THE FINER THE MORE YOU COULD ERASTRA) PUT IT IN THE ERASTRA, POURED IN WATER JUST ENOUGH TO MAKE A KIND OF SOUP AND KEEP IT AT ABOUT THAT CONSISTENCY BY ADDING MORE WATER OR GRINDING LONGER. YOU HAD A LITTLE GATE ON ONE SIDE OF THE ERASTRA WHERE YOU WOULD DRAW OFF THE TOP PULP OR SOUP. I FORGOT TO STATE THAT YOU ADDED QUICK-SILVER BY HAVING A PIECE OF BUCKSKIN TIED OVER THE NECK OF A SODA WATER BOTTLE AND SHAKING IT OVER THE PULP AS YOU WOULD A PEPPER BOX; THE QUICK-SILVER WOULD COME OUT AS A SPRAY THROUGH THE BUCKSHIN, GATHERING THE GOLD IN THE PULP AND GO TO THE BOTTOM WHERE IT WOULD SETTLE IN THE CRACKS AND ON TOP OF THE CLAY. YOU DID NOT NEED TO CLEAN UP FOR MONTHS, FOR IF THE ORE WAS VERY RICH

10 - JOE DEERING

SIMPLY ADD MORE QUICK. AFTER RUNNING IT WITH BURROS FOR SOME TIME, WE FOUND OURSELVES ADDRESSING EACH OTHER WITH ALL THE PROFANE LANGUAGE WE EVER HEARD, AND REALLY HAD INVENTED QUITE A VOCABULARY OF NEW CUSS WORDS. WE KNEW THEY WERE CUSS WORDS BECAUSE WE USED THEM TALKING TO THE BURROS. WE WERE GETTING THIN AND DID NOT ENJOY OUR FOOD; AND WE GOT TO THROWING DRILLS AND OTHER TOOLS AT THE BURRO HITCHED TO THE ERASTRA, CUSSING WHEN WE MISSED AND CUSSING WHEN WE HIT OUR MARK. OUR ERASTRA WAS BUILT ON A LITTLE PROMONTORY PROJECTING OUT INTO SALT RIVER, AND MANY TIMES WE WOULD HAVE TO GO INTO SIX FEET OF WATER TO RECOVER OUR TOOLS. ALL NICE ENOUGH IN SUMMER, BUT WAIT UNTIL WINTER! I PROPOSED TO JOHN THAT WE BUILD A WATER WHEEL (UNDERSHOT) AND RUN THE ERASTRA BY IT, AND HE AGREED RIGHT AWAY, BUT REMARKED THAT IT WAS GOING TO BE SOME JOB TO GET LUMBER IN THERE. I TOLD JOHN THAT I WOULD GO AND GET BOODY, WHO WAS OUR FOREMAN ON THE BARK AND CRISWELL COW RANCH, AND TRY AND HIRE SOME YOUNG FELLOWS WHO COULD SWIM, AND WE WOULD PACK THE LUMBER 12 FEET LONG TO THE RIVER JUST OPPOSITE THE COFFIN RANCH AND ABOUT EIGHT MILES ABOVE THE MINE, AND THERE SAW IT IN TWO. WE WOULD FIGURE ON MAKING A WHEEL SIX FEET WIDE AND 12 FEET IN DIAMETER, AND WE WOULD MAKE THE LUMBER INTO THREE RAFTS, BOUND WELL TOGETHER, WITH A FIFTY-FOOT HALF INCH ROPE ATTACHED AND DRAGGING FROM EACH RAFT, AS THERE WERE VERY LARGE BOULDERS IN ALMOST EVERY RIFFLE AND MANY RAPIDS, ONE FOR BOODY, ONE FOR THE FELLOW I HIRED AND ONE FOR MYSELF. JOHN SAID "THAT IS ALL RIGHT, BUT YOU WANT TO BE SURE THAT YOU DON'T LET THE RAFTS GET PLASTERED UP AGAINST A BOULDER IN THE RAPIDS, ALL HELL COULDN'T GET IT OFF AND YOU MIGHT GET YOUR HEAD BUMPED PRETTY HARD." I SAID, "JOHN, I WILL GET BACK JUST AS SOON AS POSSIBLE", SO I STRUCK OUT AFOOT, GOT TO THE RANCH THAT NIGHT, AND THE NEXT MORNING GOT A HORSE AND RODE INTO PHOENIX THAT DAY, ORDERED MY SUPPLIES, HUNTED UP A FREIGHTER. HE LOADED AND STRUCK OUT THE NEXT DAY, AND THE THIRD DAY ARRIVED AT THE RANCH. IN THE MEANTIME, BOODY GOT THE HORSES WE WANTED TO PACK AND THE SADDLE HORSES. WE WENT OVER THE PACK OUTFITS AND WERE READY TO LOAD THE NEXT DAY AND START FOR THE RIVER. WE GOT ABOUT FOURTEEN MILES THE FIRST DAY, BUT MADE THE RIVER THE NEXT, SPENT THE NEXT DAY IN

11 - JOE DEERING

GETTING OUR RAFTS READY AND CACHING OUR SUPPLIES OF GRUB, ETC. TO WHERE JOHN COULD TAKE THE BURROS AND GET THEM. AS WE HAD NO PLACE TO LEAVE THE HORSES, WE BROUGHT ALONG A MEXICAN BOY WHO WAS WORKING ON THE COW RANCH WITH BOODY TO TAKE THE HORSES BACK. WE STARTED DOWN THE RIVER WITH THE THREE RAFTS, WITH NOTHING ON BUT OUR OVERALLS. IN THE QUIET WATERS WE WOULD EITHER TOW THEM ALONG WITH THE ROPE, OR POLE THEM. WE WERE AS NEAR TOGETHER AS WAS POSSIBLE AND WE WERE GETTING ALONG FINE WHEN JUST AHEAD WE SAW OUR HIRED MAN'S RAFT PLASTER AGAINST A BIG BOULDER IN THE CENTER OF A RAPIDS; HE WENT OVERBOARD, AND FOR QUITE A WHILE HE DID NOT COME UP. WE COULD DO NOTHING, AS WE WERE IN THE RAPIDS, BUT PRETTY SOON HE CAME UP AND SWAM OUT. IT DEVELOPED THAT WHEN HE WAS THROWN FROM THE RAFT, THE ROPE WRAPPED AROUND HIM AND THE CURRENT HELD HIM UNDER. AS EACH OF US HAD A KNIFE, WHILE UNDER THE WATER, HE REACHED FOR HIS KNIFE, OPENED IT AND CUT THE ROPE. WE FOOLED AROUND ALL THE REST OF THE DAY TRYING TO GET IT LOOSE AND FINALLY SUCCEEDED. WE CAMPED ON A SANDBAR ALL NIGHT AND IT WAS PRETTY COLD AS WE HAD NO FIRE. NEXT MORNING, WE WENT ON AND HAD NO FURTHER TROUBLE. WE COULD HEAR JOHN TALKING TO HIS BURRO LONG BEFORE WE COULD SEE HIM. WE SOON BUILT THE WATER WHEEL, HITCHED IT UP TO THE ERASTRA, BUT IT WOULD NOT PULL IT. WE THEN COMMENCED TO BUILD A DAM ACROSS THE RIVER AND FINALLY GOT HER GOING. THE CABLE SLIPPED OFF MANY TIMES, BUT WE FINALLY GOT IT REGULATED AND SOON MADE A CLEAN UP. WE RECOVERED ABOUT EIGHT HUNDRED DOLLARS IN GOLD, BUT THE MINE WAS STILL IN DEBT, SO JOHN AND I KEPT IT GRINDING UNTIL WE HAD ERASTRAED ALL OF THE RICH ORE THAT WE HAD TAKEN OUT; AND IN SEVERAL CLEANUPS, WE RECOVERED ABOUT EIGHT HUNDRED DOLLARS MORE. WE WERE A LITTLE AHEAD, PROVIDED NEITHER OF US RECEIVED WAGES. I WENT TO TOWN FOR MORE SUPPLIES, AND UPON MY RETURN THE WATER WHEEL WAS GONE, AS THE RIVER HAD RAISED AND SWEPT THE WHEEL AWAY. SO JOHN AND I CONTINUED TO WORK AT THE MINES. UP TO THAT TIME, WE HAD NOT HAD A SINGLE VISITOR, BUT IN A FEW DAYS JIM GOODWIN OF TEMPLE DROPPED IN ON US. STAYED OVER NIGHT AND WENT HIS WAY. IN A FEW DAYS A YOUNG MAN CAME TO THE CAMP JUST BEFORE DINNER, AND I ASKED HIM TO EAT. HE STAYED WITH US ABOUT A WEEK, AND ONE NIGHT HE TOLD ME THAT HE WAS GOING TO LEAVE

12 - JOE DEERING

THE NEXT DAY, AND I PROBABLY WOULD NOT SEE HIM AGAIN; THAT THERE HAD BEEN A BIG MISTAKE MADE OR HE WOULD HAVE BEEN DOWN IN MEXICO BEFORE THIS; THAT HIS NAME WAS KING MASSEY; THAT THEY WERE LOOKING FORHIM; THAT ON A CERTAIN NIGHT, ABOUT A WEEK BEFORE (HE TOLD ME THE NIGHT) THAT HE AND ANOTHER FELLOW, WHOSE NAME HE DID NOT TELL ME, WERE LYING UNDER THE HIGHLAND BRIDGE ON THE GOLDFIELD ROAD, - THE DITCH WAS DRY, - TO HOLD UP BILL KIMBALL, WHO, THEY WERE INFORMED, WAS TO BRING TO MESA A BAR OF GOLD BULLION, AND THEN THEY INTENDED TO SKIP TO MEXICO. BUT THEIR INFORMATION WAS THE BUNK, AS THEY WAITED THERE UNTIL DAYLIGHT AND KIMBALL DID NOT SHOW UP. IT WAS COMMONLY KNOWN THAT KIMBALL BROUGHT THE BULLION FROM THE MINE, BUT WHEN, VERY FEW KNEW. KIMBALL HAD THE BOARDING-HOUSE, SALOON AND POST OFFICE AT THE MINE, AND THE OWNER OF THE MINE WHO WAS CHARLEY HALL, WAS A GREAT FRIEND OF KIMBALL'S, AND AS KIMBALL HAD TO GO UP TO THE MINE ALMOST EVERY DAY, IT WAS VERY DIFFICULT TO TELL WHEN HE HAD THE BULLION. DURING HOT WEATHER, HE ALMOST ALWAYS DROVE AT NIGHT. I HAVE NEVER SEEN KING MASSEY SINCE HE LEFT THE NEXT MORNING. IN A FEW DAYS, I WENT TO MESA, AND STOPPED ALL NIGHT AT THE KIMBALL HOTEL. IN THE MORNING AT BREAKFAST, THERE WAS NO ONE IN THE DINING-ROOM BUT KIMBALL AND MYSELF, AND HE SAID, "JIM, I MADE A FOOL OF MYSELF THE OTHER NIGHT WHEN I BROUGHT THE LAST BAR OF BULLION DOWN". "HOW SO?" "WELL, I HAD ABOUT TEN THOUSAND DOLLAR BAR AND AS I APPROACHED THE HIGHLAND CANAL, I GOT A HUNCH THAT I HAD BETTER GO AROUND. IT WAS ABOUT TWO O'CLOCK IN THE MORNING, AND I PAID NO ATTENTION TO IT, BUT IT PERSISTED, AND FINALLY I TURNED OUT OF THE ROAD, DROVE ACROSS THE DESERT TO THE ROAD A MILE SOUTH OF HERE, AND THEN CAME UP TO MESA. IT WAS BRIGHT MOONLIGHT AND I JUST FIGURE THAT I AM GETTING OLD AND PANICKY. I ASKED HIM THE NIGHT, AND IT WAS THE NIGHT THAT MASSEY TOLD ME ABOUT. I DID NOT TELL BILL WHAT I KNEW FOR OBVIOUS REASONS, BUT IT WAS CERTAINLY A QUEER HUNCH. THE GOLDFIELD ROAD IS NOW THE APACHE TRAIL, AND IN PACKING IN ANY LUMBER FOR THE MINE, WE TRAVELED ALONG WHERE THE APACHE TRAIL NOW IS, FROM THE TOP OF THE HILL AT FISH CREEK, UP FOR ABOUT FIVE MILES, THEN BORE OFF TO THE NORTH TO SALT RIVER. OUR MINE WAS SITUATED

13 - JOE DEERING

ON SALT RIVER, JUST AT THE EDGE OF THE WATER AND THE HORSEMESA MOUNTAIN LOOMS ABOVE IN ALMOST A STRAIGHT CLIFF FOR OVER TWO THOUSAND FEET. THE HORSE MESA DAM IS JUST BELOW ABOUT A HALF MILE AND THE MINE IS NOW FLOODED. I WISH TO CALL ATTENTION TO THE DISTANCE BETWEEN WHERE JOHN WAS PROSPECTING AND WHERE WE HAD EVERY REASON TO BELIEVE THE LOST DUTCHMAN IS. THIS MINE THAT JOHN DISCOVERED WHILE HUNTING THE DUTCHMAN IS AT LEAST TWELVE MILES OFF THE COURSE, AND I THINK HE FINALLY HUNTED FURTHER AWAY THAN THAT. YOU SAY, "HOW FOOLISH". BUT REMEMBER JOHN HAD BEEN HUNTING FAITHFULLY FOR YEARS, ALL ALONE, WITH HIS THREE BURROS; HE WOULD NOT SEE A HUMAN BEING FOR MONTHS AT A TIME. THE NATURAL CONCLUSION, AFTER THOROUGHLY LOOKING OVER THE CLIFFS AND CRAGS, WHERE HIS INFORMATION LED HIM TO BELIEVE IT EXISTED, THAT HE GRADUALLY EXPANDED HIS AREA OF SEARCH, UNTIL HE REALLY THREW ALL LOGIC TO THE WINDS AND JUST WENT. AT TIMES I HAPPENED TO SEE JOHN AND HIS BURROS COMING DOWN THE TRAIL TO THE HOUSE, AND HEAR THE BURROS BRAYING AS THEY ALWAYS DID WHEN THEY SAW THE GRANARY WHICH REPRESENTED BARLEY TO THEM; AND PERHAPS JOHN HAD FORGOTTEN HIS MOTHER TONGUE AND WAS ALSO BRAYING.
TO CONTINUE, JOHN AND I DRIFTED ON THE VEIN UNTIL WE CAME TO THE END; THEN SUNK A WINZE ABOUT THIRTY-FIVE FEET AND CAME TO THE BOTTOM. THE VEIN BECAME NARROWER AS WE SUNK, AND THE VALUES LESS. JOHN SAID TO ME ONE DAY, "JIM, YOU CAN HAVE MY INTEREST IN THIS MINE; IF I HAVE ANY INTEREST. I AM GOING BACK AND HUNT SOME MORE FOR THE DUTCHMAN". IN A SHORT TIME AFTER THAT I GAVE TO CRISWELL, MY PARTNER, AND ELMER BOODY, FOREMAN OF THE COW RANCH, MY INTEREST, AND I WENT BACK TO THE COW RANCH. CRISWELL AND BOODY KEPT THE ASSESSMENT WORK UP FOR A NUMBER OF YEARS AND FINALLY LET IT GO BACK TO THE GOVERNMENT.

JOHN WOULD HUNT AWHILE, THEN GO TO WORK UNTIL HE MADE ANOTHER STAKE, AND OUT FOR THE DUTCHMAN AGAIN, ALTHOUGH WHEN JOHN FIRST STARTED HUNTING, HE HAD NEVER HEARD OF THE DUTCHMAN, BUT WAS HUNTING ENTIRELY ON WHAT DEERING HAD TOLD HIM. ONE TIME, NOT MANY YEARS AGO, SIMS ELY AND A PROSPECTOR NAMED WRIGHT AND MYSELF WERE RUNNING DOWN SOME STORY ABOUT THE DUTCHMAN, AND WE STOPPED AT TORTILLA STATION, ON THE APACHE TRAIL. WE FOUND JOHN CHUNING THERE AS STATION TENDER, WHERE HE HAD BEEN WORKING FOR ABOUT A YEAR FOR THE STAGE COMPANY RUNNING

14 - JOE DEERING

STAGES FROM MESA TO ROOSEVELT DAM, OVER THE APACHE TRAIL. JOHN WAS GLAD TO SEE US, AND SAID THAT HE SUPPOSED WE WERE OUT AFTER THE DUTCHMAN. WE SAID WE WERE. HE ALSO SAID THAT HE HAD SENT IN HIS RESIGNATION TO THE HEAD OFFICE IN PHOENIX; THAT HE WAS GOING TO LEAVE JUST AS SOON AS THEY SENT A MAN TO RELIEVE HIM, BUT THE DARN SHEEP MEN HAD STOLEN HIS BURROS AS THEY DROVE THROUGH IN THE SPRING, AND HE HAD SENT TO GLOBE FOR SOME MORE; THAT HE WAS ALL READY TO PACK UP AND STRIKE OUT FOR THE DUTCHMAN, EXCEPT THAT HE HAD NOTHING TO PACK ON. WE WERE GONE JUST A FEW DAYS, AND UPON OUR RETURN WE STOPPED TO SEE JOHN; HE SAID THAT HE HAD BOUGHT FOUR BURROS IN GLOBE AND WAS WAITING TO BE RELIEVED, BUT THAT HE WAS NOT FEELING VERY WELL. HE SAID THIS TIME HE SURE HAD THE MINE AND THERE WAS NO MISTAKE, AND COULD HE REACH ME AT THE ARIZONA CLUB. I SAID, "DON'T GO OFF ALONE JOHN, IF YOU ARE NOT FEELING WELL". HE SAID HE WOULDN'T, BUT THAT HE WAS SURE HE WOULD FEEL ALL O.K. IN A FEW DAYS. A SHORT TIME AFTER, I HEARD THAT JOHN HAD DIED AT THE TORILLA STATION. HE NEVER GOT AWAY FROM HIS JOB.

I WISH TO SAY IN CONCLUSION THAT BOTH JOHN AND I FOUND MANY SANDALS ALONG THE MONUMENTED TRAIL, SUCH SANDALS AS THE PEONS OF MEXICO WORE, WHICH WERE MADE OUT OF THE CENTURY PLANT FIBER; ALSO, AROUND THE CAVES NEAR THE WATER HOLES WERE MANY PICKS AND SHOVELS, ALL OLD AND WORN. THE HANDLES WOULD FALL APART IF THEY WERE PICKED UP. THE FINDING OF THE SANDALS IS ALMOST ABSOLUTE PROOF OF THERE HAVING BEEN LARGE PARTIES OF PEONS IN THERE. IT IS ALMOST CERTAIN THEY WERE HEADED BY A DON, AS THE SUPERSTITION MOUNTAINS WAS THE APACHE STRONGHOLD, AND THE LAST MOUNTAINS THAT THE APACHES GAVE UP. IN FACT, THE LAST APACHES WERE NOT DRIVEN OUT UNTIL IN 1884.

PHYSICAL EVIDENCES

JACOB WALSH SAID THAT THE TRAIL OVER THE MOUNTAIN TO THE MINE WAS MONUMENTED WITH TWO SMALL ROCKS. THAT TRAIL AND THOSE MONUMENTS I HAVE FOUND OVER THE MOUNTAIN AND DOWN TO A CERTAIN POINT; THEN THE TRAIL AS WELL AS THE MONUMENTS BECAME DOUBTFUL.

DEARING TOLD CHUNING THAT UPON RETURNING FROM THE MINE TO HIS CAMP THE FIRST TRIP, THAT THERE WAS A WILLOW TREE GROWING ALONGSIDE THE TRAIL AND THAT WHILE RESTING IN ITS SHADE HE CUT A CRUSS x IN THE TREE TRUNK WITH A SMALL HAND AX THAT HE WAS CARRYING. THAT TREE, TRAIL, AND CROSS HAVE BEEN FOUND.

OLD JAKE ALSO SAID THAT THE INCLINED PITS IN THE MINE WERE LINED WITH MESQUITE WOOD; IT MUST HAVE TAKEN AN IMMENSE NUMBER OF MESQUITE TREES, AS EVERY TRIP THEY MADE IT TOOK MORE LAGGING, AND ONE CAN PLAINLY SEE THAT THE TREES HAVE BEEN CUT WITH LONG INTERVALS OF TIME BETWEEN CUTTINGS. THESE, I CERTAINLY HAVE FOUND, AS THERE WAS NO OTHER CONCEIVABLE PLACE THE WOOD COULD HAVE BEEN TAKEN.

I HAVE COME TO THE CONCLUSION THAT THE MEXICANS HAVE HAD THREE DIFFERENT TRAILS GOING TO THEIR CAMP IN MARSH VALLEY; THE FIRST TRAIL RAN ALONG THE EAST SIDE OF BLUFF SPRING MOUNTAIN, ABOUT HALF WAY UP THE MOUNTAIN, OR AS HIGH AS THE BLUFF WOULD PERMIT THEM TO MAKE A TRAIL. IT WAS A VERY DIFFICULT TRAIL TO UPKEEP AS THERE WERE MANY STEEP CANYONS CROSSING IT, AND IN RAINY TIMES, BOULDERS WEIGHING MANY TONS WOULD RUSH DOWN AND TEAR A TRAIL ALL TO PIECES; JUDGING FROM THE GROWTH OF THE TREES IN SOME OF THE SHORT STRETCHES OF THIS TRAIL THAT CAN STILL BE SEEN, IT WAS ABANDONED A GOOD MANY YEARS AGO. THEIR NEXT TRAIL WAS THE ONE DEERING FOUND AND FOLLOWED, WENT UP HAVALINA CANYON OVER INTO THE HORSE COUNTRY, DOWN PETERS MESA TO THE CHARLESBOIS TRAIL AND THENCE TO MARSH VALLEY (THEIR CAMP), A ROUNDABOUT WAY BUT ALWAYS HIGH UP AND FAIRLY FREE FROM FAVORABLE SPOTS FOR INDIAN AMBUSCADES.

THE THIRD, AND THEIR LAST TRAIL WAS NOT FOUND UNTIL A FEW YEARS AGO BY THE PRESENT OWNER OF MY OLD COW RANCH, GUS BARKLEY, THE MOST UNLIKELY PLACE FOR A TRAIL I EVER SAW. IT WENT UP THE SOUTHEAST SIDE OF BLUFF SPRING MOUNTAIN, AND WORKED ITS WAY AROUND THE BLUFFS TO THE TOP OF THE MOUNTAIN, A GREAT MANY HUNDREDS OF HORSES

PHYSICAL EVIDENCES

AND MULES HAVING BEEN DRIVEN OVER IT. AS YET, I HAVE NOT FOUND POSITIVELY WHERE THIS TRAIL COMES DOWN OFF THE MOUNTAIN, BUT FEEL CERTAIN THAT IT WINDS UP IN MARSH VALLEY. THIS TRAIL WAS TRAVELLED BY ALL THE LATE EXPEDITIONS TO THE MINE, AND IS ALMOST APACHE PROOF, AND THAT THEY MADE THIS TRAIL UP THIS MOUNTAIN AND THEN (WITHIN THREE OR FOUR MILES) SEVERAL THOUSAND FEET DOWN INTO MARSH VALLEY, THEIR CAMP, IS PROOF TO ME THAT THE APACHE INDIANS WERE GIVING THEM CONSIDERABLE CONCERN.

ABOUT FOUR YEARS AGO, THERE WAS AN OLD WELL DISCOVERED IN THE UPPER END OF MARSH VALLEY, NEARLY FILLED TO THE LEVEL OF THE GROUND; IT WAS CLEANED OUT BY A COUPLE OF PROSPECTORS AND PROVED TO BE ABOUT FIFTEEN FEET DEEP AND VERY NICELY WALLED WITH ROCK. IT IS MY OPINION THAT THIS WELL WAS DUG BY THE PERALTA EXPEDITIONS AND USED SOLELY FOR DOMESTIC PURPOSES. WHEN I FIRST VISITED MARSH VALLEY IN 1892, I THOUGHT IT WELL NAMED, AS THERE WERE TWO OR THREE NICE SPRINGS FLOWING FROM UNDER THE BANK ON THE WEST SIDE OF THE CANYON, AND THE WHOLE LITTLE VALLEY WAS GREEN AND MARSHY. NO DOUBT THE WATER THERE WOULD BECOME POLLUTED BY THE NUMEROUS HORSES AND MULES ACCOMPANYING THE DIFFERENT EXPEDITIONS; HENCE, THE WELL.

THERE HAVE BEEN FOUND A NUMBER OF MEXICAN WOOD IRONS AROUND MARSH VALLEY; A WOOD IRON IS A BAR OF IRON ABOUT ONE INCH IN DIAMETER, HEATED AND BENT SO THAT IT WILL FIT OVER A PACK SADDLE. TWO ARE USED ON EACH SADDLE AND ARE VERY CONVENIENT IN PACKING LONG WOOD, AS ONE MAN CAN DO THE PACKING. ALSO, NUMEROUS PEONS SANDALS, BOTH WORN AND NEW HAVE BEEN FOUND, BUT ALWAYS IN A CAVE WHERE THEY WERE PERFECTLY DRY. AS THEY ARE MADE FROM THE CENTURY PLANT, MOISTURE WILL SOON ROT THEM. THESE SANDALS ARE WORN ONLY BY MEXICAN PEONS.

NUMEROUS SHOVELS, PICKS, ETC. HAVE BEEN FOUND, WITH THE HANDLES ROTTED AWAY; AT LEAST, THEY FALL APART WHEN HANDLED. BOTH PICKS AND SHOVELS SHOWED A GREAT DEAL OF WEAR. THERE WERE A NUMBER OF SHEETS OF IRON PUNCTURED APPARENTLY WITH A PIECE OF STEEL LIKE A HARROW-TOOTH (SQUARE), EVIDENTLY USED FOR SCREENING DIRT OR GRAVEL, AND A NUMBER OF SHEETS OF IRON NOT PUNCTURED. ALL WERE FOUND IN DRY CAVES SURROUNDING MARSH VALLEY, AND ALL TO THE SOUTH, OR ALONG THE TRAIL WHICH THE MEXICANS WOULD TRAVEL IN COMING AND GOING TO AND FROM MEXICO FROM THEIR CAMP AT MARSH VALLEY.

PHYSICAL EVIDENCES

GEORGE SCHOLEY FOUND A CROSS, AND GAVE IT TO ME, MADE OF MESQUITE WOOD IN A CAVE ABOVE MARSH CREEK. IT WAS SET UP IN A CAIRN OF STONE, AND IT WAS BOUND TOGETHER WITH RAWHIDE, AND HAD EVERY APPEARANCE OF BEING ERECTED OVER A GRAVE, EVIDENTLY A MEXICAN OF MORE THAN PEON IMPORTANCE.

I ALSO SAW THE RACE TRACK THAT OLD JAKE MENTIONED; IT WAS TWO PARALLEL TRACKS ABOUT ONE HUNDRED AND FIFTY YARDS LONG AND LOCATED IN MARSH VALLEY, ABOUT THE ONLY PLACE IN THE SUPERSTITION MOUNTAINS THAT I KNOW OF THAT A TRACK OF THIS KIND COULD BE LAID OFF.

DEERING TOLD JOHN CHUNING THAT ON HIS SECOND TRIP TO THE MINE, HE DID NOT RETURN TO HIS CAMP THE SAME WAY HE DID THE FIRST TRIP, BUT CAME DOWN OFF THE BIG MOUNTAIN INTO A RINCON, AND THERE HE PUT UP FOUR SMALL MONUMENTS. "OH, NOT MUCH MONUMENTS, -- JUST A LONG SLIM STONE STUCK ON END AND FOUR OR FIVE ROCKS LAID AROUND THE BOTTOM," AND FROM THERE HE WENT UP THE BIG CANYON TO CAMP, AND "JOHN, WHEN I WORK THAT MINE, I THINK I WILL GO OUT THE OTHER WAY". JOHN ASKED HIM WHY HE PUT UP THE MONUMENTS. HE SAID THAT IF ANY PROSPECTOR SHOULD HAPPEN TO SEE THE MONUMENT THAT HE BUILT AT THE MINE, AND THEN SHOULD HAPPEN TO RUN ACROSS THOSE DOWN THERE, HE WOULD THINK THAT THE SAME CRAZY PROSPECTOR HAD BEEN PUTTING UP HIS MONUMENTS DOWN THERE, OR IF HE HAPPENED TO SEE THESE FIRST, HE WOULD PAY NO ATTENTION, IF HE HAPPENED TO RUN ACROSS THE ONE HE BUILT AT THE MINE. THOSE MONUMENTS HAVE BEEN FOUND IN THE RINCON.

I AM PROUD TO SAY THAT I CAN OFFER AMPLE LIVING PROOF OF ALL THE ABOVE PHYSICAL FACTS PERTAINING TO THE LOST DUTCHMAN MINE.

Spanish arrastra

JOE DEERING INQUEST

Sep 27, 1885 A.D, SUNDAY
Silver King, ARIZONA

Territory of Arizona }
County of Pinal }

In the matter of the inquestition upon the body of Joseph Deering Deceased,
Before B.J. WhiteSIDE J.P. Acting as Coron

We the undersigned jouriors summoned to appear before B.J. Whiteside Justice of the Peace Acting as coroner, of Pinal County, Arizona, at Silver King on The 27th day of September AD 1885 To inquire into the cause of the death of Joseph Deering, having been duly sworn according to law, and having made such inquisition, AFTER INSPECTING the body, and hearing the Testimony address upon our oauth, EACH and all do say, That we find the deceasel was named Joseph Deering, that he was a Native of Ireland, age about Thirty years; That he came to his Death on the 27th day of September A.D. 1885 by an accident occuring in the Silver King Mine in this County and territory, at about the hour of 2 O'clock AM. SUNDAY SEPTEMBER 27 1885 AND That The Same was accidental and No one to blame therefor,

All of which we duly certify by this
inquesition, in writting by us signed
this 27 day of September AD 1885

Kenneth Mackenzie
Thomas Ashton
William E. Truman
Jos. McCount
P. Stillman
T.E. Thompson
D.B. McNeil

* * *

Jim Bark told this story of Apache Jack:

In the days following the Indian wars in the Superstitions, it was not uncommon for Apaches to take jobs working for the "white eyes." Bark had a friend named George Scholey, a miner by trade, who lived in Mayer, Arizona. Scholey made friends with an aging Indian who helped him do assessment work for sundry nonresident mine owners. One day, as they sweated profusely, while leaning on shovels after finishing a job, George said, "Jack, heap work, little money." Jack responded, "Why you no get mine?" Knowing Indians never revealed that kind of information, George remained stoic except to say he had tried but never could find one.

The next day as they worked, Jack said, "George, maybe we go hunt for deer sometime." Scholey, cognizant of the workings of an Indian's mind, replied quite casually, "All right." A few days later Scholey ran into Jim Bark; he relayed the information. Bark was a treasure hunter. He encouraged George to go on the hunt and offered to foot the bill.

Later they would learn the hunt would take place on the far side of the Superstitions. Their journey would take them to Fort McDowell on the Verde River and then to the east. Apache Jack never discussed their destination, but Bark and Scholey were certain they would head into the Superstitions. Eventually they learned the hunt would take place on the far side of the Salt River in the area of the famous Apache Trail.

Shortly before they left, Jack said, "Medicine man no like for me to go. He mad." The pair left in George's car and headed for First Water on the Bark Ranch east of Goldfield. From that point they would travel on foot, only returning to First Water for provisions. They spent two weeks hunting, killed a deer and dressed it out.

Eventually, in true Apache fashion, Jack suggested they would have to climb a peak known as Black Mountain. George never mentioned the mine. After a long hike, the pair topped out on the summit that stands between La Barge Canyon on the south and Peter's Canyon on the north. (The two canyons converge at Mormon Flat.) They were headed east. George noticed Jack was lagging behind. The Indian became fidgety then, suggesting firmly they return to camp, while appearing extremely nervous.

George was angry. He had come a long way and had spent a good deal of time and money. At this point he asked, "Where is that mine you told me about?" The Indian actually shook while waving his arms in a great arc. The sweep included Tortilla Mountain and Weaver's Needle. Turning, Jack rushed away.

(Later, Bark recalled that Sims Ely and a prospecting friend had been to that spot the year before and had found 40-odd Mexican firebeds, which are dug-out impressions filled with scrub and wood and left to burn to provide a warm sleeping area.) The Indian hadn't become edgy until they were overlooking the plateau containing the firebeds.

Jack did not regain his normal composure for at least three miles. Later, he sat down to rest. Apparently he was ashamed of his refusal to talk, so he began to explain. He pointed to a place where two crazy white men had worked. They had taken out gold but were later committed to the mental institution in Phoenix. "Over there, big fight one time." Then Jack proceeded to tell a remarkable story. "A lot of Mexicans worked mine. The Apaches and Mexicans have no trouble, long time. Apaches stay near, have good food to eat. One time Mexican do bad things to Apache squaw. Bad medicine and Apaches fight."

There were only two escape routes the Mexicans could take. These led south. The Apaches had all the passes well guarded. Soon the Mexicans found themselves cut off from any escape routes to Mexico. They were being herded toward a western abutment of the Superstition wall.

Apache Jack was a young brave about 12 years of age at the time. He lived in an Indian village near the military camp of Pickett's Post, near the ancient mill town of Pinal. The Indians had been moved there to aid the military in caring for them and in keeping an eye on them.

A runner brought word of trouble in the Superstitions. The braves slipped away to aid their brothers, and Jack went with them.

After three days most of the Mexicans were dead. The last of them had been brought down on the exact spot where the two men, who went insane, later excavated. (It seems relevant to state here that those men probably found the gold that had been dumped off the Mexicans' mules as they fought to escape. The mules were no doubt eaten, the Mexican panniers slashed and the gold dumped on the ground to be covered by the wind and sands of time. The prospectors found the gold, assumed it was from that area and kept digging. Eventually they couldn't find the source and went mad. They were put into an asylum where they later died. Their names were Silverlock and Malm.)

Jack also told Scholey of the plan the Indians devised up on that mesa. It accounted for the fact it was exactly 40 years before anyone found the treasure again. Another group of Mexicans had molested some Apache women; a decision was made. The Apaches decided to bring in the squaws to cover the mine. Removing all traces of the pit, they filled the hole with anything they could find that was movable. The squaws worked one entire winter. The surface was smoothed efficiently. All evidence of Jacob's mine vanished.

Apache Jack had sealed his fate. He had come within a hair of telling his white friend where the mine was. This Indian was without cunning and was slow thinking. Leadership in their tribe was gained through fear and intimidation. Although Jack had not divulged the exact location, he had gone farther than the medicine man would allow. The Indians had been told their silence was under pain of death or insanity. Being insane was worse than death. They believed insane people were possessed of evil spirits and could never gain entrance in the "Happy Hunting Ground."

Jack told his white friend on the way back to his home, "George, if me get sick, I send for you. You come quick. Then I tell you things." Scholey, not realizing the seriousness of the matter, offhandedly reassured his friend. A week later the incident occurred. A breathless Apache, who had just run 14 miles to Scholey's house from the settlement, gasped out, "Apache Jack sick. Mr. Scholey, come quick." Scholey was working miles away on an assessment job. His wife didn't know how to reach him and had no transportation. Scholey returned a week later and went to the Indian village. He was too late – Apache Jack was dead.

Scholey knew a few of the braves and questioned them as to Jack's symptoms. He was forever convinced Jack had died of strychnine poisoning and that the medicine man was the murderer. Before George Scholey left the village a blind squaw asked to talk to him. She was a great friend of Apache Jack and mourned his death.

In a strange encounter, she explained she had worked for whites who taught her to speak pretty good English, and she also was familiar with the situation that killed Apache Jack. It turned out she was one of the squaws who was brought to cover Jacob's mine. Her words were, "Three big rocks, like wickiup." Then she traced a sort of triangle in the dirt with her walking stick. "You go by them – maybe you go down on a rope." Just then a big buck came by and kicked her on the leg. It was a warning. George Scholey's interview concluded.

Chapter 10

The Massacre Site

For many years Dutchman hunters and historians have argued about the story of the Mexican massacre. In fact, most of them deny the story. It seems odd it would be such a bone of contention. But, as we have stated before, this book offers proven answers to many longstanding questions involving Waltz, Weiss and the lore surrounding the mine.

The Spaniards and later the Mexicans were prolific miners – geniuses at locating the minerals and, once they had, knowing how to make the drift entrance so small it could be easily hidden. They antagonized the Apaches, and as time went on, fierce battles took place in the Superstitions. We believe near Waltz's mine was a large encampment where Mexican peons mined and milled ore. It is presumed these are the miners who were all slaughtered at one of the entrances to the Superstitions.

The story is brutal. It is, however, verification of the fact the peons were there to work a mine, and they obviously left carrying ore and never arrived at their destination.

Following are the diaries of Joe Green, William Edwards (soldiers in the 1st Arizona Volunteers) and Matt Cavaness, a respected Arizona pioneer, regarding this famous massacre in the Superstitions. This should end speculation that there was never a Mexican massacre.

"This account is taken directly from the journal of Sgt. Wm. Edwards who himself was a member of the 1st Arizona Volunteers Company 'F' from November 1865 through November 1866.

"Wm. Edwards grandson (Benjamin Edwards) born at Sawmill, AT. at the turn of the century was taken by his grandfather and shown the exact places where he and trooper Joe Green discovered the bones of the Mexicans and later their camps. Ben Edwards, who is related to my family (Roberts) took me into the mountains and showed me these places as well as other discoveries that have been made over the past century.

"When Ben Edwards passed away I inherited all of his and his grandfathers things, including the items that were discovered, some more than a century ago.

"Here is the original account, exactly as it was handed down to me." – Mathew Roberts

Massacre site where the skeletons of the Mexicans were found.

Wm. Edwards, Joe Green and the "Massacre Site"

Early in the year of 1866 a column of the 1st Arizona Volunteers was racing north toward the Salt River, they stopped to rest and water their horses at a cool, shady place of running water called "Bracketts" water (Weekes Wash). The troopers were chasing a band of renegade Apache who had killed some miners near Pinal. The Apache were attempting to elude the cavalry by melting back into their stronghold in the Salt River Mountains (today known as the Superstition Mountains). While the troopers rested in the shade of some Cottonwood trees and watered their mounts, the Lieutenant sent two men out to higher ground to watch for a pair of indian scouts who were overdue in reporting back to the column. One of those men was Sgt. Wm. Edwards and the other man was Joseph M. Green a scout for the 1st Arizona Vol's. Edwards and Green rode out along the northwest base what is known today as Superstition Mountain. They came upon a series of rises to get a better view of the surrounding area, as they came up along a long low ridge they reached a hilltop and there they made a most startling discovery. There before them, scattered across the hilltop, lay a grisly sight. The skeletal remains of about twenty-five people lay strewn across the ground. Green and Edwards dismounted and walked among the remains looking for clues as to who the people were and how they came to die there. It appeared to the two soldiers that the bones had been laying there for at least two years and maybe more. No clothing or personal effects could be seen to aid in identifying the remains. How these people died was unmistakable to Edwards and Green, pieces of broken arrows were present and bore the distinct workmanship of the Apache. Several of the victims skulls were crushed, undoubtedly by a blow from a war club. Joseph Green at first thought the dead were rival indians of the Apache, perhaps they were Maricopa's or more likely Pima, indians who fought and were killed in one of the many battles the tribes fought in those days. Wm. Edwards however noticed something that told him differently, there were no clothes to be found anywhere, Apaches would strip the clothes from their Anglo and Mexican victims as well as take all their personal effects but not as a rule the rival indians they would kill in battle. A rival indian might possess an item that held great medicine (power) and would work against anyone who killed it's owner and tried to use that item. But most importantly, Edwards noticed one of the skulls bore a tarnished silver tooth. No indian would have had any dental work done on them, to Edwards the bones had to be that of Mexicans as no large group of Anglos were reported missing in the Territory, or had been reported in the past to his knowledge. Wm.Edwards kept his observations to himself and presently he and trooper Green returned to their detachment and informed their Lieutenant of the discovery. The Lieutenant hastily investigated the scene and quickly came to the same conclusion as trooper Green. Being in a hurry to overtake the Apache renegades before they slipped away into the twisting Salt River Mountains, the Lieutenant ordered his men into column and called "forward-ho" into those mountains, leaving the scene exactly as they had found it. Wm.Edwards kept quiet about his suspicions, according to his journal the column followed the renegades into the Four Peaks country crossing the Black River (today known as the Salt) at the "rocky shallows" (Mormon Flat). Two days later the column chased the Apaches west to the Verde River where they lost their trail. The overdue indian scouts were never seen or heard from again. Edwards and the detachment searched for two more days to try and pick up the renegades trail but gave up after a driving rainstorm and the column turned back south to Ft. McDowell following the Verde River.

The story may have ended there, an insignificant event in the history of an emerging Territory had it not been for the keen eye and observant nature of trooper Wm. Edwards. Edwards was about to finish his enlistment with the cavalry and he planned to stay in the Territory and do some prospecting in hopes of striking it rich and making his fortune. The scene he had come across only made him think, why would a large group of Mexicans get themselves killed way out in the middle of Apache country, miles from a route to anywhere ? The only answer was, they were miners, Edwards had heard stories about Mexicans who came up from Sonora to work hidden Spanish mines lost to them forever by the Gadsden treaty that made much of what used to be Mexico, now a part of the Arizona Territory. These Mexicans would slip into mountains, work their mines, then cover them over and return to Sonora with their gold and silver. After his enlistment was up, Edwards went back to the "massacre" site, he planned to begin prospecting in the Salt River Mountains (Superstitions) then continue northwest into the Bradshaw's and end up in Prescott. But there at the massacre site he made an even more startling discovery that changed his prospecting plans and altered not only his life, but the lives of his son and grandson, not to mention a good number of other people who were to become acquainted with them.

The bones lay there exactly as he had seen them last, bleaching in the Arizona Sun. Edwards began to search around for more evidence and he didn't have to go far, less than 300 yards from the hilltop he found another skeleton, same as the others, no clothing or effects. A distance from that skeleton he found another, Edwards began to realize the massacre on the hilltop was the final stand of a running battle that started somewhere else in the mountains. By backtracking where the skeletons lay, and following the general contour of the land, Edwards was able to guess from which direction the Mexicans had came. As Edwards followed the trail of bones and skeletal remains he began to find bits and pieces of personal effects of the victims, a piece of sandal, a shred of clothing, then he happened upon the most startling discovery yet. In the words of his own journal he writes : "Come down off a little rise and see off on my right what looks like a white thing half hid under a leather bush (known today as JoJoba). Is another skeleton, only this one is whole, got the clothes and all with it. Pick on through the bones and find a knife, rusty and stained, a little medal, (a medallion) too rusty to read and a small poke bag under the mess. In the bag is about a pound of gold quill (concentrated gold ore). It sure looks rich and I know for sure they were working a mine around here somewhere about."

Edwards reasoned this Mexican must have been badly wounded in the skirmish and crawled under the bush to hide then died from his wounds there. In the confusion of the chase he must have been overlooked by the Apache, as his were the only remains found that were unmolested. Edwards continued to backtrack along the grisly trail and found more bits and pieces of the battle until finally he came upon a lightly worn trail that eventually led him down into a large canyon at a point where that canyon made a sharp bend and widened out considerably. Tall trees abounded in that area of the canyon, Cottonwood, Sycamore and Aspen grew tall there, several springs flowed from the west side of the canyon at it's bend and the canyon bottom was covered in tall grass four to six feet in height. The canyon seemed an oasis in the wilderness to Edwards. Just up out of the west side of the canyon was a flat area with a large stand of Mesquite. On that flat Edwards discovered the remains of a large encampment. Everything in the camp had been destroyed or torn up. It was the unmistakable work of the Apache. The camp had obviously been the Mexicans base of operations in the area. A closer examination of the camp and surrounding area turned up broken mining equipment and tools as well as the grisly spectacle of more skeletal remains. Edwards rested in that camp for several days contemplating the things he had seen. He came to the conclusion the Mexicans were mining somewhere in the area and had been attacked by the Apaches. They made a run for it but only got a few miles before they were cut off and forced to make a stand. It bothered Edwards that he could not evidence of an "arrastra" anywhere near this camp. He did find a few small pieces of hand cobbed gold concentrate at the camp but not the mine or the place where the ore was being concentrated. Edwards explored the area over the next few day's, following the different trails that came and went. One trail took him down canyon where he discovered a well dug into the east bank of the canyon bottom. The well was about twenty feet deep and lined with rock so not to cave in on itself. Near this well a fairly worn trail led up into a steep side canyon that appeared to be a box. Edwards followed this trail and it led him up the side canyon and out onto a large flat area high above and overlooking the main canyon. It was up on this flat area that Edwards made his most significant discovery yet. There in a rocky clearing, surrounded by thick Mesquite, Ironwood and Laurel trees was the Mexicans work camp, the place where they broke up the ore for the trip back to Mexico. A crude arrastra could be seen as well as mounds of crushed rock. A large pole, fifteen feet high, had been broken off at ground level, it had been used as the fulcrum for a crude but effective crushing stone. Edwards picked up several pieces of quartz with free gold attached, identical to the gold found in the dead Mexicans poke sack. Edwards knew the mine had to be close by, several trails criss-crossed the flat mesa top so he began to follow them to see where they might lead. Edwards could see there had been a sizable herd of animals on this mesa at some time in the past, horses mostly but also cattle. He felt he was closing in on the source of the Mexicans gold, persistence paid off when he decided to follow a path that led into a steep rocky canyon running North and South. In that canyon he began to find fragments of mined rock here and there. There seemed to be no large "dump site" and Edwards soon discovered why. The gange (waste rock) from the mine had been dumped into a seam that ran parallel to a fault which ran the length of the canyon. Edwards found the place in that canyon where the mine should have been but could see no mine opening, he realized the mines entrance must have been covered over. He began to dig and probe where he guessed the rock had been disturbed. This looked easier than it actually proved to be. After pulling down and digging away tons of rock expecting any moment to find the opening, Edwards was met with more and more rock. Edwards persisted in trying to locate the opening until one afternoon he happened to glance up along the eastern ridge of the canyon, what he saw froze him in fear. A lone Apache was making his way along that ridge, looking intently into the canyon below. Edwards was certain he had not been seen but sound carries far in those canyons and he was sure he had been heard as he dug and tossed rock aside. He watched as the Apache sat for hours intently looking for the source of the sound. As darkness fell the Apache moved back over the ridge and disappeared. Edwards waited until the moon set then cautiously made his way out of the canyon and eased back to his camp in the Mesquite thicket. An uneasy chill ran through him as he realized his horse was gone, without going any further he slipped away in the darkness and abandoned his camp and outfit. He made his way in the dark, moving only at night, afraid to go out in the open in the daylight. After a couple nights of traveling, hiding out by day, he reached Fort McDowell, tired and sore but alive. His close brush with the Apache left him wiser and considerably more cautious. After resting up and re-supplying himself at McDowell, Edwards set out once again for the mine. Once again though he ran afoul with his deadly enemy the Apache. This time he had barely entered the mountains (Superstitions) when he encountered a group of indians about a half mile distant. He watched as the group split apart and he knew they were going to try and cut off his escape routes and drive him deeper into the mountains. He quickly wheeled around and charged back down the trail he had entered the mountains on. Then, guessing the Apache had already doubled back on him left the trail and rode thru washes and arroyos, cutting across a rough stretch of desert. When he reached a high point he looked back to see that indeed the Apache had gone around to cut him off, when the two groups of indians met at the spot where he left the trail they howled with anger for letting him slip away. Edwards years as a soldier and scout had taught him a very valuable lesson of survival in this hostile land, to stay alive, one had to stay alert every minute and to always think like your enemy. This second close brush with death discouraged Edwards from further searching at that time. He thought it better to wait a while and let things cool down a bit between the Army and the Apache. He traveled up to Prescott in the Bradshaw range where he met up with a friend who was prospecting with a group of men around the (present day) Walker area. For several months he knocked around the hills with this group having only mild success. After a spell he tired of living in the mountains, eating mostly beans and occasionally fighting indians. His thoughts drifted back to the hidden mine, if only he could go back there he knew he would find the mine and secure his fortune in no time at all. Edwards left the group of prospectors and went back to the Prescott area. He found work in a nearby mine as a carpenter and timber man. In the spring he purchased an outfit and a couple horses with the money he had saved up. Then early one March day he headed back toward the mountains and the Mexicans mine. Edwards hoped to slip into the mountains undetected and find the source of the Mexicans gold, take out a sizable amount then slip back out before he was noticed by the Apaches. But once again fate stepped in and altered his plan. Camped along the Verde River near it's confluence with the Salt, a violent storm roared through his camp late one night, Edwards was drenched, then in the icy cold wind became ill and caught pneumonia. The pneumonia he contracted that night plagued him the rest of his life, that disease and a later fall from a horse that severely injured his back would effectively end his search for the Mexicans mine. Wm. Edwards returned to Prescott when he was well enough to travel again, he didn't give up on his dream of finding the mine, instead he had to delay it while he recovered and regrouped. Edwards took a job at the Vulture mine when he had recovered sufficiently, all the while the mine was foremost in his thoughts. After all he reasoned, if he had found the Mexicans camp, surely someone else eventually would find it also. ------------------

Excerpt from the *Matt Cavaness Memoirs*

Joe Green was Matt Cavaness's longtime friend who came to Arizona back in 1863 with the Peeples/Weaver expedition. Green appears on the 1860 Tulare County, California, census on the same page with George Roberts, A.H. Peeples, Oscar Hutton and Henry Wickenburg. Green was a trooper with William Edwards when they discovered what was to later become known as the "Massacre Grounds."

" During the time that I was on the ranch I seen Joe Green that I hadn't seen since he left us at Wickenberg about 1865. He had been working for the government (Army) all this time. When we met he said to me, " I got something to show you Matt and I think it's all right." I went to his house and he took out a piece of ore about two pounds out of his canteen and handed it to me. I took it and looked at it and all on one side of it was covered with gold. Just a beauty to look at. I asked him where he got that piece of ore? He stepped to the door and pointed out easterly from McDowell some twenty miles. "Now Matt, I tell you what I have thought the best for us both is for you to say what day you can go with me." We can ride there in one night. You locate the first claim, three hundred feet and I (Green) will take up the other claims on each end and take in Major Alexander and Capt. Carr and in fact all the commissioned officers in the post as I have promised to locate them in anything I find. And I tell them you found the mine and was waiting for me to come in before you located it. Of course they put some men out there to work and they have to have an escort of soldiers. Then you can put some men out there and I think we make a stake. I didn't see very much of the mine as it was in a dangerous place and I was quite a distance from the command and felt like I better keep moving. He (Green) was out hunting for deer and at the same time he was looking for mineral also. He (Green) was out scouting all the time only when the soldiers would come in for supplies. Now this was Sunday morning and I say to Green, " I am the only man on the ranch that can drive that ten mule team and haul in the hay to the post and we have to deliver that hay every day." Well, he said, you set the time and we go. I says, how will it suit you for us to go next Saturday evening? He says, that's all right, we go. We take a little lunch and a little grain and we get Sunday morning and put our horses where they can't be seen and we stake off the claims that day that day and come home that night. (Green) "We got to be on the lookout and it's right out in the Indians country and if they should see us we have a fight sure enough." I says (Cavaness) I don't care for that. He laugh at me and says, I been in lots of Indian fights Matt, since we parted in Wickenberg and I tell you it's no fun to fight Indians. I (Green) know just how to go to the mine in the night and we hardly be seen in the daytime and it's in country where no one is liable to be. (Green) Be sure and not let anyone know that I brought this piece of gold. If it does turn out to be something good, we wan't to get hold of it. And you bet I say's, I got some money too. So we had all plans arranged for next Saturday night. Monday morning about ten in comes (Ft. McDowell) two Indians and want to have a talk with the commanding officer. But a few day's before that they had him and had a powwow with Major Alexander and he had given them some rations and blankets. They camped close to the post (Ft. McDowell) a few nights and one morning they was all gone. So the guard at the post reported this to the General and he gave orders to put them in the guardhouse. He say's, I learn them to be good. I give them food and everything and it didn't last ten day's and they go to stealing and killing again. So in the jail they go put them in a cell where they couldn't see out. There's one guard in the whole house and about Wednesday morning they bring some coffee and hardtack for the Indians and they open the cells and let them eat in a room in the center of the jail with the guard sittin by the door. Those two Indians look at the food and all at once makes a leap and grabs an ax that set in the corner of the room for cuttin firewood. Mr. Indian comes down overhand with that ax and and hit the Sergeant on top of the head it glances and cuts off one ear and broke the collar bone. Outside he run, the other one grabs the guard guns and runs back in the cell and close the door. All the guards run and commence shooting at the Indian running across the plaza. Now he was running and Lt. Frasha runs out of his room after him and shoots him in the back killing him dead. The bugle blows and the whole post was out on the Plaza and Capt. Carr reports to Major Alexander that one Indian is in his cell with a pistol and a rifle with a bayonet on it. Major Alexander issues orders to take him out of that cell so Capt. Carr took charge and marches everybody around the guard house. They could look in and see the Indian armed to the teeth in his cell. If they runs in he will kill somebody for sure, so they cut holes in the outside wall and commence shooting inside the cell. Well, Joe Green is standing outside one hole and he shoots through the hole at the indian inside, things get quiet so Joe looks through the hole and the Indian stuck the bayonet through the hole at him. It didn't hit him in the face but hit him in the left breast and put it most way through him. We pack him off to the post hospital and Dr. Gwyther tries to patch Joe up but he die that night. Just before sundown they finally kill that indian in the cell but he shoot another soldier before they do. I never had a better friend than Joe Green and I know he found that mine. When I got through with haying for the Army I had too much money to go looking for a mine, but since then I have gone looking for it several times but didn't find it. I know it's there and it will be found some day by someone. ——————

Matt Cavaness Jr.

The following items have been discovered in the area known today as Peter's Mesa by William Edwards, Ben Edwards and Mathew Roberts.

1. An Iron strongbox, 14" x 10" x 8". Inside the strongbox was a fully loaded Patterson model Colt 5 shot revolver. A homemade butcher type knife with a wooden handle. A tool for re-loading the revolver, a second cylinder for the revolver also fully loaded, 10 lead bullets, some crumbling paper and a small leather pouch with 7.75 ounces of hand cobbed gold concentrate. The revolver was rusted solid and appeared to have been in the strongbox at that location for at least half a century. The strongbox was also rusted shut and had to be pryed open. The papers were unreadable as they crumbled into little pieces when handled. The revolver was found to be an 1846 model, the serial number was only partially readable. The strongbox was found inside a small hole dug into solid rock at the base of a tall rock formation. The entrance had been walled up and filled in to hide it's existence. The hole was not natural and had been neatly mined out to hold the strongbox. Ben believed from the size of the hole in comparison to the size of the strongbox that the hole had not been dug with the strongbox in mind. Something else was cached in that hole before the strong-box had been hidden there. Ben found tiny pieces of white to grayish quartz in the dirt at the bottom of the hole when he sifted it out. The size of the cache hole is about 2 feet x 3 feet x 3 ½ feet.

2. A short handle miners pick, a short handle shovel, a 5 pound hammer with a broken handle that had been wrapped with rawhide and a cloth strip, 8 hand forged drill bits from 8 to 16 inches in length, several iron wedges, two well worn sharpening stones, 50 feet of rope which had rotted badly, a crude hand made miners bar and an army canteen that held about a gallon of water. The canteen was dated as one that was used by the army prior to the Civil war 1861-1865. The miners pick was forged in St.Louis, Mo. identified by the forging stamp on the head. That forge burnt down in 1851. None of the other items could be successfully dated. The items were all found together in a natural cave whose dimensions are 4 feet x 3 feet by 2 feet.

3. Three sticks made of a wood not native to Arizona, two of them were cut exactly to 33 inches in length and the third was exactly 3 times as long at 99 inches. They were found in a cave with several clay pottery jars which probably were used to hold water. Also found in this cave were the tattered remains of an old overcoat.

4. An empty "tin" , a metal container used in the old days to pack seafood or canned meat. The tin was of the type used between the end of the Civil war and the 1870's. The container was large enough to hold 5-10 pounds and was sealed with lead. Some of this lead was still attached to the containers lid. The "tin" was found empty and half buried in soft dirt under a prickly pear just a few yards away from the site of a very old camp.

5. The partial remains of two separate skeletons were found at (site # 5 and # 6) both skeletons had been "buried" in shallow graves and the bones either uncovered by rain and elements or dug up by animals. Both skeletons were found at the base of a ledge. Bits and scraps of leather were the only remains of clothing. No weapons, money (coins) or personal effects were found with either remains so it is assumed the bodies were intentionally buried. It was impossible to tell how the people died or who they were, the bones had been in the ground, a very long time. Both skeletal remains are now buried at site # 27.

6. An old worn out pair of boots, the type miners wore in the 1860's. Hob nail style. The laces were untanned strips of cowhide or possibly deerhide. The boots were found together and as if they were placed where they were found in waiting for future use. Found at site # 23. They were behind some rocks just off a seldom used trail at the crest of a canyon. There was no right or left boot, both boots appeared to be exactly the same but impossible to tell if made as right or left.

7. The fully clothed remains of a skeleton found under a leather bush (jojoba) by Wm. Edwards of the 1st Arizona Volunteers F Company in November of 1866. With the skeletal remains were items of clothing, Mexican style, a knife, a rusted medal on a chain (medallion) two pieces of flint and a small untanned cowhide bag containing about a pound of very rich highly concentrated gold ore. It seemed to Edwards the man had crawled under the bush to hide and died while there.

8. At a site known to some as the "salt flat" on Peters Mesa, a bare area of rock sits on a rise. Here was found a work camp for the breaking down of ore by hand (hand cobbing). Ore would be brought to this place and workers would bust the quartz away from the rock to concentrate it into it's highest quality form so it could be transported more easily to an arrastra or refining site. At this site numerous small pieces of quartz with gold attached have been found over the years and some may still exist to this day in the area. The last piece discovered there was found in 1989.

9. A Spanish or Mexican drift mine dug along a fault line on Peters Mesa. The drift angles back into the hanging wall (westerly) and slopes downward at a 5 degree angle for about 60 feet where it dips sharply at a 45 degree angle for another 80 feet where it widens out on a lode vein. The entrance to this drift is concealed to be no more than two foot square and the addit had been completely concealed at one time. The waste rock for the drift was carried down canyon and dumped in a fault seem about 200 meters from the drift. Samples from this mine do not match well with the gold pieces discovered on the surface of the Mesa. discovered by Wm. Edwards in late 1866 or very early 1867.

10. Discovered walled up in a small natural hole in the rock face of a ledge was 19.75 lbs. of quartz/gold concentrate. The quartz appears to have been broken down by hand into smaller pieces of a more concentrated (valuable) form, eliminating weight and reducing the size of the overall load. The gold content has been assayed and the quartz/gold analyzed for mineral content. These findings are recorded in my files as (assay report number 26) and (Crissmon and Nichols analysis report no. 26). The ore concentrated ratio ranged from 55 to 70 percent Au. gold content with native Ag. (silver) present (and visible) in the ore at (+ or –) 1.3 percent.
Au. Gold content established at 11.455 lbs. at 996 fine.
Ag. Native Silver content established at 0.144 lbs. at 997 fine.
Location of the discovery site, photo of the ore and site, and the account thereof is listed in my files under (Site #20).

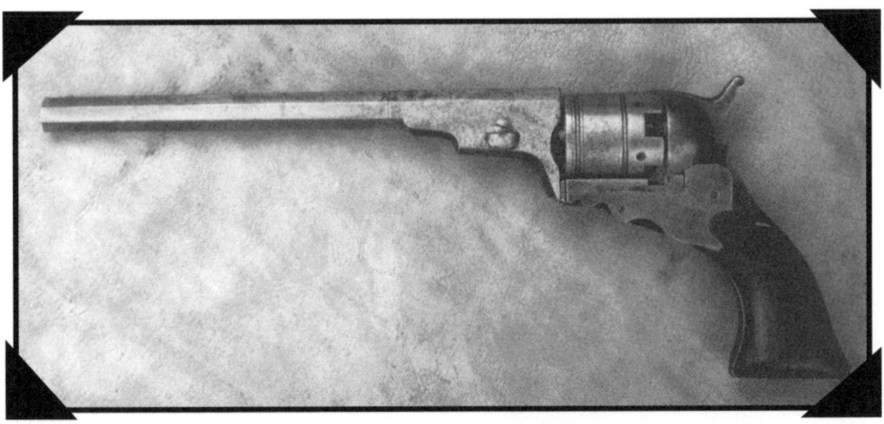

Photograph of the pistol (Paterson model Colt) found on Peter's Mesa.

Chapter 11

The Tragedy of Adolph Ruth

The story of Adolph Ruth – one of the more contemporary stories involving the lost mine – has been contemplated for years by true Dutchman hunters. Because of new evidence, this book will present a totally different theory regarding the outcome of the story.

Fate is an interesting thing. One often wonders why people take the routes they do, and when the result is fatal, we all question why and how it happened.

In this case it started because of a political matter. To digress we need to go back to 1913 when General Carranza, a former governor of a Mexican state, led a rebellion against General Huerta, an upstart, whose efforts resulted in the assassination of Mexico's President Madero. The Americans supported Carranza in his bid to retake the country; however, the general needed money badly. His plan to recoup funding included selling cattle to Texas. Being aware of our strict sanitary codes, he hired an American, Dr. Erwin Ruth, to inspect the cattle. Ruth was a veterinarian of stature in the U.S. and Mexico.

During his tenure, he was riding south in Mexican territory when he was met by a squad of soldiers escorting a carefully guarded prisoner. Ruth immediately recognized the captive as a prominent insurrectionist named Gonzales. Erwin asked if he could speak to the accused. His request was granted. Gonzales told Ruth he was headed for the firing squad and asked the doctor for a favor. Ruth agreed.

Gonzales wanted Ruth to go to his home and escort his wife and daughter to El Paso to find a safe haven. In return for the favor, his wife would give the doctor papers and two maps that were hidden there. Gonzales said the maps were to two gold mines in the United States. Mrs. Gonzales's maiden name was Peralta. Ruth kept his word. Later, after he arrived home in Washington, D.C., he showed the maps to his father, Adolph. The old man's interest was definitely peaked. He suggested they search for the lost mines; although, the family was dead set against the venture.

Adolph Ruth (circa 1930)

Erwin loved his father and wanted to please him. He was, however, a very busy man. Several years would pass before they were actually able to leave on their first adventure.

They failed to locate the California mine, and during their quest, Adolph got separated from his son, slipped from a ledge and broke his leg. A doctor was summoned who operated on the man and inserted a steel plate. The accident left the elder Ruth with a limp, as one leg was shorter than the other. From that time on he used a cane.

Shortly, Adolph declared his intentions to go to Arizona to hunt for the other lost gold mine. He had succumbed to an often fatal disease: **gold fever**. His son was too busy to accompany him and said so. No fear or family objection could dissuade him. Adolph made plans to leave after convincing a friend to join him. The entire matter sounded exciting and very appealing, and the pair left for the West in May 1931.

Their journey was pleasant until they reached the outskirts of Phoenix, where they stopped at a gas station to ask directions to the Bark ranch. It was over 50 miles from town to the base of the Superstition Mountains. The desert heat was intense; the remoteness was unsettling; and the roughness of the upcoming country soured his companion, who immediately called it quits. Ruth, however, was determined to go on alone. The pair parted company after Ruth received directions to the cattle ranch.

Jim Bark had sold out to Tex Barkley, who now called the place the Half Circle U Ranch (later it became known as the Quarter Circle U). There were several men on hand when the old man arrived. Tex was there, a cowboy or two and some old prospectors, one of whom was Brownie Holmes. Ruth spoke candidly about his intentions. Because he was a greenhorn prospector, he even stated he had a map and boasted they'd be surprised at how quickly he would go right to the mine. Up until now, most folks would have suggested that statement cost him his life. It certainly was bad judgment. Although the cowboys seemed indifferent, the prospectors exhibited an intense but furtive interest.

Bark's Ranch was sold to Tex Barkley, and he called it the Half Circle U Ranch – later known as the Quarter Circle U (circa 1962).

Ruth asked if there were a sharp peak anywhere near the ranch. Tex was quick to reply, "Why, yes, that'd be Weaver's Needle." The rancher explained it had been named for Pauline Weaver, a famous mountain man who had been a friend to trappers, Indians and whites alike and whose entire life had been part of Arizona's history. [*The rock also looks like a needle formerly used by weavers. (*Barnes' Arizona Place Names, *1972).* – Ed.]

Ruth appeared excited. He questioned the rancher at length about the terrain, camping spots and the peak itself. Barkley was a realist. He shook his head slowly before explaining that no one would go into that country this late in the year. He further stated it was not safe for anyone to go into the Superstitions alone at any time.

We can only presume that Ruth was completely unaware of the stories surrounding the search for the mine or the curse. Given his reaction to the information from someone who obviously knew, one could predict it wouldn't have mattered anyway. The man, it seemed, was intractable. "Never mind that," Ruth replied. "Is there any water near that peak?"

Reluctantly, Barkley admitted there was always water in West Boulder Canyon, less than two miles from Weaver's Needle bearing a little northwest. Seeing Ruth's grit, he added that the best route was the trail up Willow Canyon, the mouth of which was only a mile from the ranch. He explained that this trail led to a summit overlooking Weaver's Needle and then turned down to it. The pool was only a quarter mile or so down that canyon.

Barkley was busy preparing for a cattle drive to the railhead where they would be shipped, but he offered to pack Ruth in himself if Ruth would wait until he had finished his business. Tex left almost immediately. As soon as he had gone, the prospectors offered to pack the greenhorn into the Superstitions (on June 13). They returned on June 15 saying they had left Ruth in camp. Barkley returned several days later. Upon learning what had happened, he went after Ruth without delay. Tex said later, he'd had a hunch. Tex Barkley was a sincere man; he must have felt some responsibility for not having stopped the old man from going into the Superstitions.

It was hot. The caliche crunched under the horses' hooves as Tex and a cowhand made their way through bone-dry washes. Once they reached the camp, Tex realized there hadn't been anyone there for more than 24 hours. Further examination of the camp revealed Ruth was wearing only light shoes when he left. Barkley was puzzled. He and the cowhand continued down the canyon, firing their guns and calling Ruth's name, but to no avail. They continued searching until dark before returning to the Half Circle U.

West Boulder Canyon is at the divide between Maricopa and Pinal counties; therefore, Barkley notified both sheriffs. By morning deputies had arrived for the search.

Although it might sound simple, the search would prove arduous. The sheer rock-walled canyons absorbed the blistering heat of the summer sun that in turn radiated out all day and long into the night. Man and beast dripped sweat. The breeze seemed to die, leaving them breathless. Rattlesnakes, the dreaded reptile of the mountains, had long since crawled into dark crevices, scorpions were imbedded in damp sand hidden from view, and all other animals had gone to higher ground. Even the birds were silent.

Weaver's Needle stands among three canyons. The canyons converge some two miles north of the spire, and at that point there is a ridge a mile long covered with dense brush. The ridge runs northwest and finally is lost in a right-angle intersection with La Barge Canyon, which runs northwest to its junction with the Salt River.

One thing is certain, the man with the "leather" face, conditioned to heat and parching sun, might be able to sustain himself under these circumstances. He would know to hide himself under deep shaded canyon walls, seek cool seepage from a spring and stay out of the noonday sun. But, it was evident from Ruth's attitude he feared little and really didn't listen to the advice given by men who had spent a lifetime in those canyons.

Ruth's family was notified. His wife was beside herself with worry, and his son left immediately for Arizona to start his own search.

It is important for the reader to realize the time frame of this incident. It was during the Great Depression. The country was reading about gloom and doom – death from suicides and the unpleasantness of the hunger of the unemployed hordes who stood in soup lines. News was scant. The story of this search in the scorched mountains east of Phoenix whetted appetites and took on international import. The publicity did many things. It enhanced the searches and brought many strangers to the area. It also reawakened the search for Jacob's gold.

At Ruth's camp Tex found the following letter:

> My dear Wife and Children,
>
> Yesterday, Saturday, June 13th, Mr. Purnell and Jack Keenan and I rode 3 burros and two carried my tent, bedding, fifty pounds of flour, 10 pounds of sugar, coffee, etc. I rode my burro until we got to this water. I didn't get off because I was afraid I could not stand on it again. Purnell and Keenan both got off their mounts and walked. It sure was rough, sometimes straight up and down. Several times my burro slid down stiff-legged and from six to ten feet.
>
> We left the ranch at 6:30 A.M. and arrived here at 10:00 A.M. They at once put up the tent and my bed. Everything else was put on the ground. I was too tired and aching to do much. I could hardly stand up. The distance we made was between six and seven miles. I went to bed about 7:00 P.M. and got up at 6:00 A.M. It was cool last night, but the flies bothered you as soon as it was daylight. The highest yesterday was 93, but now, at 2 o'clock it is 94.
>
> I am sitting under a big willow tree, water in front of me, water behind me. I made a shelf of sticks, wired them together and tied them with the axe. I put the bacon, chocolate, prunes, sugar, etc., on it.
>
> I'll gradually put things ship-shape. Sometime tomorrow morning I'll prospect some. I left my car at the Barkley Ranch.
>
> Love A. Ruth

Ruth's son learned later that the cowboys, Keenan and Purnell, came back to Barkley's ranch to use the old man's automobile to go joy riding about the countryside. They also took their lady friends to the picture show in Phoenix, 56 miles away.

Erwin Ruth claimed the pair returned on June 18th and found Adolph's camp abandoned. Sims Ely reported that Tex Barkley was the first one to view the abandoned camp. Whatever the truth may be, the search lasted 45 days and then was ended because of the July heat. Temperatures near or exceeding 120

L.F. Purnell (left) and Jack Keenan were cowboy prospectors who packed Mr. Ruth over the southern pass to West Boulder Canyon.

degrees and lack of available water are tough conditions for man and beast.

Ruth's son tried to hire pilots to aid in the search; he reasoned that a body would be easy to spot from the air. Of course, it was claimed pilots feared flying in those canyons because of wild air currents that were swift and uncertain. No aviator could control his airplane with certainty, and it would be suicidal to attempt it. Most of the pilots who took the job stayed at about 7,000 feet, which denied any view of a body lying in the canyons or on ridges below.

Brownie Holmes, son of the man who sat at Waltz's deathbed and who claimed his gold, was part of the first party of searchers. He advised Erwin Ruth that he and a pilot had tried to fly the canyons and had nearly crashed. They had been forced to abandon the flight.

Finally, a pilot named Goldtrap consented to attempt flying at dawn when the air currents were at their lowest force. He was adamant he would fly for only two hours after sunup. His photographs were enlarged to 100 times their original size. Everyone was hoping for a clue, but none was found. The pilot was later quoted, "The profusion of pinnacles and spires reminded

W.A. "Tex" Barkley was a veteran cattleman of the Superstition Mountains. It was from his ranch that Adolph Ruth set out on his ill-fated treasure search in 1931.

me of headstones in a graveyard. The only trouble was the shadows were too deep and long for good pictures."

Later Goldtrap said, "A man's crazy to fly over that mountain. It's no place for an airplane. There wasn't a spot as big as my hat that I could set the plane down in. If we had smacked, the ship would have been strewn over the tops of peaks to the bottoms of canyons. It wouldn't be necessary to go after us if we had crashed; there wouldn't have been anything left of us to find."

After the aerial survey plus a $300 reward and still no results, the 45-day search was called off. The Ruth family finally concluded that all that could be done had been done.

Barkley and Maricopa County Deputy Sheriff Jeff D. Adams persisted. They were, after all, two of the cleverest mountain men. In one shale-covered canyon, they found small pieces of tissue paper and tinfoil such as Ruth used to wrap around food cubes for his emergency pack. There were no footprints – the land was too rocky for that. All their efforts failed. Even Adams's hunting dog, a bloodhound named Blood, had no success, just blistered feet from running over scorching rocks. At the ranch house, his feet were doctored with mutton tallow to relieve the burns.

Reward for Missing Prospector Increased

July 10 1931 P. 1 col 4

The reward for the finding of A. Ruth, amateur prospector lost in the Superstitions in search of the Lost Dutchman mine, has been increased to $300, the sheriff's office receiving a letter from a nephew of the missing man saying that he would offer $200 in addition to the $100 reward offered by the wife of the lost prospector. Although the search for Mr. Ruth continues, no success has as yet been reported, according to Deputy Sheriff Chester McGee, who is supervising the hunt.

Articles from local Arizona newspapers (July 10 and 21, 1931).

P. 3 THE ARIZONA DAILY STAR JULY 21, 1931

col 4-5

Heat In Superstitions Ends Efforts to Find Prospector

Deputy Sheriff and Cattleman Unable to Stand Summer Temperatures as Rattlesnakes Hamper Search Among Canyons and Peaks of Range

FIRST WATER RANCH, Ariz., July 20.—(AP)—The furnace-like heat of the Superstition mountains brought an end today to formal efforts to solve the disappearance of Adolph Ruth, amateur prospector of Washington, D. C.

The 66-year-old man entered the desolate range June 13 to hunt for the "Lost Dutchman" gold mine. He has not been seen since. The last entry in his diary was dated June 15. His camp appeared unmolested. The major portion of his supplies, his boots and his field glasses were found there.

Deputy Sheriff Jeff Adams and W. A. Barkley, cattleman, who went into the inner mountain fastnesses last Saturday with the intention of remaining a week to explore the deep ravines and caves at the junction of the east and west Boulder canyons, sent up a smoke signal of distress and a pack train was sent in for them today. They came out parched and almost exhausted, without having found any substantiation of clues they found on a previous trip. These included Ruth's handkerchief, a paper wrapping from his pipecleaners, some tinfoil and a freshly built mining claim monument.

Their exploration of caves and the deeper ravines, they said, was hampered by rattlesnakes which had taken refuge there from the fierce heat radiating from the broken boulders and volcanic walls of the rubble-choked canyons.

Adams and Barkley, each of whom has spent the greater portion of his life in the Superstition country, said the summer temperature of the range, in the section they believe Ruth's body is, makes it almost impossible for a human being to remain there long enough to conduct a thorough search.

Dr. Erwin C. Ruth, the missing man's son, said he would abandon the search probably until next fall, when, if no further clues by that time have been found, he will organize an expedition to comb the section thoroughly in a final effort to solve the mystery. He said he planned to return to Washington tomorrow.

Naturally, the newspaper resurrected the issue of the curse and claimed Ruth was just one more victim.

On December 7, 1931, the *Arizona Republic* newspaper sponsored an archaeological expedition into the Superstition Mountains to gather information about the reported sightings of many unexplored Indian ruins. The group consisted of E.D. Newcomer, *Phoenix Gazette* staff photographer; Brownie

Brownie Holmes poses for news photographer E.D. Newcomer after locating Ruth's skull.

The remains of Adolph Ruth; photo taken January 6, 1932, by David Newcomer.

Holmes, guide for the expedition; Harvey L. Mott, *Arizona Republic* literary editor; Richie Lewis, owner of the horses and dogs used; and Odd S. Halseth, archaeologist.

They were in the mountains for five days before anything of importance occurred. On December 12, one of Richie's dogs Music located the skull of Adolph Ruth under a palo verde tree in Needle Canyon. A recent rain had obliterated any tracks near the find. Brownie Holmes picked up the skull to examine it. He noted two holes near the temporal region; then commented to the others that this was Ruth's skull and that he appeared to have been shot. Brownie had been at the Half Circle U Ranch when Ruth arrived. He had time to watch the man and observe his features. He said the forehead and shape of the skull looked just like that of Adolph Ruth. E.D. Newcomer, the photographer, asked Holmes to cradle the skull in his hands so he could take a picture.

Odd S. Halseth was anxious to oversee the examination. It was silent for a long time while the professional fingered the object. Finally, he stated the skull was that of a dead Indian, and it was very old.

For those of us who were not present, it is essential to know one bit of conclusive evidence: Portions of skin tissue still clung to the bone, and there were blowflies buzzing around. Although it is very dry in the desert, one would imagine the flesh would eventually reach a state where blowflies would no longer be attracted. Holmes's statement regarding the identity of the skull would later be proven correct.

W.A. "Tex" Barkley placing the skeleton remains of Adolph Ruth in a gunny sack for the trip out of the Superstition Mountains.

Of course, by now, a controversy had developed among the members of the expedition over what they should do. Some wanted to continue the trek. Naturally, the newsmen wanted to return to town. They had a new story – a front page story. Remember, these fellows were from two different newspapers. Oh, what a dilemma!

At last a decision was made. The group would continue on to Charlebois Spring and spend the night. Once they arrived, Holmes and Lewis set about fixing up their camp while Halseth expressed concern over the security of the find. It wasn't unusual for coyotes to carry off bones from any creature. To prevent that from happening, he used bailing wire to hang the eerie skull in a nearby sycamore tree.

A later report stated it was damp and cold that night of December 12, 1931. Picture if you will, the campfire and the group huddled around it while Ruth's skull danced overhead as the trees soughed.

The party left in the morning around 10:00 A.M., reaching First Water by 3:00 P.M. It was 7 miles to the ranch and another 56 to Phoenix. Since it was midnight when they arrived back in Phoenix, it was too late to have a pathologist examine the skull. Within two days, three doctors had viewed it. All agreed it looked like Ruth. Halseth was adamant. He shipped the artifact off to the National Museum of Anthropology in Washington, D.C., to have it examined by an expert. Unfortunately, Halseth was out of funds. Offering the matter to the museum as an experiment, he hoped for a positive outcome. The Science Service Bureau declined his proposal; however, by that time the December 13, 1931, edition of the *Arizona Republic* bore a bold headline that read:"SKULL BELIEVED THAT OF MISSING PROSPECTOR FOUND IN MOUNTAINS."

There are always men of science who are captivated by a puzzle and who will willingly invest their time and efforts in solving it. In this particular case it was Dr. Ales Hrdlicka, anthropologist and pathologist for the National Museum. His answer came by wire. "Skull unquestionably that of aged white man recently shot possibly."

The newspaper had gambled and won. The scoop set off a chain reaction in newspapers all over the United States as the entire matter was rehashed. Of course, this set new searches

SKULL BELIEVED THAT OF MISSING PROSPECTOR FOUND IN MOUNTAINS

Sunday Morning, December 13, 1931

POSSIBLE MURDER OF A. RUTH, AGED EASTERNER, SEEN

Grisly Evidence Discovered Deep In Stormswept Canyons Of Superstition Mountains By Expedition On Archeology

BROUGHT TO CITY

Prehistoric Village Is Surveyed During Trip Arranged By Newspapers And Heads Of Municipal Department

EVIDENCE which may point to the murder, deep in the canyons of the Superstition mountains of Dr. A. Ruth, aged searcher for a fabled mine who has been missing for six months, was brought out of the storm-swept hills yesterday by an expedition sent out a week ago.

The Arizona Republic, in co-operation with the archeological commission of the City of Phoenix, sent into the Superstitions a party of five men to survey and check reported sites of prehistoric villages.

Yesterday they returned, bringing with them the major portion of a skull which many bits of evidence point to having been that of the veteran government employe who entered the Superstitions in May to hunt for the fabled Lost Dutchman mine.

Holes In Skull

Holes in the skull, which gave indication through fragments of flesh attached of having been relatively recent, pointed to the logical conclusion that a high-powered bullet had entered the head from the left side.

During the many weeks of search for Dr. Ruth, officers never strongly considered the possibility that he had met foul play. They operated almost exclusively on the assumption that he had wandered from his base camp and become confused in the labyrinth-like canyons of the forbidding range.

Dr. Ruth, however, naively told many persons that he was taking with him into the mountains an ancient map of the Pedro Peralta, or Lost Dutchman mine which reputedly produced hundreds of thousands of dollars worth of almost pure gold before its last possessor died without revealing its location.

Murder Is Possibility

Finding of the skull by the archeological expedition, however, led last night to consideration of the possibility he may have been killed for the map he was known to possess.

The skull was found accidentally by the party, within three miles of the point at which Dr. Ruth was last seen—the permanent camp he established with the intention of remaining for 10 days to two weeks. His camp was found later by

(Continued On Page 5, Col. 3)

Skull Found In Mountains

(Continued From Page One)

searchers to have been left just as he walked away from it.

Lying on its left side, exposing the larger hole in the right, the skull was found beneath a palo verde tree in one of the deep, unnamed canyons which cut the Superstitions into thousands of segments.

Every evidence points to its having been dragged there by a coyote, a fox, or one of the many other wild creatures which somehow find sustenance in the somber mountains.

Animal Tracks Found

Despite rains which for days had deluged the mountains, turning rocky slopes into quagmires and dry washes into roaring streams, several indistinct animal tracks were discerned near the skull.

The immediate area was searched for other parts of the skeleton from which came the skull without avail. The possibility exists, however, that it is within a radius of 1,000 yards, a distance beyond which an animal would be unlikely to drag it.

If it were, days of careful searching would be necessary, in all probability, to find it, so torn and rough is the country.

On a nearby hill, an old mine prospect, abandoned years ago, gave evidence by new marks of having been worked within a relatively late period compared with the original working.

Brought To Phoenix

The skull was carefully packed over many miles of treacherous mountain trail to bring it to Phoenix late yesterday.

It was submitted immediately to the study of two pathologists and a dental surgeon.

Dr. Orville Harry Brown pointed out that many characteristics of the skull agreed with characteristics evident in photographs of Dr. Ruth in the files of the Arizona Republic.

He pointed out, for example, that the skull had a marked protuberance on the left forehead, as Dr. Ruth's forehead apparently also had.

He indicated from preliminary measurements that the left side of the skull, from nose line to left cheekbone, was wider than the right, as Dr. Ruth's appeared to be from his photograph, and that one ear was slightly lower than the other.

Dr. James J. Lasalle agreed with the forehead indication, but pointed to an apparent difference in the length of the nose. Dr. Brown disagreed with this point of view to some extent and it was agreed that accurate measurement was not possible.

Dr. Claude M. Moore, dental surgeon, said the upper jaw showed that the person whose skull it was had either worn false teeth or had been for many years without teeth. The upper jaw was worn smooth and gave no indication of tooth sockets.

He said it was impossible to determine which of the two instances had been the case.

Men who knew Dr. Ruth and were members of the Arizona Republic archeology commission expedition recalled that Dr. Ruth had had false teeth.

Dr. Moore was of the opinion the skull was relatively recent. Dr. Lasalle said he thought it could be either old or recent, but that in any instance it was a peculiarly shaped skull.

Odd S. Halseth, archeologist for the archeological commission of the city of Phoenix and for the expedition, said he could not determine definitely whether the skull was that of a white man or Indian, but would begin minute measurements today with a view to checking them against anthropological tables for final decision.

"It gives many indications," he said, "of being a skull of comparatively recent date. I have seen and excavated skulls hundreds of years old, however, bearing an equivalent amount of skin and in somewhat the same condition, dependent upon the climate. The skull is not, however, definitely Indian in type and may readily be that of a white man and of Dr. Ruth.

It was agreed that definite identification, or determination whether the skull is of recent or old date, would be impossible until further portions of the skeleton from which it came are discovered.

Prehistoric Village Surveyed

The discovery was a sudden and abrupt climax to a week of intensely interesting archeological research during which a rumored prehistoric village was discovered and surveyed carefully.

It was a week of thrilling adventure steeped in the mystery and majesty of the Superstitions. It was, too, a week of alternate brilliant sunshine and dismal, soggy, apparently endless driving rains.

The archeological investigations of the party were conducted by Mr. Halseth.

Members of the party included E. D. Newcomer, Phoenix Gazette staff photographer, and Harvey L. Mott, staff writer of the Arizona Republic.

Guides for the party were George "Brownie" Holmes, veteran explorer and guide in the Superstitions, and Richie Lewis, famous rider and now operator of the Tempe Riding academy.

An 11-horse pack train was required to transport the party over the almost phantom trails of the mountains which brood over the east end of the Salt River valley.

Their experiences form a fascinating serial of modern adventure in an ancient stronghold of the West. It will be published exclusively in the Arizona Republic beginning Tuesday morning.

Article from
The Arizona Republic
(December 13, 1931).

SHERIFF SENDS MAN ON TRAIL OF PROSPECTOR

Deputy To Seek New Clues To Adolph Ruth In Mountains

Sheriff J. R. McFadden today ordered his veteran deputy, Jeff Adams, to make another attack upon the forbidding wilderness of the Superstition mountain range in an effort to solve the mystery of the disappearance of Adolph Ruth, 66-year-old searcher for the Lost Dutchman mine, and former government employe, who dropped from sight last June.

The discovery of a skull, said to closely resemble in certain aspects the head of Mr. Ruth, by a party of archeologists, again brought attention to the amateur prospector.

Deputy Adams will leave tomorrow morning to begin a search in the area adjacent to where the skull was found in an attempt to find other bones of the skeleton.

The search will be made not only to attempt to definitely clear up the fate of Mr. Ruth, but also to discover whether the Washington gold seeker met with foul play, Sheriff McFadden said.

FOUL PLAY INDICATED

Deputy Adams and W. A. "Tex" Barkley, Mesa cattleman whose stock range includes the Superstition mountain, made futile searches for Mr. Ruth during July and came to the conclusion at taht time that he had been murdered.

They believed that someone followed Mr. Ruth into the mountains when he began his hunt for the legendary mine and murdered him either for the map he was known to have possessed of the location of the treasure trove or else waited until he discovered it and then killed him to seal the secret for themselves.

Identification of the skull as that of Mr. Ruth would substantiate the murder theory, as there are jagged holes in both sides of the skull, it is reported, believed to have been made by a bullet.

Odd S. Halseth, leader of the expedition sent into the Superstition mountains by the archaeological commission of the city of Phoenix and the Arizona Republic, that found the skull, is preparing to send it to the National Museum in Washington, D. C., to determine whether it is that of an Indian or a white man.

UNDER GOVERNMENT PERMIT

Mr. Halseth explained that he is exploring the Superstition mountains under a special department of agriculture permit and anything he finds of a possible archeological value he regards as government property.

If it is not the skull of an Indian, then he has no further interest in the matter but will aid the authorities in an investigation if necessary, he declared.

Whether it is the skull of Mr Ruth, an Indian or any one else, authorities here were willing to accept the conclusion that the man was another victim of the Superstition mountain.

The will-o-the-wisp phantom of a gold mine of fabulous wealth has accounted for many deaths of known authenticity. Mr. Ruth is believed to be at least the ninth victim on the fabled golden altar since the mine was re-located after it had been operated by its Mexican discoverers.

Many additional names might be added to the death list, it is believed, if there is any truth in the legends of the Lost Dutchman mine.

LORE INSPIRED NAME

A death phantom is ascribed to the mountains by others than the prospectors and veterans of the wilderness. The Indians have long regarded the section with superstitious dread and it is from their lore the frowning cliffs and peaks derived their name.

Deputy Adams, a veteran peace officer and travelor of Arizona's mountains and deserts, is loath to comment on the lost mine or the traditions of the mountain. He is familiar with the hardships and dangers that may be encountered.

It is his opinion that Mr. Ruth was a victim of the mountain's legend, even though he hilds the thery that he was murdered through human greed. If the skull was not that of Mr. Ruth, ther the mountain claimed still another victim. Through his renewed search tomorrow he hopes to clear up the possible double mystery.

Article from* The Phoenix Gazette *(December 14, 1931).

into motion. By January 8, 1932, the skeletal remains of Adolph Ruth were found by Tex Barkley in a small tributary canyon.

The intrinsic facts of the case are fascinating; the lengthy search captures interest; the controversy bears discussion; the find is certainly worth mentioning. But there is one more significant matter to be discussed.

The Ruth brothers were persistent in their quest. They pursued the matter with the National Museum and were finally told by Dr. Hrdlicka that the bullets were fired at such an angle the wounds could not have been self-inflicted. (Twelve years later, Sims Ely interviewed Dr. Hrdlicka, who by then had retired. He told Sims he remembered every detail of the case and added two more details – the caliber of the fatal bullet was a .44 or .45, and the shot had been fired from an old model army revolver.) Authorities in Phoenix refused to take action. They even refused to hold an inquest. The brothers were furious at local law enforcement for not finding their father's "killer," and they were angry that the remains had not been located until Tex Barkley found them.

Deputy Sheriff Jeff Adams accompanied Tex Barkley on that final search. On a brushy ridge overlooking West Boulder Canyon, a considerable distance from where the skull was discovered, they found the dismembered skeleton of the elder Ruth. Among the scattered bones was found a silver surgical plate. In the pockets of the trousers were some trinkets that belonged to the man, and in the coat pocket were papers in his handwriting.

Ruth was a precise man who kept a daily record of his movements and who wrote detailed letters. His notebook, which was with the remains, stated the following (quoted from Sims Ely's book *The Lost Dutchman Mine*, 1953): "It lies within an imaginary circle whose diameter is not more than 5 miles, and whose center is marked by the Weaver Needle, about 2,500' high – among a confusion of lesser peaks and mountainous masses of basaltic rock.

"The first gorge on the south side from the west end of the range they found a monumented trail which led them northward over a lofty ridge, thence downward past Sombrero Butte (now known as Weaver's Needle), into a long canyon running north, and finally to a tributary canyon very deep and rocky and densely wooded with a continuous thicket of scrub oak."

The description ended here, but lower down on the page was a single item: "*Veni. Vidi. Vici.*" Then, as if hastily jotted down, in pencil rather than ink as the rest of the page, Adolph's last words (in two lines):

"I came I saw I conquered

about 200 feet across from cave."

Deputy Sheriff Jeff Adams (on the left with his dog Blood) accompanied Tex Barkley on the final search for Adolph Ruth.

Now, it seems expedient to pause for some questions and a look at some of the clues. In the notebook was an interesting sentence: "The first gorge on the south side from the west end of the range they found a monumented trail which led them northward over a lofty ridge." Why did Sims write "they found" and "which led them"? Was Sims taking license with the exact words? That's doubtful. Something as important as the old man's notebook would have been copied exactly by anyone so captivated by the mystery. If Ruth did say that, why did he use *they* and *them*? One can only speculate about the use of words. It does seem more likely that Ruth had papers that Gonzales said accompanied the maps and that those papers were written by the Peraltas, probably describing the trail they took.

It seems relevant that "*Veni, Vidi, Vici*" was written by the early Spanish miners. We know for certain Ruth was alone in his quest.

It has been rumored that the doctor who put the plate in Ruth's leg had told him if he fell on it again, it would kill him. Also the notebook found in his pant's pocket had blood stains on it.

In this chapter, the reader will find a letter that has come to light. It was sent from Maricopa County Deputy Sheriff Sgt. Jeff

Office of
SHERIFF OF MARICOPA COUNTY

Phoenix, Arizona

January 25, 1932.

Hon. Carl Hayden,
United States Senate,
Washington, D. C.

Dear Sir:-

Replying to your letter of January 21st respecting a proposed inquest to be held upon the body of Adolph Ruth I have this to say.

The portion of the body found by us as nearly as I could judge from the geological maps was found within the boundary of Pinal County although the boundary line is very close to where the body was found. What remained for us to discover was only the skeleton minus the head which is now at Washington D.C. The flesh had all been stripped from the bones by either coyotes or wild hogs and the bones were scattered hundreds of feet apart on the mountainside.

We found intact all of his papers including the map or directions to be taken to find the Lost Dutchman Mine which Mr. Ruth was supposed to be trying to locate. Afetr finding and assembling these bones we followed the directions given to reach the alleged Lost Dutchman Mine. This trip took us two days of very hard labor and following these directions we came to the place pointed out in the instructions and found no evidence of any human being ever having been there at any time in the past.

So far as the officers here are concerned we have no reason to believe that Mr. Ruth came to his death by the criminal act of any person. So far as we can ascertain he had never been dispossessed of any single thing that he had taken with him.

These bones were found on the wall of a little canyon that comes very sharply down the mountainside and runs into Needle Canyon and about three quarters of a mile from where the head was found and the head was found on a mesa about two hundred yards from the Needle Canyon and my conclusion is that in the floods

This letter was sent to Senator Carl Hayden from County Deputy Sheriff Sgt. Jeff Adams in January 1932.

(J.D.) Adams to Senator Carl Hayden in January 1932. It is self-explanatory and leads to a new theory regarding Ruth's demise. In the letter, Sgt. Adams clearly states that when they found the old man's remains, all his papers and a map were found at the sight. If he had been murdered, it would surely have been for the map and the directions to the mine.

Jeff Adams later became sheriff of Maricopa County. We know

Hon. Carl Hayden,
January 25, 1932,
Page 2

that have washed down both canyons the head was rolled down and carried to near the place where it was found by the floods as other of his bones were found in the same canyon and that it is more than probable that the fractures in the skull were produced by being rolled by the floodwaters against the rocks in either of these canyons because it is evident that all of these canyons have carried heavy floods at different times since the disappearance of Mr. Ruth.

The remains were found over six miles from the place where his camp was established over a very rough and rugged country. Mr. Ruth left his camp provided with a one quart thermos bottle which was found empty, at the season of the year when there was no water to be found in the mountains and you know that a crippled man walking six miles over rugged xxxxxxy rough mountainous country for six miles under the burning heat of an Arizona sun in the month of June would absolutely perish of thirst.

The authorities in this county, and the County Attorney agrees with them that his death did not occur under such circumstances as afford a reasonable ground to suspect that it had been occasioned by the act of another by criminal means. We do not know of a single thing that can be developed beyond what is now developed by an inquest and both the County Attorney and myself are of the opinion that it would be absolutely a waste of money to hold an inquest upon that portion of the remains now here.

The only portion of the skeleton that even is claimed to show evidence of violence is in possession of deceased's son in Washington, that is the skull, and in fx view of the fact that he was in possession of everything he ought to have including the directions to the Lost Dutchman Mine we do not feet that there is anything more that we could do.

Whether or not an inquest is to be held is a matter, of course, to be decided ultimately by the coroner and not by the Sheriff or the County Attorney. The County Attorney has already advised the coroner that an inquest is an unnecessary expenditure of money in this case.

Very truly yours,

(Sgd) J. D. Adams

DEPUTY SHERIFF.

he never found the Lost Dutchman Mine. The letter also exonerates Tex Barkley, who many thought had killed Ruth for the map because he also looked for the mine. Others loudly accused the two cowboys, Keenan and Purnell, of doing the old man in. But still, this strange story took some new turns.

Bob Corbin became the first Republican county attorney of Maricopa County in January 1965. At the same time, the sheriff of Maricopa County was a Democrat named Cal Boyce, who had been in office for many years. Coincidentally, they both had an avid interest in chasing lost mines. Their mutual hobby brought them together occasionally. Eventually, they became great friends and shared their secrets over coffee.

Cal was getting old and readily admitted he could no longer prospect in the mountains. He suggested that he give his younger friend information about some mines he thought were worth seeking. Later, Bob stopped at Cal's home when he learned his friend was ill. During the visit, they got to talking about the Adolph Ruth death in the Superstitions. Cal told Bob that Jeff Adams had given him the map he found on Ruth's remains. After seeing the look on Bob's face, Cal smiled and pointed to his book-lined wall at the end of the room. "It's over there in one of those books. I'll look for it and give it to you, if you like." Bob just grinned.

Within several weeks Bob Corbin learned his friend Cal had died. He didn't think it would be courteous to disturb the family at such a sad time. Cal had forgotten his promise because he was so ill. Once again the Peralta map had disappeared.

In the chapter about Walter Gassler, a modern-day treasure hunter whose death in the Superstitions opened a new mystery involving Jacob's gold, we learn one more clue in the Adolph Ruth saga, but perhaps the most tragic part of this saga is that the Ruth brothers spent two decades trying to prove there was a killer. They wanted someone to pay for their father's death.

The Story of Dr. Ruth
(from Bark's Notes)

THE STORY OF DR. RUTH FROM BARK'S NOTES

IN MAY 1931 THERE ARRIVED, AT THE HALF CIRCLE U RANCH, SITUATED ABOUT FIFTY MILES EAST OF PHOENIX, ARIZONA, ON THE EDGE OF THE SUPERSTITION MOUNTAINS, AN ANCIENT MODEL "T" FORD CAR WITH TWO MEN IN IT. ONE OF THE MEN WAS KNOWN TO THOSE AT THE RANCH AS THE MAN RUNNING THE FILLING STATION AT THE JUNCTION OF THE FLORENCE AND SUPERIOR HIGHWAY; THE OTHER OCCUPANT OF THE CAR, IT DEVELOPED, WAS A DR. RUTH, WHOSE HOME WAS IN WASHINGTON, D.C. WHENCE HE HAD JUST COME WITH ANOTHER MAN WITH WHOM HE HAD BECOME ACQUAINTED IN WASHINGTON.

THIS NEW ACQUAINTANCE WANTED TO GO TO WASHINGTON STATE UNIVERSITY AND AGREED TO DRIVE THE FORD AS FAR ON THE WAY AS THE DOCTOR WISHED TO GO. THE DOCTOR SAID THAT HE WANTED TO GO AS NEAR THE SUPERSTITION MOUNTAINS AS HE COULD.

THEY HAD STOPPED AT THE ABOVE MENTIONED FILLING STATION, MADE INQUIRIES AS TO THE ROAD TO THE HALF CIRCLE U RANCH AND WHAT KIND OF A ROAD IT WAS. THEY WERE TOLD THAT IT WAS A VERY DIFFICULT ROAD TO DESCRIBE, AS REALLY IT WAS ONLY A DIRT TRACK AND THERE WERE MANY WOOD ROADS AND ALL JUST ALIKE; BUT THAT EVIDENTLY THE DOCTOR WANTED TO GO TO THE RANCH OWNED BY TEX BARKLEY, KNOWN AS THE HALF CIRCLE U, THE ONLY ONE NEAR THE SUPERSTITION MOUNTAINS.

UPON HEARING THE ABOVE, THE DRIVER OF THE MODEL "T" SAID HE WOULD GO NO FURTHER, AS HE HAD HAD ENOUGH TROUBLE TRYING TO KEEP THE "T" GOING WHILE ON THE PAVED HIGHWAY. HE WOULD STAY THERE AND TAKE THE BUS WHEN IT CAME ALONG, TO PHOENIX OR FURTHER, IF HE COULD.

AT THIS POINT IN THE CONVERSATION THE MAN AT THE STATION SPOKE UP, SAYING HE WOULD DRIVE THE FORD OVER TO THE HALF CIRCLE U RANCH IF THE DOCTOR WANTED HIM TO. DR. RUTH HAD A COMPLETE CAMPING OUTFIT WITH HIM. HE TOLD BARKLEY THAT HE WANTED TO GO OVER THE MOUNTAIN AND WANTED TO KNOW IF HE WOULD PACK HIM OVER.
TEX WAS BUSY AT THIS TIME, ROUNDING UP AND GATHERING HIS CATTLE FOR THE MARKET, ALSO BRANDING ALL UNBRANDED STOCK. TEX TOLD ME THAT IT WAS VERY HOT, AND AS THE DOCTOR WAS ALONG IN YEARS, AROUND SIXTY, AND QUITE LAME IN HIS RIGHT LEG, HE WAS

2 - DR. RUTH

AFRAID TO TAKE THE OLD MAN OVER THE MOUNTAIN AND LEAVE HIM ALONE.

FEELING SATISFIED THAT THE OLD MAN COULD NOT TRAVEL ANY DISTANCE AFOOT, HE TOLD HIM THAT JUST THEN HE WAS VERY BUSY, BUT IF HE WOULD WAIT UNTIL RODEO WAS THROUGH HE WOULD TAKE HIM OVER; BUT IN THE MEANTIME, HE WAS WELCOME TO STAY AT THE RANCH WITHOUT COST. TEX TOLD ME THAT BY THAT TIME HE HOPED THE OLD MAN WOULD REALIZE HOW HOT IT WAS AND THAT WATER WAS VERY SCARCE, AND PERHAPS, GIVE IT UP UNTIL COOLER WEATHER, WHEN TEX COULD KEEP AN EYE ON HIM.

THE OLD FELLOW STAYED AT THE RANCH UNTIL SOME TIME IN JUNE. THE WEATHER WAS GETTING HOTTER ALL THE TIME. ONE DAY TEX AND HIS COW PUNCHERS LEFT THE RANCH WITH A HERD OF BEEF CATTLE TO BE SHIPPED OVER THE MAGMA RAILROAD TO PHOENIX, SAID RAILROAD BEING ABOUT TWENTY MILES DISTANT FROM THE HALF CIRCLE U RANCH, THIS JOURNEY REQUIRING ABOUT THREE DAYS TO MAKE THE ROUND TRIP.

WHEN TEX LEFT THE RANCH WITH THE SHIPMENT OF CATTLE, HE LEFT THE DOCTOR AND A COUPLE OF PROSPECTORS THERE. UPON HIS RETURN TEX'S FIRST INQUIRY WAS, "WHERE IS THE DOCTOR?" THE PROSPECTORS REPLIED THEY HAD PACKED THE OLD FELLOW AND HIS OUTFIT OVER THE MOUNTAIN AS HE HAD ASKED THEM TO DO SO AND THEY HAD ESTABLISHED HIM AND HIS CAMP ON WEST BOULDER. TEX TOLD THEM THAT THEY SHOULD HAVE HAD MORE SENSE THAN TO TAKE AN OLD, CRIPPLED MAN LIKE THE DOCTOR AND LEAVE HIM ALONE OVER THERE. HE ORDERED THEM TO RIDE RIGHT BACK AND BRING HIM TO THE HOUSE.

THEY LEFT FOR OVER THE MOUNTAIN AND WHEN THEY RETURNED, THEY SAID THEY COULD NOT FIND THE OLD MAN. HIS CAMP WAS JUST AS THEY HAD LEFT IT, BUT NO SIGN OF HIM. THEY SAID HIS PROSPECTING BOOTS WERE AT HIS CAMP, BUT HIS CANTEEN AND PERSONAL BELONGINGS WERE GONE. THEY ALSO CLAIMED THEY TRIED TO TRAIL HIM BY HIS TRACKS BUT IT WAS IMPOSSIBLE. THEY WAITED AT HIS CAMP FOR SOME TIME, HOPING HE WOULD MAKE HIS APPEARANCE.

TEX BARKELY THEN TOOK UP THE SEARCH FOR THE DOCTOR, BUT WITHOUT SUCCESS. FINALLY, TEX MOVED THE DOCTOR'S BELONGINGS OVER TO THE HOUSE AT THE HALF CIRCLE U. RANCH.

3 - DR. RUTH

THE DOCTOR WAS AN EDUCATED MAN, VERY NEAT IN HIS HABITS, ALWAYS NEAT IN APPEARANCE, AND HAD A PLACE FOR EVERYTHING AND EVERYTHING IN ITS PLACE. HE WAS ALWAYS NICELY SHAVED AND HIS HAIR COMBED; QUITE A CONTRAST FROM THE AVERAGE COW PUNCHER AND PROSPECTOR.

IT WAS GENERALLY KNOWN AT THE HALF CIRCLE U RANCH THAT THE DOCTOR HAD SOME MAPS (TWO) PERTAINING TO THE LOST DUTCHMAN MINE, SUPPOSED TO BE LOCATED IN THE SUPERSTITION MOUNTAINS.

HE WAS VERY CERTAIN HE COULD FIND THE MINE FROM THE DIRECTIONS AND MAPS IN HIS POSSESSION. ALTHOUGH HE VERY SELDOM MENTIONED THE MINE, OR MAPS, HE AT ONE TIME TOLD A GIRL THAT THE PEOPLE AROUND THERE WOULD BE SURPRISED TO SEE HOW STRAIGHT HE WOULD GO TO THE MINES WHEN ONCE HE WAS OVER THERE IN CAMP. NOT MUCH ATTENTION WAS PAID TO THAT BY THE COWBOYS AS THEY HAD COME IN CONTACT WITH MANY PROSPECTORS WHO HAD MAPS OF THE LOST DUTCHMAN, AND WHO WERE GOING TO FIND IT VERY SHORTLY. IN FACT, PRACTICALLY EVERY COW PUNCHER THAT HAD WORKED AT THE HALF CIRCLE U RANCH HAD A MAP AT ONE TIME OR ANOTHER, SHOWING EXACTLY WHERE THE MINE AND FABULOUS AMOUNT OF GOLD WAS LOST.

SOME COW PUNCHERS PAID AS HIGH AS TEN DOLLARS FOR THEIR MAPS AND THE MORE FORTUNATE ONES OBTAINED THEIR MAPS FOR A SMALLER FIGURE; BUT THEY ARE ALL EQUALLY VALUABLE.

YOU SELL A COW PUNCHER A MAP, SAY FOR FIVE DOLLARS, AND IF HE FAILS TO FIND A MINE WORTH A MILLION HE NEVER INVESTS IN ANY MORE MAPS, BUT IN SPITE OF THIS, HE IS JUST AS CONFIDENT AND WILL BET AND LOSE HIS MONEY ON ANY GAME OF NO CHANCE.

SEVERAL MONTHS LATER A SCIENTIFIC EXPEDITION EXPLORING IN THE SUPERSTITION MOUNTAINS FOUND A HUMAN SKULL WITH TWO HOLES IN IT. THIS FIND WAS MADE ABOUT TWO MILES NORTH OF THE NEEDLE. THEY TOOK IT TO PHOENIX AND TURNED IT OVER TO THE AUTHORITIES THERE; IT WAS SENT TO DR. RUTH'S FAMILY LIVING IN WASHINGTON, D.C. THE SKULL WAS IDENTIFIED AS THAT OF DR. RUTH. THE TWO HOLES WERE IDENTIFIED BY A GOVERNMENT EXPERT AT WASHINGTON AS MADE BY A GUNSHOT, BUT NOBODY WAS FOUND.

- - -

4 - DR. RUTH

THE FIND OF THE SKULL INTENSIFIED THE SEARCH IMMENSELY. EVEN AIRPLANES WERE EMPLOYED IN THE HUNT. FINALLY, PINAL AND MARICOPA COUNTIES EACH APPOINTED TWO DEPUTIES, AND TOGETHER WITH TEX BARKLEY, MADE A SYSTEMATIC SEARCH.

ABOUT SIX MONTHS AFTER RUTH'S DISAPPEARANCE, THEY FOUND THE SKELETON ON THE SIDE OF A SMALL CANYON ABOUT ONE-HALF MILE FROM WHERE THE SKULL WAS FOUND. IT WAS IN A VERY ROUGH COUNTRY.

A FEW SILVER DOLLARS, HIS FOUNTAIN PEN, CANTEEN, COMB AND VARIOUS PERSONAL BELONGINGS WERE FOUND ON THE REMAINS, BUT NO MAPS OF THE MINE, AND BOTH SHOES WERE MISSING. THEY FOUND A SILVER PLATE WHICH WAS INSERTED AROUND THE BONE IN HIS RIGHT LEG, MADE NECESSARY WHEN A FEW YEARS BEFORE HE HAD BROKEN IT. THIS CAUSED THE LEG TO BE SEVERAL INCHES SHORTER THAN HIS LEFT, MAKING IT QUITE DIFFICULT FOR HIM TO GET AROUND IN THE MOUNTAINS OR TO WALK IN ROUGH COUNTRY.

THE SKELETON AND ALL OF HIS PERSONAL BELONGINGS WERE BOXED BY THE SHERIFF'S OFFICE IN PHOENIX AND EXPRESSED TO HIS FAMILY IN WASHINGTON, D.C. UPON THEIR RECEIPT HIS SON, ERWIN RUTH, LEFT WASHINGTON FOR THE SUPERSTITION MOUNTAINS. UPON HIS ARRIVAL HE TOLD THE FOLLOWING STORY ABOUT THE MAPS AND LOST MINE; THAT HE, ERWIN RUTH, WAS EMPLOYED BY THE GOVERNMENT AS AN INSPECTOR OF LIVESTOCK, BEING STATIONED ON THE MEXICAN BORDER. ONE DAY WHILE RIDING SOUTH ON MEXICAN TERRITORY HE WAS MET BY A SQUAD OF MEXICAN SOLDIERS. IN THEIR CENTER WAS A MEXICAN PRISONER WITH HIS HANDS TIED AND MOST THOROUGHLY GUARDED.

RUTH IMMEDIATELY RECOGNIZED THE PRISIONER AS A MAN KNOWN AS GONZALES WHO HAD BEEN A PROMINENT INSURRECTO. AS RUTH WAS WELL KNOWN IN THAT PART OF MEXICO, HE ASKED THE OFFICER IN CHARGE IF HE COULD SPEAK TO GONZALES. HE WAS GIVEN THE PRIVELEGE, SO HE ASKED GONZALES WHAT THEY WERE GOING TO DO WITH HIM. HE SAID THAT HE SUPPOSED THAT THEY WERE GOING TO STAND HIM AGAINST AN ADOBE WALL AND SHOOT HIM. GONZALES ASKED RUTH IF HE WOULD DO HIM A FAVOR. RUTH REPLIED, "CERTAINLY". GONZALES ASKED IF HE KNEW WHERE HE LIVED AND RUTH REPLIED IN THE AFFIRMATIVE. GONZALES THEN ASKED HIM TO GO THERE AND TELL HIS WIFE, WHOSE NAME WAS PERALTA BEFORE SHE MARRIED, THAT IN A CERTAIN PLACE IN THE HOUSE SHE WOULD FIND SOME PAPERS,

5 - DR. RUTH

AMONG THESE WERE THE TWO MAPS OF THE MINE LOCATED IN THE SUPERSTITION MOUNTAINS THAT WE HAD TALKED ABOUT MANY TIMES. SHE WAS TO TAKE THE PAPERS AND GIVE THE MAPS OF THE MINE WHICH DESCRIBED THE LOCATION IN THE SUPERSTITION MOUNTAINS IN ARIZONA TO ME; ALSO, FOR HER AND THEIR LITTLE DAUGHTER TO GO WITH ME AS I WAS GOING TO TRY TO GET THEM INTO THE U.S. AT EL PASO. I SUCCEEDED IN CARRYING OUT HIS WISHES, AND LEFT MRS. GONZALES AND HER DAUGHTER, WHO WAS ABOUT EIGHT YEARS OLD, IN EL PASO.

WHEN I RETURNED TO WASHINGTON, I GAVE THE MAPS TO MY FATHER AS HE WAS CONSIDERED THE MINING MAN OF THE FAMILY. HE WAS ALL FOR STARTING FOR ARIZONA IMMEDIATELY, BUT THE REST OF THE FAMILY WERE ALL AGAINST THE PROPOSITION. DAD BIDED HIS TIME AND STARTED FOR ARIZONA AND THE SUPERSTITION MOUNTAINS AT THE FIRST OPPORTUNITY.

I FORGOT TO MENTION THAT WHEN THEY FOUND THE BODY OF RUTH, THEY ALSO FOUND A BLANK CHECK BOOK. IN THE FLAP, NEATLY FOLDED AND TUCKED THEREIN WAS A LEAF FROM THE NOTE BOOK BELONGING TO RUTH. I HAVE A PHOTOSTATIC COPY OF THIS NOTE AND IT HAD SEVERAL QUITE LARGE BLOOD STAINS ON IT. THE WRITING WAS AS FOLLOWS:

> IT LIES WITHIN AN IMAGINARY CIRCLE, WHOSE DIAMETER IS NOT MORE THAN FIVE MILES, AND WHOSE CENTER IS MARKED BY THE WEAVER NEEDLE ABOUT TWENTY FIVE HUNDRED FEET HIGH - AMONG A CONFUSION OF LESSER PEAKS AND MOUNTAINOUS MASSES OF BASALTIC ROCK.
>
> THE FIRST GORGE ON THE SOUTH SIDE FROM THE WEST END OF THE RANGE THEY FOUND A MONUMENTED TRAIL WHICH LED THEM NORTHWARD OVER A LOFTY RIDGE, THENCE DOWNWARD PAST SOMBRERO BUTTE INTO A LONG CANYON RUNNING NORTH, AND FINALLY TO A TRIBUTARY CANYON VERY DEEP AND ROCKY AND DENSELY WOODED WITH A CONTINUOUS THICKET OF SCRUB OAK.
>
> VENI. VIDI. VICI.

IN PENCIL I CAME I SAW I CONQUERED
ABOUT 200 FEET ACROSS FROM CAVE.

THIS NOTE WAS ALSO SENT TO HIS PEOPLE IN WASHINGTON AND IDENTIFIED AS IN THE DOCTOR'S HAND WRITING.

HIS SON, WHO IS STILL EMPLOYED BY THE GOVERNMENT IN WASHINGTON, THOUGHT THE LATIN WAS NOT WRITTEN AT THE SAME TIME AS THE BODY OF THE NOTE, AS IT HAD A STIFFER APPEARANCE, AND THE LEAD PENCIL WRITING WAS STILL AN AFTER THOUGHT, AND ALL WAS

6 - DR. RUTH

WRITTEN AFTER THE DOCTOR HAD REASON TO BELIEVE IN THE NECESSITY OF SECRECY. AFTER THE FINDING OF THIS NOTE, THE HUNTING PARTY OF FIVE WENT INTO A HUDDLE AND THEN ASKED BARKLEY IF HE KNEW OF SUCH A CAVE. TEX REPLIED THAT HE KNEW OF A CAVE UP THERE, WHETHER IT WAS THE RIGHT ONE, OR NOT HE COULD NOT SAY, SO THEY GATHERED UP THE BONES AND BELONGINGS OF RUTH AND TOOK THEM TO CAMP. EACH GOT AN EMPTY GRAIN SACK AND RETURNED TO EXPLORE THE CAVE, AND INCIDENTALLY, TO GET A SCAK OF GOLD ORE OR NUGGETS, WHICHEVER IT MIGHT BE. TEX LED THEM TO A CAVE, AND IT CERTAINLY WAS A ROUGH TRIP, BUT THEY FOUND NO MINE OR GOLD.

NOW IT IS UP TO ANYONE WHO READS THIS TO DRAW HIS OWN CONCLUSIONS AND DO HIS OWN DEDUCTING. PERHAPS THE OLD DOCTOR WAS ON THE VERGE OF DISCOVERING THE LONG LOST MINE. WHAT HAPPENED, HOW HE DIED, AND WHAT HE KNEW WILL PROBABLY NEVER BE KNOWN OR MADE PUBLIC.

Letters to Erwin C. Ruth from Emilio (Peralta) Drigo (circa 1932)

May 2, 1932
Chicago - Ill

Mr. Irwin Ruth,
Dear friend -

While visiting the family in Mexico last month we were descussing the story of your fathers death. One friend of my uncle send him news papers from Arizona.

My uncle asked me write to you. He had discussed this matter with Mercedes Gonzales my distant cousin who know you when her mother give you papers of Arizona gold mine. Mercedes and my uncle think you should have that mine if you want as oweing to promise of my uncle Gonzales and account your father loose his life trying to find the place.

For some reason of her's Mercedes would not write this letter. Maybe or not you know why but if you kike her address, I can send it.

I come from Spain where I am born. My people leave Mexico many years ago so I speak English poorly therefore I have my girl write this who speaks English better. She is from Tampico Mexico with English father.

I hesitate writing to you because I learn you and your people are with Government and I have been in this country too long and do not want to leave yet. Six years I am here. Iwant you not to advertise my name or to mention it to officials. The famklies in Mexico say you were gentleman of kind heart and could be trusted, so I take this chance.

If you like my uncle to help you I will tell you where he lives. He told me about twenty five years ago he made journey to those Superstitions Mountain range with my cousin and showed him the very place gold was removed. He said place had been shaken up or flooded some but he was sure he know the place. He is 86 years but was there when just growing into man hood while his father and brothers took out gold.

He was old man when he showed my cousin the place and they talked seriously then to work again for gold but give up later, they think undertaking to big and costly.

My uncle said his father never wanted American to have the old mine but now he thinkgs your family deserve it if you can afford to work it practicly.

My uncle beleive that much gold still there and can be found there if you could dig for enough to reach this old mine where they worked on it years ago.

Below I give you my address and will be glad to get your answer.

Yours truly

/s/Emilio (Peralta) Drigo

P. S. My uncle say to tell you he has his fathers maps with true directions still but as he do not need directions can give you better as old ones are too correct and too mathematical.

E.D.

The above is a true copy: Erwin C Ruth

Erwin C. Ruth

Chicago, Ill. Aug. 7, 1932

Mr. E. C. Ruth
Dear Sir:

I received a letter from my uncle, and he would be glad to meet you in Phoenix and talk with you. He says if you are sincere in o perating the old mine he will give you old necessary information and maps frely, but he wants to be assured that you wil not sall the maps and papers to others. Yars ago my uncle give a man part of the information with which to raise fonds for operating porposes. The were to shere equaly. Uncle would deliver the couplete directions and derect explorations when the man had everything redy. but this man only sold the information so uncle become disguste.

I am sorry this is the best I culd do. I advaise you to see my uncle. also visit him and Mercedes. because she has much influense withe him and take my advaise if you want recuperate the old mine. Work fast because it is no time to lose in this case. and if you do, well be easy for you to get mor information.

Sincerely yours

/s/E. Peralta

The above is a true copy:

Erwin C. Ruth

Chapter 12

Walter Gassler's Search

Adolph Ruth's death brought forth a stream of newspaper articles throughout the United States, and later as far away as Australia, Europe and Mexico. But it was the article in the *Oakland Tribune* in 1932 that caught the eye of Walter Gassler. So intrigued was the man, he headed for the Bancroft Library at Berkeley. After his first bit of research was complete, Walter spent every weekend and vacation day in the stacks. The librarians came to know him and offered their help.

In reading his diary, one can see his quest became a research project leading him and the librarians all the way back to Father Juniper Sierra and Father Kino of Jesuit pioneer fame.

The project lasted two years. In 1934 Walt decided he knew all he wanted to know. Coincidentally, he received an offer to become a chef at the famous Arizona Biltmore hotel. When Walter arrived at the hotel, he learned a friend had already been offered the position, and not wanting to replace him, Walter declined the offer. Believing this to be an omen, he left for the Superstitions.

He arrived at Apache Junction, then consisting of a restaurant, gas station, four white cabins and an abandoned drug store. Pitching a tent next to the empty building, he camped and asked a lot of questions. In due time Walter learned of the Barkley ranch and its personable owner.

The next move was obvious. Once Walter arrived, he was given access to all the information from a lot of sources. All the reporters, who surfaced during the Ruth matter and enjoyed the Barkley hospitality, had probably sent their own articles to the ranch. Tex had gathered material from prospecting friends,

Walter Gassler in the Superstition Mountains (circa 1936).

locals and libraries. Walt had high regard for the Barkley collection, but he also mentions how disappointed he was to find so much conflicting evidence regarding Jacob and his mine.

Gassler seemed confused as he wrote of his doubts that Jacob and his partner had come upon the Mexicans who showed them the mine and offered to share it.

> What a ridiculous story. Think! The Superstition Mountains, wild Apaches, claim jumpers—all kinds of wild animals. Wouldn't you have your guns loaded and ready at all times, right handy beside you for fear of your life? I am sure you would have your gun ready when strangers would approach you till they were out of sight. I can hardy understand how Herman fell for that story. [Here he refers to Herman Petrasch with whom he had made friends.] Also, you notice the story says they walked not climbed and all the other stories say it is in a very steep place.
>
> As to the stories of the Peralta Mines which Waltz claims he found, he said that the mine existed in a canyon that runs north and south. Its peculiarities were potholes, hundreds of them, shortly after you leave Tortilla Creek and follow it up maybe a mile—you can't miss them and it will identify that you are in the right canyon. You follow up Peter's Canyon shortly after you enter Tortilla Creek. You will find the potholes as it enters the canyon properly. There, the story continues, the mine is about 1½ miles from a cave at the foot of a high bluff and a two-room house with a sod roof in this cave. This was the headquarters for the Peralta who worked the mine above. Which brings you up Pistol Canyon and Peter's Mesa.
>
> Here you must remember that Rhinehardt and Herman, but especially Rhinehardt, spent a lot of time in Peter's Canyon to find that mine. That made me suspect that Petrasch knew more than he told and he must have heard it in Waltz's ramblings when he was dying, but just never got hold of the whole story. Of course, there probably were no names at that time for those canyons. I gave a map, colored topographical, I brought from San Francisco where I bought it and was made before Tortilla Flat was put under water by the dam. I think I recall it was called

the Ray Triangle. Tex was crazy about it. So when I quit looking for the mine, I gave it to him. It probably still is in the family.

What made a story about a cave and a tunnel 200 feet across, that map that is floating around—don't say it is a Superstition Peralta Map and it might have some features of it alike. It could easily have come from all together some other part of the country. Especially since Ruth had several maps of different places and different mines with him. I am positive his Superstition map disappeared when he died, but it sure gave a lot of writers a bone to chew on. And look at all the stories that sprung up, everything around Weaver's Needle. Nobody was ever able to find the mine on those wild goose chases.

[Walt went on to say that Tex had told him when he was a youngster chasing cows back there around Peter's Canyon, that the cave was still there and had a few rafters and poles hanging there.] But there is still more to come.

The cave was very convenient to Peralta as right around the bend of the bluff there was and still is, a permanent water hole at the foot of a 10 to 20-foot drop of the creek. It is probably 6 to 8 feet wide. Maybe 8 to 10-foot deep and always has water in it.

[He then talked about the shape of the mine and the ledge and how you could walk and face LeBarge Canyon, which has a magnificent sight of the Needle.] It is like looking at a picture gallery and at the end is this grand sombrero, no wonder, if this should be the location, they named it the Sombrero Mine.

Walt had deduced there was more than one mine. After all, Waltz claimed he covered it. "It is my opinion the Apaches covered it and Waltz never found it, so let's concentrate on Waltz."

Now, Walter talked to Tex's wife, Gertrude Barkley, who said she had known Jacob Waltz when she was a little girl. She remembered the German coming through the ranch and always bringing candy for the children. She claimed he was gentle, a man who would never harm anyone. She begged Walt to give it up and not become a "Dutchman-a-holic."

Walt had finally married and had a little girl. Mrs. Barkley was probably trying to save his marriage.

Mrs. Barkley also told him of the night Adolph Ruth went into the mountains. She said he had on a trench coat. It was drizzling when he came to the ranch. He showed her a manila envelope and said, "I'll show you how easy it is to find that mine. I'll be gone one or two days." She further stated he started to show her a map and then reconsidered, saying he would show it to her when he got back.

After Ruth left, Mrs. Barkley became anxious while awaiting her husband's return. When Tex did get back and learned of the trip, he immediately saddled up and went to look for Ruth, telling his wife he was afraid for the man's life.

Walter said he believed Waltz had found two skeletons of Mexicans who had not been able to get away from the Apaches. The Indians had already torn down the rock house to keep anyone else from finding it. "Those Mexicans wouldn't come back, they knew the Indians would kill them." But, Walt did believe the Apaches chased Waltz away from the mine, and he did leave his grub and mules. Gassler further believed that while Jacob Waltz was making his getaway, he came across the remains of the house, the charcoal pits and a pile of worked gold. That is when old Jacob fabricated his saga.

At this point in the notebook, Gassler wrote something that bears repeating. It attests to the fact there were divided camps in the search. It also allows the reader to see how easily information was transposed in order to confuse and mystify the searchers.

> "Old Holmes asked Tex one time if he knew of a grove of trees on a mountainside and if he did to let him know. Of course Tex knew the place he was referring to, but he told me (Walt) that he never cottoned to Holmes and never had any further contact with him. That was the time Holmes also told Tex the Dutchman still had several pouches of gold hidden back there. This also must have come from his deathbed murmurings."

Walt had a friend, a mining engineer by the name of Tom Reis whose experience consisted of 40 years in Chile, Bolivia, Colombia and Mexico. They came together in Arizona, at which time Walt convinced the engineer to come to the Superstitions for a look at the area.

Walt had found two pieces of quartz in the wash below the tree

Superstition Mountains, Arizona.

grove. Tom Reis said it would assay out at $20 a ton. They both admitted it wasn't worth fooling with, but it did prove there was gold in the area.

Tex and Reis hit it off immediately, so much so that Tex allowed the engineer to ride his favorite horse into the mountains, and he normally didn't let anyone ride that horse. Gassler never learned to ride, so he made the trip on foot.

Leading the way, Gassler followed the creek bed to the top of Miner's Needle. It was really rough. When they got to Bluff Springs, it was filled with water even though it was August. Eventually, Tom asked, "Are you sure you know where you are going?" Cow trails leading here and there were the only signs there had been life in there. The growth was wild and difficult to move through. When they got to Charlebois Spring, there was water, and Tom seemed a trifle more believing.

In those days you could still see the sheet metal shack, corral, some hedges and wildflowers put there by Charlebois himself. Walt was certain that all of that had been wiped out by the big fire that occurred back there.

Now, they were moving toward Peter's Mesa in dense mesquite and oak brush. It was so high Walt had to elevate his arms to walk through it. Walt said there was a path, but it was barely visible, and at every step he could hear the whir of two or three rattlers being disturbed. Walt said, "Tom looked as if he really thought I was lost."

Finally they got there. Walt expressed real joy as Tom gave detailed information on the charcoal pits and grinding stones nearby. The mining man explained there must have been 60 or 70 people working there for at least six months. But Tom definitely disagreed with Walt's assumption that the grove was the place where they mined. He took Walter over the crest of a hill facing south in Pistol Canyon saying that in his experience this is where it would have been.

When they arrived at the Barkley ranch that night, Tom told Tex, "If anyone would have told me that Walt Gassler could have made that trip on foot, I would have called him a liar."

Then, Tom talked to Tex about the huge field of stumps they found from the ironwood and mesquite trees that Tom admitted proved there was a big production going on back there at one time.

Now, Walter Gassler's notes broke. His next heading was "Ruth's goal and Waltz's dream fairy tale."

I think it is impossible that one or even two men could ever recover this mine. When you think that 60 to 70 men worked there and what a hole that must have been—well. Then, the Apaches covered it. One man could probably work two or three years and just scratch the surface. I feel silly now that I thought I could do it alone. Just listen to what Tex told.

An elderly Indian woman Tex knew well told him at one time she remembers when she was a little girl her mother took her with her and a small tribe of Apaches into the mountains to cover up a mine. She said they had a camp on top of small mountain, on the side below it was a big cone-shaped mine with a pole in the middle and chain attached to it so one could grade the chain and walk around the small path down to the pit of the mine. Close to the mine there were three large stones (boulders), tepee shaped. She said they threw the tools and chains down into the pit and then, started to cover the mine. It took six moons and

> the women did all the work. The men hunted and made mescal up at the camp. Tex said you could still see the place where they made the mescal. Then, the old Indian said they planted young trees of mountain laurel and cottonwood over the area to make sure it wouldn't be found again. They also blended the surroundings to make it look natural. That solves part of the Puzzle. I also think they tore part of the cabin down. They didn't seem to care about the charcoal pits. They probably used the grinding stones to grind up their corn.

Walt, at this point, described what Sims Ely had said in his book about the Indian woman who was a friend of Apache Jack's. He said, "The tools and chains are probably buried too deep to be reached by a metal detector."

Here Walter Gassler complimented himself:

> It still amazes me to think, how, as a complete greenhorn from San Francisco, I could have found that mine the very first time out.
>
> Tex explained to me at the ranch all the problems. He warned me not to drink any stale water, to watch out for snakes, and to avoid any human beings back there as much as possible. So I took off. I found Charleybois Springs and then went up from there. It was pure luck. Somehow I got to Peter's Mesa then, went on top of that ridge, went to the northern end down off it and kind of circled around on the bottom southerly and came to the bottom of a wash running south. Then I saw beside the wash a big pile of rocks, kind of square looking and thought it was odd. Glancing up on the hillside above, I saw a grove of trees and three big boulders. And, I knew I'd made it. Further investigation led me to the charcoal pits with the grindstones laying around. My heart jumped out of my mouth; hell, I was at the Lost Dutchman. But, the task proved too much in the end. I had to give it up. Oh, I dug alright but to no end. There was one hole already there and I thought Tex, maybe, had already tried it.
>
> You must remember I could only prospect in summer. I had

to work in wintertime. No water up there and it was two miles to Charleybois Spring and a long way to the watering hole in Peter's Canyon.

Mrs. Barkley told me later that Tex had said, "that greenhorn will never find it." And, when she found out what he had done she really scolded him. When I did find it, he sure became a different man.

Then, I went up to Bluff Springs and found that cave. And, it is a funny cave, not very big. You can crawl comfortably in there. It is about 15 to 20 feet deep and has a chimney hole up through it, straight up. You can crawl up it or climb a big boulder above and see down in. The top blends right in with the plateau. You would never see it is there.

On the bottom there is a large, thick bush growing absolutely innocent. The reason I noticed it was I was sitting nearby and thought I saw a dark blotch. It couldn't have been a shadow as the sun was at the wrong place. So I ambled over to it and poked it with my walking cane, which I always have with me, and I went right through. I almost fell down. I poked some more in case there were rattlers in there. But, I got no response and crawled in. Another funny thing about the cave is if you put anything light down like papers or leaves or light sticks at the entrance and the top is open, it sucks them right up and out. Maybe the Indians used it for smoke signals or the Mexicans. When I got back to the ranch and told Tex he was dumbfounded. In fact, he saddled right up and went out there. After that, we were intimate friends.

All the stories about Dutchman, Ruth, and the Peraltas I got from him. In the late evenings after supper we would go out and sit awhile on that hill rise as you came around the bend in the river. We would roll a smoke and talk a little while. I wish I had known then what I know now; we might have solved the puzzle while he was still alive. I still feel bad today that I couldn't have done better for him and his belief in me.

Now I know those stones I found in the wash must have been discarded stones from that rock house.

Tex also told me about a cave on Horse Mesa. I looked for it but never found it.

There was an old prospector in Tortilla Flat; they called him Monument Smith. I got to know him real well. He wanted to know about that cave. I told him all that Tex said. Monument wanted me to question Tex more but I said, "If Tex tells you something and that's all he tells you, you'd better leave it alone." Then, there was Anderson, bad character. Run into him once and he asked me. Told him, "Don't know." Billy Martin was there fur trapping in the winter and guiding. Met him after he got rid of his gang and was cleaning up his camp down below Charleybois. He asked me if I saw his burros. I said yes, they were just above Charleybois and above the mesquite flats. I told him to be careful—I just passed two rattlers in there. He said, "The hell with them, I'm not going through that. They'll come back, I'd rather wait." We walked to his camp and talked. I really liked him. Before I left he gave me a bottle of old Taylor, said he don't drink. I didn't either, but I gave it to Tex; he grinned from ear to ear.

I told Tex about all the people who wanted to know about that cave. Tex said, "I was up on Horse Mesa early one morning and I followed a bunch of wild horses, they led me right in there. Nobody would've ever thought it'd be in there. But, one time I was tramping all the way down LeBarge to Canyon Lake, I had a real powerful pair of binoculars. When I was by that willow marsh I saw up on the cliffs what looked like a cave, I guess, I filed it away in my memory." Then, Tex said you could look 1,500 feet straight down in the river. I thought he meant the Salt. But, when I read Barnard's book where he describes that cave from where he could see the Weaver's Needle, I realized that Horse Mesa to Barkley was Malpais Mountain Flat and there is that cave. Wild horses from up there even today I suspect. When I was up there in 34 to 42 some 50 to 60 horses with a white stallion came every morning. So, the mystery is finally solved. I have not been there yet. I want to see if Barkley left any kind of hint there for me. He was always playing jokes on me.

He has destroyed or dismembered all the grindstones, has scattered them and turned them upside down so you never realize what they are. And, since the pits are all covered you don't know where they would all come from. He must have even tried to tumble the boulders down the slope with a lasso since the middle one is broken off about a third of the way down. I am sure he would figure I would know and understand. I sure do, Tex.

One evening he revealed to me that originally only two boulders were up there by that mine and the Peraltas found another one just like it and put it there for a marker. I

Weaver's Needle.

asked him, "How do you know?" He grinned and said, "Greenhorn, if you look under them there stones you will see that two of them have no growth of any kind coming from underneath while the third one has all kinds of them. So, the first two are natural and the other was put there."

Then came the time when I told Tex, I just could not keep up. Either I would have to give up prospecting or my family. He was shook up some; then, that evening, he told me about Ruth. He said he knew he was dead. He said he looked for a couple of days and then, got serious about tracking. It had rained back there before Ruth went in and he found his cane prints. He said he followed up to Peter's Mesa and said Ruth found the landmark. He said he must have been tired and when he saw the grove of trees on the slope he headed for them.

He sat on a protruding rock, rested and looked around and saw the three boulders in front of him and knew he had come to the end of his search. He must have taken out his notebook and wrote, "VENI, VIDI, VICI." In the meantime, whoever was his guide got impatient and told him, "Let's go." Ruth realized he would die when the guide knew the location and he would die if he did not. Then, he must have realized what he had done by writing those words which meant, "I came, I saw, I conquered," so he added "two hundred feet across from the cave" which would surely be misleading.

That is my guessing. Tex claims the guide put the gun to the side of Ruth's head and told him to get going. He must have gotten excited and pulled the trigger, took the map and disappeared. Tex found the body, got a mule and Tom Dickens, a cow puncher; they put his body in burlap and transported it to Black Top Mesa hoping it would be found there. Again, I asked him why. He explained that a few years ago somebody claimed they found the Lost Dutchman and about 200 people came stampeding back there and scattered his "Cattle to Kingdom come—clear up to four peaks." It took about 60 days and extra cowhands to recover all the cattle and he said, "*not ever again*." I sympathized fully with him.

I also figured Tex was baiting me. Well I horsed around another week and I guess, the last day, I prepared to go into the hills about 8 A.M. on Saturday. He came out with me to the road leading to Miner's Needle. He looked me straight in the eye, then I saw he came to a decision about something. He said, "Wait a minute, Walt, I will show you positive proof that the mine is there." He strode towards the house, stepped on the veranda just as Bill, his son, and Brownie Holmes came around the house corner up from the smithy, tipsy already. Tex saw them, turned around, came back and said, "I don't want them to see it. I'll show you some other time."

What was it? I'm sure it was the map. But, I figured it could have come from the notebook or gold he had found up there. Wish I knew what it was.

That was the last time I ever saw Tex.

I went away to Tennessee for a number of years. When I came back I visited with Mrs. Barkley when I had time again.

I started to collect information again and came to all these conclusions. I also remembered something else he told me that makes me think he told me the truth about finding Ruth's body up there on Peter's Mesa.

He told me when he found the body it was laying on its side, not the way a body falls forward, he said the killer had the pistol actually right to his head—flush. Then, he pulled the trigger so that it knocked Ruth sideways and he died that way and laid that way. That's what makes me think he actually found the body shortly after death. He also said to me that when Brownie found the skull and he took in the searching party, he took them right to the bottom where he knew the body was. Brownie said, "You fellows go up this way and I'll go up over these rocks." Tex said Brownie knew they could not miss the body and they didn't.

Still another time Tex told me about an Indian that owed him a favor. He said the Indian told him about a gold mine and would take him there. They went up to what is now called Peter's Canyon, clear up to the cave and the creek obstruction around the bed from there. As they approached this place, Tex said the Indian actually froze, stared up the canyon like he seen a ghost, grabbed Tex and told him "up there, take right hand canyon." He fled and Tex never saw him again.

You will also notice that whenever they claim they found the mine or gold it was always coming up from the north going south. I was forever trying to find them potholes in that elusive canyon but always workin' up from the south I never found it. Then I ran across that old map at the Phoenix Library that showed the old Spanish trail that branches off from Tortilla Creek and goes into Sheep Mountain, Geronimo's Head to Malpais Mountain into Peter's Canyon up to Peter's Mesa and Charleybois Spring, then finally, I went to Tortilla Flat up Tortilla Creek to find that trail, missed it somehow, but came on Peter's Creek and followed it up the canyon to the cave and made my camp there. Then I was going up to Peter's Mesa, but next morning it was black and cloudy and a heavy storm forecast and I did not want to get trapped in that canyon so I beat a hasty retreat. And then on the way out of that canyon and a short distance down Tortilla Creek on a slight bend of the river, I saw the start of that old trail. It was quite visible coming that way. While on the way up you miss it if you aren't careful. Don't even know if it's possible today. Maybe Tom Kollenborn can find it on his horse.

That trail should lead you within a few hundred feet of that cave on LeBarge Canyon side of the mountain. There you can look 1,500 feet down to the river. You should come out a short way south of the other cave and the fall according to Barnard's story; Waltz stayed at the cave on LeBarge side of the mountain. He watered his burro on the old water hole that would surely be that large pothole which always has water and is only minutes away from that cave. This trail gave Waltz also the option of going the way of Tortilla Flat to Phoenix or the other way south by Charleybois and Bluff Springs or Whiskey Springs. He could alternate and confuse his followers.

Surely there must have been some certain marker or sign—something to identify the exact or close location to the mine. Just to say 2½ to 5 miles from the Weaver's Needle just would not have been enough for anybody, for a general direction. Yes, but since so many places back there look so much alike and identical, it just would have been better identification and that might have been the oral secret and final clue to the map Ruth was carrying. Ruth's son said he had some personal clues not on the map. Will this all bring the search to an end or are there some more loose ends? Maybe we will find out more in the LeBarge facing cave?"

So end Walter Gassler's notes. Walter never said anything about dates. He disappeared from the Superstitions for many years but later became the curse's most current victim.

Chapter 13

The Curse Claims Another Victim

Arizona's Department of Law in the state's capitol complex sprawls along an entire city block, just a stone's throw away from the decaying cemetery where Jacob Waltz is buried. The man who headed that department in 1983, Attorney General Bob Corbin, had searched for the Lost Dutchman's Mine since his arrival in Arizona in 1957. His interest in the mine was well known through television documentaries, articles in magazines, newspapers, radio talk shows, speeches and one episode of the television favorite, "Unsolved Mysteries." The star of that show, Robert Stack, told Bob the Superstition episode was the most popular show they had ever done, and it had been rerun more than any others of their shows.

Bob Corbin's office was filled with antique oak furniture and private memorabilia depicting the Old West. At exactly noon on December 12, 1983, his secretary ushered in his next appointment. The visitor was Walter Gassler.

Over a period of years, Corbin had been inundated with requests from reporters, photographers, movie producers, businessmen, prospectors and a few characters (whose mentality was questionable) to take them into the range. It was also not uncommon to receive phone calls from around the world from people who claimed they knew where the mine was. Since the attorney general enjoyed talking about the mine more than he did eating, he often scheduled those appointments during lunch time.

Walter Gassler, he observed, showed his age on a wrinkled, honest face, but his weathered body moved with rigid determination. In his hand he carried an orange, line-ruled notebook. After introduction, for they had never met, Mr. Gassler

Bob Corbin (left) in the Superstitions being interviewed by radio announcer George Scott (right). (Circa 1968)

proceeded to tell the lawyer he had looked for the Dutchman since 1931. He was close to finding the mine, he had said with complete sincerity, and really felt he needed the security of some trustworthy companions. He further stated he knew Corbin and his prospecting companion Tom Kollenborn were good friends, and he would like them both to share his experience and the gold.

Corbin was noted for being down to earth and never openly argumentative. Besides, he told Tom later, he liked the old man. However, it was a fact that hundreds of people over the years had made the same statement.

Gassler was no fool; he had anticipated that reaction and being prepared for the eventuality, had brought with him a diary of his experiences in the Superstitions. To further plead his case, he presented the notebook to the attorney general saying, "I know everyone tells you they know, but please read these and after you have, if you think I don't know what I'm talking about, I won't trouble you again."

A tentative trip had been discussed but, for one reason or another, never came about.

Bob later said, "The man seemed astute and after having read his notes, I have no doubt his opinions were sound. But, whether he has actually located it still has to be determined."

Bob's friend Tom knew of Gassler and knew he was a friend of Tex Barkley. In true western fashion, if one had a cowboy friend who accepted someone, so would others. However, Tex Barkley was now dead.

In May 1984 Gassler called Tom to ask if he knew where he (Gassler) could get hold of Bob Corbin? It was explained that Bob was enjoying a brief vacation at his cabin in Prescott and there were no phone lines on the attorney general's mountain. Gassler seemed upset. He said he had finally figured out the puzzle and asked if one of the two could accompany him into the mountains on the following day. Tom regretted his refusal but explained he had business and couldn't get away. "Besides," he told Bob later, "how many times have we heard that same story over the years?"

On Thursday of that same week, Tom called Bob Corbin to tell him Walter Gassler's body had just been found on the trail above Charlebois Springs. There was to be a coroner's inquest, but it looked as if the man had died of a heart attack. Of course, both men were shocked and felt deep remorse they hadn't been able to share what was to be Walter Gassler's last journey into the Superstitions.

That weekend a man appeared at Tom's home. He introduced himself as Walter Gassler's son and was invited in. After pleasantries were exchanged, the man brought out some rich gold ore. He said it was in his father's backpack when he was found on the trail. Tom was shocked. However, he maintained his dour facade while he examined the ore and, later, said it looked exactly like the Dutchman's ore that came from under his deathbed. The man listened patiently, and after a bit said, "I understand that my father gave his notes to Bob Corbin. Is that true?" Tom nodded, saying little, in true cowboy fashion. It was apparent the man wanted those notes back. Tom suggested he call the attorney general at his office in the state capitol. The matter was left right there.

A week later Tom called Bob to see if he had heard from Gassler's son. Bob said no, but he would gladly give the notes back to Walter's family; he felt that was only fair.

Again, fate seems to work in strange ways.

Tom is invited regularly to give slide presentations of the Superstitions and to relate stories of the Dutchman and others. There are exotic tales of some 68 murders and wonderful explanations of the characters who lived and died in those canyons for myriad reasons. The show is well accepted and has been shown throughout the U.S. and Europe. This particular slide show was in Florence, Arizona, the county seat of Pinal County, home of the Silver King Mine, Tom Weedin and the place where the swamper was last seen.

Once the show concluded and Tom was busy collecting his materials, he was approached by a man who introduced himself as Walter Gassler's son. The fact that Tom is a stoic probably kept him from gaping. The person who stood in front of him bore no resemblance to the man who had recently come to his home to discuss the ore. This man was polite and well dressed; however, he proceeded to make the same request about the diary and was given the same answer. This time, Tom went right to the phone to call Bob Corbin.

Two days later Bob received a phone call from a man claiming to be Walter Gassler's son. Bob said he was busy and would return the call at a later date. He asked his secretary to get a current phone number. This time the number given was the same as the one Walter had in his diary. It appeared this was the right son. Corbin said he was also impressed with the fact that the man said, "I don't want to take the notes away from you; my father wanted you to have them. But, if you could Xerox them and then give them to me for our family momentoes, I'd appreciate it." Of course, the attorney general complied.

The pair met for lunch. During the course of the meal, the son was politely questioned as to his father's effects that were brought into town by the sheriff's office. Bob said, "Ron, did you find your dad's backpack?"

Ron looked puzzled and answered rather hesitantly. "Why, no. It was listed on the paper which recorded his effects, but when I picked them up it wasn't there." Corbin believed the man to be sincere.

Had Walter gone in alone, discovered Jacob's secret, extracted some ore, hurried away in fear and had a heart attack on the trail? The inquest confirmed he had died of a heart attack. He was in his 80s and surely a shock of that magnitude might have killed a younger man. Or, was he followed by some unknown person who tried to force him to take them back to the mine? Or, is it a simple fact that someone found the pack and removed the ore? Was the gold taken from the pack on the body? Or did someone in the sheriff's office take it in the excitement of the moment? Whatever happened remains a mystery.

Tom Kollenborn identified the gold. He has a degree in geology and has looked at a lot of specimens – he is not a novice. He stated the gold was almost identical to the gold that came from under Jacob's deathbed.

All the above facts cannot be disputed. Walter did go into the mountains on that last quest; he did say he knew where the mine was; and he died in possession of valuable gold ore.

The following is an affidavit from former Attorney General Bob Corbin:

AFFIDAVIT

This is to certify that I am Bob Corbin, Attorney General of Arizona. I have held that office since 1979.

My hobby is searching for the Lost Dutchman Mine in the Superstition Mountains which I have done since 1957.

Approximately three years ago I had the opportunity to be shown ore samples, documents and jewelry made from said ore which is owned by an Arizona businessman. This individual obtained these valuable items from Brownie Holmes, son of Dick Holmes, who was present in the home of Julia Thomas on October 25, 1891 when Jacob Waltz died.

I have in my possession an hour long tape made before Brownie's death which states that he gave these items to said businessman. He further states that the Dutchman gave these items to his father which were stored in a candle box under Jacob's deathbed. I have seen the report on said ore which was assayed at Goldman's store by Dick Holmes. This report states that the ore assayed out at $110,000 a ton. Gold was selling for $20.64 an ounce in 1891.

With the items were shipping papers indicating some of the ore had been sent to a jewelry company in San Francisco in 1892. A request accompanied it directing the jeweler to make a matchbox, ring, tie pin and cuff links from the ore. I have also seen the papers shipping this jewelry back to Phoenix.

I had permission to photograph the matchbox which will be used in this book. I have seen the ring made from this ore, but I have not seen the tie pin or the cuff links. This is the first time the Dutchman's Gold has ever been photographed.

I was told by the individual who owns the assay report, jewelry and ore that this ore had been sent to the University of Arizona, School of Mines which has samples of gold ore from every known Arizona gold mine to see if the ore could be matched up. I was further told by this individual that the University of Arizona, School of Mines had informed him that this ore came from an unknown source.

The reason that I cannot name the individual who has the ore, jewelry and documents is that I gave my word I would never divulge his name without his permission if he would show the ore to me which, as stated above, he did.

Further affiant sayeth not.

Bob Corbin
BOB CORBIN

SUBSCRIBED AND SWORN to before me this 10th day of April, 1985.

Kathleen M. Copus
Notary Public

My Commission Expires:
11-9-88

BC:kmc

Conclusion

For over a century, hordes of prospectors have sought the secret to Jacob Waltz's lost gold mine in the dreaded Superstition Mountains. Reams of books have found their way into print regarding that mystery; unfortunately, many of those works were touted as truth when in fact there was no truth. Writers enlarged on theory or historical fact to add romance to the printed page. No one could prove the truth. Often when one cannot verify a fact, it is almost a foregone conclusion that an explanation will follow in the form of what later becomes a myth. Repeat it often – it becomes legend.

Of those whose religion includes chasing lost treasure there are legions. They honestly believe. When one talks with them, it is obvious how dedicated they are to their perspective and nothing or no one will change their minds. Of course, it is enviable for people to have a fantasy especially about gold so rich as to create the ultimate fantasy. The Lost Dutchman's ore is worth $1,600,000 to the ton on today's gold market. Instead of wasting their lives sitting in front of an electronic marvel, treasure hunters are steadfastly pursuing an elusive dream that keeps their blood churning – searching for bonanza tantalizes.

Naturally, following the trails into one of God's most magnificent natural creations is in itself a healthy endeavor and an arduous task. The Superstition Mountain Range has extraordinary beauty. However beautiful, it is equally dangerous, and going into those treacherous canyons takes stamina, along with an experienced rock horse, and very sturdy legs.

This particular title, my third book on the Superstitions, has for this author been a labor of love. Just as most of my peers have done, I've searched for the truth for many long years. Usually I don't express an opinion. Readers experience pure joy at trying to figure out the goal and the explanation of the material presented in a book. But, all authors have an opinion. I am no exception. In this case I have traveled this road before; however, prior to this writing, I was unable to locate the exact truth regarding this subject. That fact caused me great consternation. Following the thread of this man's life, and what transpired after his death, usually brought me to frustrating dead ends.

In the fall of 2000, when I became privy to the pioneer journals of four families and read their notes, I became ecstatic. Just reading the material gave me a thrill, because the knowledge contained in the notes had otherwise eluded me. Now, it became apparent I could document Jacob Waltz from birth to the grave.

Ordinarily, when I pondered the questions that always surfaced, there was only speculation – until now. In *The Bible on the Lost Dutchman Gold Mine and Jacob Waltz*, the answers are documented. And, yes, there was a kernel of truth in every legend. That has been said many times. But at this writing Jacob becomes a real entity as he moves through history with familiar names. We are all now able to enjoy his story as he lived it, along with the trials and tribulations of his compadres, the tireless pioneers whose incredible efforts opened the great American West and settled the place called Arizona.

I am not a Dutchman hunter; I am, however, married to one. My husband, Bob Corbin, has searched for nearly 45 years. I considered his search one of the great privileges afforded in his life. And, as the years progressed and so many people literally begged him to take them into the Superstition Mountains, I realized how fortunate he was to have had such an avocation. I also have been fortunate enough to have gone into the range to examine its stark beauty and to sleep in the shadow of Weaver's Needle by a waning campfire while listening to the stories of those stalwarts who still search for Jacob's hidden gold.

Imagine the difference between what we knew before and what we now know? Think of those who have spent a lifetime searching, some for months and years at a time. But, perhaps more importantly, think of those who have actually located the Lost Dutchman's Mine and then paid the ultimate price. Was it fate?

The Indians, Ghost Horse in particular who vowed to avenge his people, put a curse on the "White Eyes" who made gold and the search for it their dedication. The promised revenge on all those people who violated their sacred sites seems certain.

Jacob said, "We never slept near the mine." He didn't elaborate. Later, a few tough men – soldiers and miners – all stated, "It was eerie there . . . we never slept there." What was meant by that statement? These people were afraid of very little, and yet they were obviously apprehensive of being near that mine at night. Did they sense a spirit presence? In all the stories about Apaches who knew of it, even those who befriended white men, there was a fear of revealing what they knew. Some came very close to giving their information up, but in the end they disappeared or died.

Years have passed, 111 years to be exact, since Jacob Waltz left this mortal coil. Can we even imagine what he would say if he knew how famous he'd become? Would any of these wonderful characters believe they had an intricate part in the most famous story of the American West? (The Lost Dutchman Mine was listed in *U.S. News and World Report*'s July issue in 2000 as one of the great lost treasures of the world.)

One can certainly visualize young Jacob Waltz along with his friend, Weiss, leaning over the railing, peering out to sea expectantly as the *Olbers* weighed anchor headed for the New World. The sounds of the rising rigging, the shouts of her crew and the slosh of ocean racing against her hull as they moved out of the harbor must have excited the pair. Leaving behind poverty, desperation, hunger and cold must have made them exuberant. There was probably not an instant when they actually believed it was happening. Considering where they had been and what they were escaping from, they must have been in almost disbelief. At last, they had hope. They were in their late 20s, strong, unschooled, hard working and eager as they took a final look at Bremen.

They would be on that ship for one month and 17 days before catching a glimpse of their new country. They would never see their homeland again. Perhaps they never gave that a thought; perhaps it didn't really matter. One would wonder when Jacob decided to hunt for gold. It seems likely he was a "redemptioner" and would be busy for a lot of years paying off his debt, but considering the state of things in Prussia, working on a farm in a good climate and having a clean bed, fresh air and good food must have been very acceptable. It was 1839.

The real adventure didn't start until Jacob joined the wagon trains at Castroville headed for California. He had filed for citizenship in 1848, and by 1850 he was on his way to the gold fields and his destiny. He had been in America for exactly 11 years.

One would wonder what transpired in Castroville as the wagon train travelers packed for the journey ahead. It was common knowledge the Indians were on the rampage. Of course, they probably thought there was safety in numbers, and for the most part that was true. The collage of immigrants, locals and adventurers from all over the world would seem unruly in a village compound filled with Conestoga wagons and animals. One would wonder at their absolute courage.

Many of them would never realize their dreams but would die along the trail. It was not uncommon to bury the dead on the road in a hasty funeral and move on immediately. Their families would never know where the body was buried, and no headstone would grace the land. But perhaps, the most poignant thought relates to farewells. In Jacob Waltz's case, his family members were there, and saying goodbye was forever. He never went back. His sister and her farmer husband went to live in Kansas. (We know Jacob sent a huge sum of money to her through Wells Fargo many years later.) The others may have remained in Castroville, but they were parted and apparently never saw each other again.

It is spectacular to read the words of A.H. Peeples, Matt Cavaness, George Roberts, Daniel E. Conner, Harwood Hinton, Rheinhart Petrasch, Bertie Roberts, Jacob Weiss, Dr. Thorn, Brownie Holmes, Jim Bark, Sims Ely, Tom Weedin, Dr. Walker, Julia Thomas, William Edwards, Joe Green, Mathew Roberts, Adolph Ruth, Erwin Ruth, Tex Barkley and Bob Corbin. These people either knew Jacob, shared his dream, walked his trail or came after and searched for his mine. Although we were privy to few words of the man himself, we came to know him vicariously. I admire and respect what I know of him.

Waltz apparently was a simple man. To those he befriended, he was genuine. His generosity seemed extraordinary for that culture. There is no question about his loyalty or his dedication to his friends, and he was known to be a tough man: fearless, aggressive, instinctive as well as sensitive. Jacob was a recluse. He was quiet, a loner of sorts, kept his own council but was available to his fellow man when needed. Look at his service to

Julia and Reiney, not only paying their debts but offering his mine to them for the future. There was his aid to the sick man, Magdeberg, who eventually died and left his entire fortune to Jacob. His dedication to his boyhood friend, Weiss, goes without saying – he mourned him until he gasped his last breath. And, he sent a fortune to his sister in Lawrence, Kansas.

At no time did history record Jacob spending the gold on himself. He farmed vegetables, raised his chickens and ducks and went into town regularly to sell them. He had a fortune in the adobe, Reiney told us that in his own words, but still Jacob worked. Probably that had been his way his entire life. Finding the mine was pure joy – not for the gold but for the thrill of it. He had lost his friend, Jacob, and more than likely he equated the gold with that loss. It might explain why he never went back for the caches. He wasn't greedy, in fact, he probably disdained the ore that had caused him to lose the only real friend he ever had. Supposition? Yes, but hardly a stretch with all this evidence. The pair were like brothers in every sense of the word. Although he didn't need or want the gold, he promised Julia and the boy they would go collect the caches in the spring.

There were those who called him coward. Imagine! Would a coward have saved a fellow miner during an Indian attack with the Peeples party? Would he have dragged the man out of the fray after his leg was shattered by an Indian missile while the arrows rained around them? Would a coward have crossed the desert against marauding savages, expecting day after day to be attacked, killed or worse? Native methods of torture defied belief; and, these people all knew it. They had often come upon the bodies of those who had been tortured. They knew what capture meant.

It is obvious Jacob never used the gold for his own aggrandizement, and it might be the reason he was allowed to keep it. The history of the goldfields records many instances where men took their ore to San Francisco or Denver and built great homes – monuments to themselves and their wealth. Then, those self same folks became the scions of society. Not so with our protagonist.

Everyone else who found the gold, or came close to finding it, died. That was decreed by the Indians. In fact, Jacob was allowed to keep it only until he decided to give it away – then death claimed him. That the mine is cursed goes without saying;

just read about the lives of those people who spent a lifetime searching. Look at the tragedy surrounding most of the names listed. The loss of life and the unparalleled circumstances of those who searched seems incredible. And, yet, they still came. They had **gold fever** – one of man's great plagues.

It is important that we recognize the pioneers. To have been privy to their own words fascinates, opening new vistas of our imagination as we peer into their devastating crossing toward the Arizona Territory. They were indomitable – a breed that will never be seen again. It is obvious they never considered their vulnerability; it was their destiny. When one reads their accounts, they are without passion or fear or even intolerance. They are matter of fact.

Matt Cavaness talks about riding for three days in search of water for the wagon train because the children were crying. He admits he had no food or water; it was summer, and he must have been exhausted. Since I have spent a lot of years in the desert, I can't imagine riding a horse for three days without food and water and keeping up the strength it would take to accomplish this feat. Matt was probably one of the greatest Indian fighters who ever lived. That he was fearless goes without saying. He recounts being surrounded by his attackers and never uttering a whimper. He just plowed on, resisting and shooting. He lived to be 83 years old, and what a life he had! The fates decreed he would not be killed when so many others were.

One of the stories about Matt that particularly interested me was about his fight with a wildcat. It was Christmas Day 1853. Matt was 10 years old. He, his sister and his dog Tige went for a walk when Tige chased the wildcat up a pecan tree. Matt wanted that wildcat's skin. He had no gun but promised his sister he would get a stick, climb the tree and send the cat off the limb so Tige could kill it. His sister remarked that the tree was very tall, and if he fell it would kill him. The limb was 60 feet up and out over the river. With stick in hand, when Matt confronted the cat, it attacked him and made a run at his face. Then it bit him on the neck and held on. In the ensuing battle, the pair tumbled into the river; whereupon, the dog leaped in and took over. Tige eventually dragged the angry wildcat up on land and killed it. It was freezing cold, and they were wet. Matt's sister found some punk, and they used Matt's knife to cut it up. They built a fire to dry his clothes and skinned the wildcat. Matt's lack of fear was apparent even then. To smite the enemy was a duty he did with aplomb. One has to wonder if his apparent fearlessness was his secret weapon?

In the Peeples's group were famous Arizona names – names most folks are familiar with but of whom little is actually known. Included were Pauline Weaver, Joe Green, Coho Young, A.H. Peeples, George Roberts, Matt Webber, Henry Wickenburg, Samuel Groupa, Oscar Hutton, Charles Hahn, Jacob Waltz, Jacob Weiss, Gideon Roberts and, of course, many others with whom we are less familiar.

The talk of being in the gold camps is enlightening. Peeples says, "It was just he (Roberts) and I and Matt Weber back then, cutting timbers when we made the strike. In a couple months the word had spread to the mining camps down along the American River and the miners came pouring in. The placer mines around the quartz lodes were very rich in coarse gold and money came flooding in as the town sprang up overnight, in six months there was two thousand men in congregation along the banks and in the canons. In the beginning everyone was finding gold, twenty to two hundred dollars a day was common." Jacob Waltz came to join them and mined alongside them.

Picture this scene. All the miners were crammed into the new makeshift town on the river. Everyone had gold. Supplies would be paid for with gold dust, much of it slipping into the cracks of the floor as it was laid to waste while they drank and measured. People questioned them. "There's more where that came from," they would state haughtily. They stood side by side in the icy water working their shovels and swinging a pick. At first it was peaceful and exciting, but as time went on too many appeared – slim pickings caused unrest. Fights and bloodshed became the norm. Again fate stepped in – the mines flooded with water, and having no pumps, it signaled the end. George Roberts had taken out about $15,000 worth of gold as had his partner. Jacob Waltz and another man worked their claim side by side with the others. They all had to leave; it was over.

They were clever and were becoming experienced. Soon there were numerous sites that became well-known mines. They were making money – lots of money – and gold was selling for $16 a troy ounce. It went up to $20.60 an ounce in due time, but most of the time it went for whatever they could get for it. Ultimately, the Waltz mine assayed out at $110,000 to the ton in the late

1800s, and Jacob is supposed to have said that it wasn't the richest ore he'd seen but was all he had left. At today's gold prices, the Lost Dutchman would be worth $1,600,000 to the ton.

The pioneers kept moving toward Prescott ignoring the Indian raids. The attacks were all too frequent, but still the pioneers prevailed. The hardships they endured seem overwhelming, but they never gave up. They were the survivors; by that time the weak ones had been culled.

One surprising fact was that people in the towns the wagon train passed through tried to blackmail them by demanding payment. The pioneers either attacked their adversaries or bypassed them, whichever was more expedient. What seems extraordinary is the fact they were having to fight Anglos, Mexicans and Indians. Tough became the norm. Life in the early West was unimaginable.

Once these people all converged on Prescott, which was a dirt path in the road near the Bradshaw Mountains, they realized the potential of the earth. They were professionals now – not a ragtag bunch of settlers but experienced miners who knew the lay of the land. It wasn't long before they located gold – in places just waiting to be extracted from the earth. Of course, they happened onto Rich Hill, as it came to be called, by accident, where they extracted $150,000 in gold nuggets in a relatively short time.

Peeples went on into a beautiful valley north of there, where he established his homestead. They named it Peeples Ranch and later Peeples Valley, a name it still bears. The others trekked into the Bradshaws to Big Bug Mesa, Turkey Creek, Lynx Creek and the Hassayampa River. Jacob Waltz is known to have filed claims at Turkey Creek, along the Lynx, at the Grose Lode and in other places. There is no record of what he actually took out; however, it is known that the Grose Lode became a successful venture for the new owners and was mined for many years, even after Jacob Waltz and Jacob Weiss sold out and left Prescott.

(As an aside the reader should note that the Indians called the Hassayampa River, Haviyamp [also spelled Haviamp in Conner's notes]. Mr. Wheelhouse of the Walker Party changed the name to a Spanish pronunciation Hassayampa. It retains that name today. Another creek that was important to the Walker Party was Cum-yam-muck [with a strong accent on the second syllable]. It was changed to Walker Creek and later called Lynx

Creek, a name it still bears, because they killed a lynx when they first camped there.)

We soon learn that Prescott, in its infancy, became the hub of mining activity. Waltz and the entire Peeples group were joined by the Walker Party that settled in and around the area where they fought Indians on a regular basis while establishing mines and camps. Naturally, there was trouble as strangers came into the area regularly seeking to make a strike. Eventually a mining district was formed as the area became more civilized. People wanted to settle down. This situation became almost intolerable as greed took over. After numerous pitched battles, rules were established that prohibited Chinese and Mexicans from mining. There was even an attempt to prohibit soldiers from staking claims. This matter caused a furor and was resolved quickly in favor of the military. Gold fever was alive and well in Prescott; however, the savages were ever present. In fact, it was not uncommon for scalps to be found hanging on the outside door of the bar in town after a raid. The Indians really believed they could terrorize these newcomers into leaving.

Arizona was still a territory. Eventually Prescott became the territorial capital and remained so until 1912 when statehood was established, and Phoenix became its capital.

Jacob Waltz and Jacob Weiss might have decided to move on since there were so many problems surrounding mining in Prescott. Strikes were fewer and farther between. Naturally, in summer there was little water for mining or drinking. Trying to keep livestock for pack animals became almost impossible as the Indian raids increased.

We have a document within this manuscript that shows Jacob Weiss buying part of the Grose Lode claim in 1863. We can document his death in Dr. Walker's infirmary in 1871. Therefore, Waltz and Weiss found the Lost Dutchman Mine sometime between those years. It is likely they located it between 1867 and 1870.(Jacob Waltz homesteaded his farm in the Salt River Valley [Phoenix] in 1868.) It seems realistic to believe they met Peralta and rode shotgun for him and his peons on the first trip before giving him their gold. Then, it is probable they went back to find the gold they put into the caches. If the story is true, that was when they saw the Mexican peons mining at the site and shot them. It wouldn't have taken all that long to accomplish two trips. In fact, that was the time Jacob said the mules had trampled all their stores except the meat that was hanging in the trees. Waltz is supposed to have left for supplies, didn't get back in time and Weiss was attacked by the Apaches and sustained what turned out to be a mortal wound.

We learned from Tom Weedin that Weiss stumbled out of the Superstitions and collapsed at a water hole. He was delirious. Pima squaws found the man and brought him to Dr. Walker's where he gave his map to the doctor. Later he died. Now the infamous map was transferred to Dr. Walker. The doctor shared his experience with his friend Tom Weedin, whose letters are included in this book as proof of this information. Walker was killed and his Indian daughter stripped of her inheritance by the court. Juana left the area and never spoke to a white man again. That map was obviously still in Dr. Walker's house and was claimed by his relatives. We learned from Sims Ely that Weedin's wife destroyed the map copy to keep her husband from going into the dreaded Superstitions. Sims claimed Weedin's daughter gave him that information after her mother's death. Both maps ultimately disappeared.

Jacob Waltz lived until 1891. He is supposed to have offered two of his friends the mine. He was old and sick and probably knew he wouldn't live very long. Although he tried many times to explain the directions, Julia and Reiney couldn't or wouldn't listen. That fact caused them irreparable harm.

(We should probably pause here to interject the information about the German translations [see page 184 in Chapter 6]. Julia and Reiney each told Reiney's brother Hermann Petrasch what Jacob had said in German. Hermann then compared the two translated statements, and when they both coincided, he considered that to be the truth. We should also add that Bertie Roberts was quoted as having said that Reiney wanted attention and was prone to exaggeration, but Hermann, the older of the two, was a stoic who rarely spoke, but when he did it was always factual.)

Julia sold the ice cream parlor to pay for the wagon, horse and supplies for their search for the mine after Jacob's death. Eventually, Jim Bark had to feed them at his ranch. They were broke. Even Reiney's family members, who were miners, came down from Colorado to search. No luck. It got so bad the brothers fought among themselves and wound up not speaking for the rest of their lives. Julia became a drunk and died of Bright's disease. Reiney committed suicide. The senior Petrasch died in an insane asylum and Hermann, the elder brother, lived in the Superstitions searching unsuccessfully until his 80s. Dick Holmes, who took the gold from under Jacob's deathbed, claimed it was a gift from Jacob as were the directions to his mine. Dick and his son Brownie spent their entire lives searching for Jacob's gold. They failed. Brownie in his declining years gave the few pieces of ore he had left to a businessman in Arizona who befriended him. A photograph of that ore appears on the cover of this book.

The Arizona School of Mines that examined the ore said it came from no known Arizona mine. That fact should end once and for all the theory that it was high graded from the Vulture. The photographs of Jacob's gold (taken by Bucky Singleton, the head photographer for Motorola) were taken to the Arizona Mineral Museum for examination in 1980. The geologist at the museum said he had never seen gold that rich in all of Arizona's recorded history.

It might be of interest for the reader to note there have been four calderas in the Superstitions – a caldera being an imploded volcano in which the sides collapse forcing the magma up and out. Some believe that occasionally a chimney of almost solid gold rises up with the magma and when cooled becomes a find as rich as the Dutchman is supposed to be. The United States Geological Survey (U.S.G.S.) identifies three calderas in the Superstitions and a fourth on the eastern edge. The three inside the range are Doggie Springs, Tortilla and the Superstition calderas. Goldfield, Mammoth, Black Queen, Bulldog and Wasp mines all sit on the outside (western) ring fracture of the Superstition caldera. It is along the outer edge of these ring fractures that gold and silver deposits are most likely to be discovered. The outer rings of all three calderas overlap each other within the interior of the Superstition Mountains. Erosion has overburdened most of these rings, but in a few places canyons have exposed the fractures, and those are the most likely places to find mineral wealth. A long, deep fault running north and south down Pistol Canyon exposes a good section of where the Doggie Springs and Tortilla calderas converge and overlap each other.

The chapter about the Peraltas seems quite explicit. These people were well known in Mexico, New Mexico, California and Arizona. Their family name was respected, and they were astute business people. They did not always fare well because of prejudice. They lost their mines in the territory; they were prohibited from owning mines when the mining district was formed and became victims after the Reevis – Baron of Arizona fiasco occurred. Later they were allowed to own mines again, but it appears it was too late, at least for Miguel, who obviously gave his rights to the Superstition Mine to the two Jacobs in return for the $60,000 in ore that he needed to pay his gambling debts. He and his father lost the Valenciana Mine in the Bradshaws because of the rules in their mining district. They laid claim to that mine in the early 1860s. (His son said after his father's suicide that Miguel was looking for the mine in the Superstitions. Miguel said in his suicide note, "I had the money; I have lost it: good bye." I believe Miguel Peralta was talking about the Lost Dutchman Gold Mine.)

The map that Ruth had was supposed to have come from a Peralta. The map that Weedin had was a copy of Dr. Walker's – also a Peralta map. (That was reputed to be a gift from a dying man – Jacob Weiss.) The map given to Bob Corbin was supposed to be a Peralta map – a copy of the original. But, something that seems to escape everyone's notice is the fact that the geography changed after an earthquake in the late 1800s in the Superstitions. The Salt River changed course naturally over the years and, eventually, dried up almost totally because dams were built. Numerous springs dried up in the Superstitions after the earthquake.

The maps that are available, some of which are in this book, are extremely vague in their drawings. We have included them because Dutchman hunters want every chance, and we would like them to have it. When possible the origin of the maps is given.

The chapter on the Indians was especially revealing because for the first time we have an explanation of their culture, their religion and their philosophy. All this information offers a reason for their wrath against men who only wanted gold. It is impossible to fathom the reasons for their methods of torture until one reads the philosophy regarding their souls and their burial locations. They had no use for the mineral that caused all their trauma, and when they obtained it by raiding, they took it to their sacred sites within the Superstitions to store in caves as a gift to their gods. It is a grievous sin to remove any gold from those sites. Mexicans first violated these unknown laws. They were captured and tortured. Indian women were violated by these interlopers, which exacerbated the entire situation and started a war that lasted almost a century.

Jim Bark's notes have become very famous. We have included a fair amount of them, especially those about Dr. Thorn, Blank, the two soldiers Joe Deering and John Chuning. The outcome of these stories is the fact that the Dutchman Mine had been found numerous times and that others of note had actually seen the gold and spoke of how rich it was. These stories validated the curse. Just read about their lives; they also offer clues as to the mine's whereabouts. The answer lies within these covers.

The Adolph Ruth story is certainly captivating. It is the stuff movies are made of. Here is an aged, crippled man who feels compelled to climb down from a burro after a hard day and be left alone in a remote mountainous place of which he has no knowledge. He readily admits in a letter to his wife that he can barely stand after dismounting. He had a steel plate in his leg and had been told by a reputable doctor that if he fell on it again, it would kill him. Yet, this invincible spirit hiked on a rocky slope toward a treasure he was convinced he could find with no difficulty. He had told everyone who would listen that he had a map. So, of course, when he was found dead, it was assumed he had been murdered.

The world press had a field day. It didn't come to light for a long time that Tex Barkley, the rancher who found Ruth's body, had moved the remains to avoid the problem of all those who might climb around the area and scatter his cows. What appears macabre is the fact that Adolph Ruth's head was missing. Brownie Holmes found it on the later search with the newsmen. Much has been made of that fact. Most people believe that animals probably carried off Adolph's skull. Perhaps that is why Deputy Jeff Adams didn't bother telling the world he had found the map on Ruth's body. He and Barkley were friends, and they probably made a pact that their secret was safe if they didn't say anything about any of it. Eventually, Adams became sheriff. There is a copy of Adam's letter to Carl Hayden in Chapter 11 detailing the fact that he (Adams) found Adolph Ruth's map and confiscated it. Barkley and Adams were Dutchman hunters, and both were completely familiar with that area of the Superstitions; but neither of them ever found the mine.

Cal Boyce, who later became sheriff of Maricopa County, tried to give Bob Corbin a copy of Ruth's map but died before that could be accomplished. The map was originally given to Ruth's son by an insurrectionist who was about to be shot in Mexico. The map came from his wife's family; her maiden name was Peralta.

Tex Barkley owned and lived on a ranch built in 1877 by Matt Cavaness after his discovery of a spring in that location. It has been referred to as: The Board House, The Cow House, Bark's Ranch, The Quarter Circle U, The Half Circle U, The Milk Ranch and Barkley's Ranch. Matt Cavaness raised milk cows and sold butter and milk to the Silver King Mine. After his divorce in 1881, his wife, Alice, and their son Aaron continued to live there. The Indians would come to the ranch for milk and eggs. They paid in gold. In 1882 Matt built a ranch house about three miles southeast of there at a spring in Milk Ranch Canyon, which was named simply The Milk Ranch. One needs an explanation because of the confusion about the location of these often named places. (A body, presumed to be one of the two soldiers, was dug from a shallow grave in that area.)

The last in this long line of visitors to the Superstitions who was infected with the lore and who honestly believed he would locate Jacob's mine was Walter Gassler. Believe it or not, he was a successful pastry chef. Walt had followed the Ruth news story with rapt attention. Not only did he read all the newspapers, but he went to the library where he enlisted the aid of the librarians to do a search of the stacks. He then spent two years researching his subject. In 1934 Walt started his quest. After arriving in Arizona, he made friends with the Barkley family and became a regular at the ranch. Eventually he brought in a mining engineer, Tom Reis, who had 40 years experience in South and Central America. Tom showed Walt where the Mexicans had been – a group of 60 or 70 of them for at least six months. These were probably the Mexicans mentioned by Apache Jack and the cavalry soldiers.

Reis and Gassler found gold-bearing quartz. The mining engineer reasoned it was proof there was gold in the area, although it was low grade. They followed all the leads and, of course, never found the mine. Walter was about to lose his wife over his obsession. He gave it up and went away to Tennessee for many years. He did say he found the stone house and the cave, which were important clues.

In 1983 Walter Gassler, now an old man, returned to Arizona and reintroduced his search for the mine. In doing so he remembered reading that Bob Corbin, who was then Attorney General of Arizona, was an avid Dutchman hunter. It seemed reasonable that since Corbin had a reputation for being completely honest, he might have found the right man to trust. Walter now believed he knew the secret to locating the mine. He made an appointment to see the attorney general and brought his copious notes along to give to this new acquaintance. His words were, "I know everyone tells you they know, but please read these and after you have, if you think I don't know what I'm talking about, I won't trouble you again."

Corbin said later he read the notebook and knew Walter had done his homework. The lawyer was impressed. A trip was planned that unfortunately never materialized.

In May 1984 Gassler called Tom Kollenborn, another Dutchman hunter and friend of Corbin's. Since he couldn't get hold of Corbin, who was vacationing at a remote cabin in the Bradshaws, Walt asked Tom if he could get away to go into the Superstitions. He said he had solved the riddle. Tom declined.

We know now Walter went in alone and was found dead on the trail near Charlebois Spring. The coroner said he died of a heart attack. Tom Kollenborn received a visitor who claimed to be Walter's son. The man produced gold ore that he said Walter had in his backpack. Tom, being familiar with the gold that was taken from under Jacob's deathbed, said the specimens were alike. The man wanted to know if Bob Corbin still had Walter Gassler's notes. Tom shook his head and said, "No, I have a copy which Bob gave me, but only Bob has a copy of the map." Tom later said, "Obviously he wanted the notes and map." Tom told him to call the attorney general in his office at the capitol. He never called.

Later a man approached Tom at a slide show he was giving in Florence, Arizona. He said he was Walter Gassler's son and wanted to know if Bob Corbin still had his father's notes? This man was a completely different person than the man who had appeared at Tom's house. This time the matter was checked. The second man turned out to be the real son; he was given a copy of the notes.

Now, of course, we are faced with a mystery. Did Walter really figure out the puzzle and come out with the gold? Was he confronted by an interloper who caused him to have a heart attack? He was old at the time of his death; perhaps the shock of finally finding the mine brought on the attack. Whoever came upon him took the gold and probably pretended to be his son, thinking Kollenborn would get him the map. Both Bob and Tom regretted not having gone in with Walter, but for years they have had calls from people all over the world who say they know where the mine is. If they went in with all of them, it would be a full-time job.

Later, an old man, who lived at a stable outside Apache Junction, said he was hiking near Charlebois Spring on the day Walter's body was found. He saw a man, who fit Tom's description of the interloper, in that area. The old man did not stay in the area and never saw Walter alive or dead. Walter's real son met with Bob Corbin to receive his copy of the notes. He said when he went to the sheriff's office to retrieve his father's possessions, the backpack was listed but was missing, as was the gold.

It seems the *curse* is at work again. Even if the reader is not superstitious, one has to admit death stalks the mine. However, people come from everywhere to hunt for it. No one seems the least bit intimidated. They have **gold fever** – a disease without a cure. If you are one of that group, we hope you have enjoyed *The Bible* in its entirety. If you actively seek the location of Jacob's mine – good luck and safe journey!

– Helen Corbin

Bibliography

Altschuer, C.W. *Chains of Command Arizona and the Army 1856–1875.* Tucson: Arizona Historical Society, 1983.

Bancroft, H.H. *History of Arizona and New Mexico 1530 – 1888.* San Francisco: 1889.

Barnes, W.C. *Arizona Place Names.* Tucson: University of Arizona Press, 1935.

Beeching, W. *Diary of a Journey to California, 1849.* Berkley: Bancroft Library, University of California.

Bieber, R.P. "From Texas to the Gold Mines." In *Southern Trails to the California Gold Fields 1849.* Glendale, Calif.: The Arthur H. Clark Library, 1937.

——. *Through Mexico to California Letters and Journal of John E. Durivage.* Glendale, Calif.: The Arthur H. Clark Library, 1937.

Corbin, H. *The Curse of the Dutchman's Gold.* Phoenix, Ariz.: Foxwest Publishing, 1990.

Donohue, J. *After Kino, Jesuit Missions in Northwestern New Spain 1711–1767.* Rome, Italy: The Jesuit Historical Institute, 1969.

Dumke, G.S. *The Mexican Gold Trail, Journey of a Forty-niner.* San Marino, Calif.: The Huntington Library, 1945.

Elliot, W.W, ed. *History of the Arizona Territory 1884.* San Francisco.

Ely, S. *The Lost Dutchman Mine.* New York: William Morrow and Co., 1953.

Farish, T.E. *History of Arizona* (volumes 1, 2, 4 and 6). Phoenix: 1915–1918.

Gerald, R.E. *Spanish Presidios of the Late Eighteenth Century in Northern New Spain.* Albuquerque: Museum of New Mexico Press, 1968.

Goetzmann, W.H. *Army Exploration in the American West 1803 – 1863.* New Haven, Conn.: Yale University Press, 1959.

Hastings, J.R. "Arizona and the West." In *People of Reason, the Colonization of Sonora to 1767,* 1961.

Hathaway, M.L. *An Album of Memories Arizona Territorial Pioneers.* Tucson: The Arizona Historical Society, 1972.

Hayden, C.T. *The Hayden Biographies of Arizona Pioneers.* Tempe: Arizona State University, 1969.

Kessel, J.L. *Mission of Sorrows Jesuit Guevavi and the Pimas 1691 – 1767.* Tucson: University of Arizona Press, 1970.

Kinniard, L. *The Frontiers of New Spain Nicolas LaFora's Description, 1766.* Berkley, Calif.: The Quivara Society, 1958.

Lacy, J.C. "Early History of Mining in Arizona Acquisition of Mineral Rights from 1539 – 1866." In *History of Mining in Arizona*, edited by M.J. Canty. Tucson: Mining Club of the Southwest Foundation, 1987.

Lockwood, F.C. *Pioneer Days in Arizona.* New York: MacMillan, 1932.
——. *Life in Old Arizona.* Los Angeles: Ward Ritchie Press, 1943.
——. *Pioneer Portraits.* Tucson: University of Arizona Press, 1968.

Manje, J.M. "Luz de Sierra Incoanita." In *Unknown Arizona and Sonora 1693 – 1721.* Tucson: Arizona Silhouettes, 1954.

Martin, D.D. *An Arizona Chronology The Territorial Years, 1846 – 1912.* Tucson: University of Arizona Press, 1963.

Martin, M.E. *From Texas to California in 1849.* Southwest Historical Quarterly, 1925 – 1926.

Martineau, J.H. *Settlements in Arizona.* Berkley: The Bancroft Library University of California, 1885.

Mattison, R.H. *Early Spanish and Mexican Settlements in Arizona.* The New Mexico Historical Review, 1946.

McCarty, K. *The Spanish Years 1776–1821.* (Monograph 4) Tucson: Arizona Historical Society, 1976.
——. *A Spanish Frontier in an Enlightened Age: Franciscan Becinninas in Sonora and Arizona 1767–1770.* Washington DC: The Academy of Franciscan History, 1981.

McClintock, J.H. *Arizona, the Youngest State.* Chicago: Chicago Press, 1916.

Miller, D. *Soldiers and Settlers, Military Supply in the Southwest Arizona Territory from 1861–1865.* Albuquerque: University of New Mexico Press.

Miller, Jos. *The Memoirs of Matt Cavaness Arizona Pioneer.* Tucson: Arizona Historical Society, 1950.

Mitchell, J.D. *Lost Mines of the Great Southwest.* Mesa, Ariz.: J.D. Mitchell & Sons, 1933.

Moore, M.L. *The Interior Provinces of New Spain, The Report of Hugh O'Conor, January 30, 1776.* Arizona and the West, 1971.

Officer, J.E. *Hispanic Arizona 1536–1856.* Tucson: University of \ Arizona Press, 1987.

Opler, M. *An Apache Life Way, Religious Institutions of the Chiricahua Apache.* New York: Cooper Square Publishers, 1965.

Parsons, G.W. *The Private Journal of George Whitwell Parsons.* Phoenix: Arizona Statewide Archival and Records Project, 1939.

Peplow, E.H. *History of Arizona, Volume 2.* New York: Lewis Historical Publishing Company, 1958.

Rockfellow, J.A. *Log of an Arizona Trailblazer.* Tucson: Arizona Silhouettes, 1955.

Schellie, D. *Vast Domain of Blood.* Western Lore Press, 1968.

Sheridan, T.E. *Arizona, a History.* Tucson: University of Arizona Press, 1995.

Southworth, J.R. *The Mines of Mexico Volume 9.* Liverpool, England: Blake and Mackenzie, 1905.

Trimble, M. *Arizona A Cavalcade of History.* Tucson: Treasure Chest Publications, 1989.

Tuck, F. *The History of Mining in Arizona.* Phoenix: Department of Mineral Resources, State of Arizona, 1963.

Villa, E.W. *Historia del Estado de Sonora.* Hermosillo, Sonora, Mexico: Editorial Sonora, 1951.

Wagoner, J.J. *A History of Mining and Cattle in Arizona and Sonora.* Hermosillo, Sonora, Mexico: Mexico Historical Review, 1952.

——. *The Arizona Territory 1863–1912.* Tucson: University of Arizona Press, 1970.

——. *Early Arizona Prehistory to the Civil War.* Tucson: University of Arizona Press, 1975.

Wyllys, R.K. *Arizona the History of the Frontier State.* Phoenix: Hobson and Herr, 1950.

Young, O.E. *The Spanish Tradition in Gold and Silver Mining.* Arizona and the West, 1965.

Acknowledgements

Robert K. Corbin, Arizona Attorney General (retired)

Mathew Roberts, historian

Gregory Davis, Director of Research & Acquisitions, Superstition Mountain Historical Society

The Cavaness family notes and diaries

The Gassler family notes

Paul Pettit, Petrasch family member, historian

Peeples family historical manuscripts and photographs

Bertie Roberts (deceased), Roberts family historian

Professor Meneray, historian, Tulane University Archives

Judge Al Flood (deceased), Justice of the Peace – Phoenix, former Don's Club president

Bucky Singleton, senior photographer, Motorola (retired)

Carl Ricketts, research assistant, New Orleans

Arizona State Historical Society

Arizona Archives, State Capitol

University of Arizona Library

Arizona State University Carl Hayden Memorial Library

Photograph Collection from *Arizona Sheriff* magazine

Tulane University Archives

Wolfe Publishing Co.:

- **Mark Harris**, publisher
- **Roberta Montgomery**, senior editor
- **Dixie Wenger**, assistant editor
- **Gerald Hudson**, art director (cover design)
- **Rebecca Pinkley**, production director

Kathy Corbin, computer assistant

Index

Bob Corbin

Former president of the NRA and Arizona's Attorney General Bob Corbin shown here on a recent trip into the Superstition Mountains in his quest for the Lost Dutchman Mine.

Corbin has been interviewed by NBC, CBS, *Newsweek, Geo, Los Angeles Times*, the *Denver Post*, A&E Television, "Unsolved Mysteries" and local radio and television stations. In 1984 the U.S. government closed the Superstitions to mineral exploration, provoking Bob Corbin's quote: "You cannot legislate away men's dreams."

Helen Corbin

"Helen Corbin distinguished herself as an incomparable researcher with her first book *The Curse of the Dutchman's Gold* on the Lost Dutchman Gold Mine in the Superstition Mountains. Now comes another fascinating book, this one on the life and times of Jacob Waltz, the Dutchman himself.

For those interested in one of the world's most mesmerizing lost treasure stories, *The Bible on the Lost Dutchman Gold Mine and Jacob Waltz* is absolutely vital reading." – Bob Early, editor of *Arizona Highways* magazine.

In July 2000 *U.S. News and World Report* named the world's greatest mysteries: Jacob Waltz's Lost Dutchman Gold Mine was listed among such notables as Noah's Ark, the Holy Grail, the Ark of the Covenant and King Solomon's Mines.

Coincidentally, author Helen Corbin, working on her third Superstition title, was given permission to use heretofore unpublished diaries from four pioneer families. The research, for the first time in history, documents Waltz from birth to the grave. Besides the manifest of the ship on which he sailed to America, there are German translations of the directions to Jacob's infamous mine given to Reiney Petrasch and Julia Thomas before the old man died. In fact, this landmark text offers historical proof of these and all major speculations that have occurred in the hundreds of books published in the 110 years following the German's death.

Author Corbin, winner of awards from the National and Arizona Press Women, is married to long-time Dutchman hunter Bob Corbin, formerly Arizona attorney general and National Rifle Association president. The Corbins reside in Prescott, Arizona.

Helen Corbin's Books

The Curse of the Dutchman's Gold – The top-selling book on the Lost Dutchman Mine, which includes maps, documents and evidence proving the existence of this cryptic lost treasure. **$16.95**

King of the Ice – The life story of an Alaskan homesteader whose heroics in the bush are legend. Don Johnson became a mentor to the Inuits near the Arctic Circle and the Aleuts on the Aleutian chain. A true adventure; a must read. **$16.95**

Senner's Gold – A diary found in an attic trunk revealed Al Senner's true story about the 1,000 pounds of stolen gold, where he hid it in the Arizona Territory and a skeleton in a shallow grave. Another true Superstition Mountain lost treasure. **$16.95**

Treasure at Battery Point – An infamous island lighthouse, just off Crescent City, California, becomes the setting for a page-turning thriller surrounding an apocalyptic search for sunken gold amid a dangerous love triangle. **$16.95**

The Bible on the Lost Dutchman Gold Mine and Jacob Waltz – The life and times of Jacob Waltz are documented from his birth to the grave with heretofore unpublished diaries from four pioneer families who knew him. Also included are German translations of directions to the mine given to Julia Thomas and Reiney Petrasch by Waltz. **$24.95**

* * *

Also available are 100 leather-bound, numbered, limited editions of ***The Bible on the Lost Dutchman Gold Mine and Jacob Waltz***. Price: **$100.00** (ppd). Books will be autographed by the author and her husband Bob Corbin.

When ordering leather-bound copies, please send check or money order to the author at HC 31 Box 308, Prescott AZ 86303. Please state clearly the name(s) of the person(s) to whom the book should be autographed.

All other Helen Corbin books should be ordered from Wolfe Publishing Co., 2625 Stearman Road, Suite A, Prescott AZ 86301; or call toll-free **1-800-899-7810**. (*Please add $5.00 for shipping and handling. Call for shipping charges on multiple books.*)